Methods in Molecular Biology™

Series Editor
John M. Walker
School of Life Sciences
University of Hertfordshire
Hatfield, Hertfordshire, AL10 9AB, UK

For further volumes:
http://www.springer.com/series/7651

Legume Genomics

Methods and Protocols

Edited by

Ray J. Rose

Environmental and Life Sciences, CILR, The University of Newcastle, Callaghan, NSW, Australia

Humana Press

Editor
Ray J. Rose
Environmental and Life Sciences,
CILR, The University of Newcastle
Callaghan, NSW, Australia

ISSN 1064-3745 ISSN 1940-6029 (electronic)
ISBN 978-1-4939-6024-8 ISBN 978-1-62703-613-9 (eBook)
DOI 10.1007/978-1-62703-613-9
Springer New York Heidelberg Dordrecht London

Printed on acid-free paper

Humana Press is a brand of Springer
Springer is part of Springer Science+Business Media (www.springer.com)

Preface

Legumes, with their nitrogen-fixing capacity, are the third largest family of flowering plants and are critical for agricultural and ecological sustainability. They are important for food, feed, human nutrition, bioenergy, and industrial purposes.

Genomics can now be readily applied to legumes with the increasing sequence information and analysis tools available. Model legumes such as Medicago and Lotus, as well as soybean and pea, have provided much of this information, which is now being applied to the legume family more broadly. This book provides a consideration of a range of methods and protocols, including available resources, used to discover new genes and to understand genes and their interactions.

The book is targeted to plant molecular biologists, molecular breeders, plant physiologists and biochemists, developmental biologists, and those interested in plant–microbe interactions. The chapters are suitable for those already in the field, those plant scientists entering the field, and graduate students. Legume genomics offers a potentially productive field for new investigators.

The first chapter provides an overview of legume genomes followed by chapters on techniques used in genome characterization and analysis. Chapters then follow on transcriptome analysis and miRNA identification and analysis. Also considered are a range of resources available for forward and reverse genetics in Medicago, Lotus, and pea. These approaches have an increasing role in identifying gene function. Molecular markers developed through genomics are also considered. A number of chapters deal with a range of transformation strategies used to investigate gene function using transformed roots, stable plant transformation, and transient expression, and there is a guide to investigating subcellular localization using fluorescent fusion proteins. There are chapters on proteomics, phosphoproteomics, and metabolomics where, increasingly, mass spectrometry developments provide new approaches for investigating the products of genes—the proteins and the metabolism they regulate and support.

Callaghan, NSW, Australia ***Ray J. Rose***

Contents

Contributors

STIG UGGERHØJ ANDERSEN • *Department of Molecular Biology and Genetics, Centre for Carbohydrate Recognition and Signalling, Aarhus University, Aarhus, Denmark*
JUSTIN E. ANDERSON • *Department of Agronomy and Plant Genetics, University of Minnesota, St. Paul, MN, USA*
SERGIO ARCIONI • *Consiglio Nazionale delle Ricerche (CNR) - Istituto di Genetica Vegetale, Perugia, Italy*
MARK A. BAKER • *Biological Sciences, Priority Research Centre in Reproductive Science, School of Environmental and Life Sciences, University of Newcastle, Callaghan, NSW, Australia*
NICHOLAS J. BALTES • *Department of Genetics, Cell Biology, and Development, University of Minnesota, Minneapolis, MN, USA; Center for Genome Engineering, University of Minnesota, Minneapolis, MN, USA*
ORNELLA CALDERINI • *Consiglio Nazionale delle Ricerche (CNR) - Istituto di Genetica Vegetale, Perugia, Italy*
STEVEN B. CANNON • *Agricultural Research Service, United States Department of Agriculture, Ames, IA, USA*
MARIA CARELLI • *Consiglio per la Ricerca e la sperimentazione in Agricoltura (C.R.A.) - Centro di ricerca per le Produzioni Foraggere e Lattiero-casearie, Lodi, Italy*
DAVID A. COLLINGS • *Biomolecular Interaction Centre, School of Biological Sciences, The University of Canterbury, Christchurch, New Zealand*
CECILIA CONTRERAS-CUBAS • *Departamento de Biología Molecular de Plantas, Instituto de Biotecnología, Universidad Nacional Autónoma de México, Cuernavaca, Mexico*
ALEJANDRA A. COVARRUBIAS • *Departamento de Biología Molecular de Plantas, Instituto de Biotecnología, Universidad Nacional Autónoma de México, Cuernavaca, Mexico*
SHAUN J. CURTIN • *Department of Agronomy and Plant Genetics, University of Minnesota, St. Paul, MN, USA*
DANIELA S. FLOSS • *Boyce Thompson Institute for Plant Research, Ithaca, NY, USA*
J. STEPHEN GANTT • *Department of Plant Biology, University of Minnesota, St. Paul, MN, USA*
ROHINI GARG • *National Institute of Plant Genome Research, New Delhi, India*
RENÉ GEURTS • *Laboratory of Molecular Biology, Wageningen University, Wageningen, The Netherlands*
NICOLAS GOFFARD • *Plant Science Division, Research School of Biology, College of Medicine, Biology and Environment, The Australian National University, Canberra, ACT, Australia; Computational Biology & Bioinformatics, IP&Science, Thomson Reuters, London, UK*
MARIA J. HARRISON • *Boyce Thompson Institute for Plant Research, Ithaca, NY, USA*
VALÉRIE F.G. HECHT • *School of Plant Science, University of Tasmania, Hobart, TAS, Australia*

ROGER HELLENS • *The New Zealand Institute for Plant & Food Research Ltd, Auckland, New Zealand*
NIJAT IMIN • *Plant Science Division, Research School of Biology, College of Medicine, Biology and Environment, The Australian National University, Canberra, ACT, Australia*
MUKESH JAIN • *National Institute of Plant Genome Research, New Delhi, India*
HANS DE JONG • *Laboratory of Genetics, Wageningen University, Wageningen, The Netherlands*
HEE-KYUNG LEE • *Plant Biology Division, The Samuel Roberts Noble Foundation, Ardmore, OK, USA*
ROBYN LEE • *Department of Biochemistry, University of Otago, Dunedin, New Zealand*
ILARIA LOSINI • *Parco Tecnologico Padano (PTP), Sassari, Italy*
RICHARD MACKNIGHT • *Department of Biochemistry, University of Otago, Dunedin, New Zealand; The New Zealand Institute for Plant & Food Research Ltd., Sandringham, New Zealand*
ANNA MAŁOLEPSZY • *Department of Molecular Biology and Genetics, Centre for Carbohydrate Recognition and Signalling, Aarhus University, Aarhus, Denmark*
DHANANJAY MANI • *Department of Agronomy and Plant Genetics, University of Minnesota, St. Paul, MN, USA*
KIRANKUMAR S. MYSORE • *Plant Biology Division, The Samuel Roberts Noble Foundation, Ardmore, OK, USA*
KIM E. NOLAN • *Biological Sciences, Australian Research Council Centre of Excellence for Integrative Legume Research, School of Environmental and Life Sciences, The University of Newcastle, Newcastle, NSW, Australia*
FRANCESCO PANARA • *Consiglio Nazionale delle Ricerche (CNR) - Istituto di Genetica Vegetale, Perugia, Italy*
KELSEY PICARD • *Department of Biochemistry, University of Otago, Dunedin, New Zealand*
PIETRO PIFFANELLI • *Parco Tecnologico Padano (PTP), Sassari, Italy*
ANDREA PORCEDDU • *Consiglio Nazionale delle Ricerche (CNR) - Istituto di Genetica Vegetale, Perugia, Italy*
AMIT RAI • *Metabolites Biology Lab, Department of Biological Sciences, National University of Singapore, Singapore, Singapore*
PASCAL RATET • *Institut des Sciences du Végétale, CNRS, Gif sur Yvette, France*
JOSÉ LUIS REYES • *Departamento de Biología Molecular de Plantas, Instituto de Biotecnología, Universidad Nacional Autónoma de México, Cuernavaca, Mexico*
RAY J. ROSE • *Environmental and Life Sciences, CILR, The University of Newcastle, Callaghan, NSW, Australia*
ALEXA M. SCHMITZ • *Boyce Thompson Institute for Plant Research, Ithaca, NY, USA*
CARLA SCOTTI • *Consiglio per la Ricerca e la sperimentazione in Agricoltura (C.R.A.) - Centro di ricerca per le Produzioni Foraggere e Lattiero-casearie, Lodi, Italy*
ISLAM SHAHIDUL • *Faculty of Science, School of Plant Biology, The University of Western Australia, Crawley, WA, Australia; The UWA Institute of Agriculture, The University of Western Australia, Crawley, WA, Australia; Department of Horticulture, Bangladesh Agricultural University, Mymensing, Bangladesh*
YOUHONG SONG • *Biological Sciences, Australian Research Council Centre of Excellence for Integrative Legume Research, School of Environmental and Life Sciences, The University of Newcastle, Newcastle, NSW, Australia*

COLBY G. STARKER • *Department of Plant Biology, University of Minnesota, St. Paul, MN, USA; Department of Genetics, Cell Biology, and DevelopmentUniversity of Minnesota, Minneapolis, MN, USA; Center for Genome Engineering, University of Minnesota, Minneapolis, MN, USA*
JENS STOUGAARD • *Department of Molecular Biology and Genetics, Centre for Carbohydrate Recognition and Signalling, Aarhus University, Aarhus, Denmark*
ROBERT M. STUPAR • *Department of Agronomy and Plant Genetics, University of Minnesota, St. Paul, MN, USA*
FRANCES C. SUSSMILCH • *School of Plant Science, University of Tasmania, Hobart, TAS, Australia*
SANJAY SWARUP • *Metabolites Biology Lab, Department of Biological Sciences, NUS Environmental Research Institute, Singapore Centre for Environmental Life Sciences Engineering, National University of Singapore, Nanyang Technological University, Singapore, Singapore*
MILLION TADEGE • *Department of Plant and Soil Sciences, Institute for Agricultural Biosciences, Oklahoma State University, Stillwater, OK, USA*
SHIVSHANKAR UMASHANKAR • *Metabolites Biology Lab, Department of Biological Sciences, NUS Environmental Research Institute, National University of Singapore, Singapore, Singapore*
DORIAN FABIAN URBAŃSKI • *Department of Molecular Biology and Genetics, Centre for Carbohydrate Recognition and Signalling, Aarhus University, Aarhus, Denmark*
DANIEL F. VOYTAS • *Department of Genetics, Cell Biology, and Development, University of Minnesota, Minneapolis, MN, USA; Center for Genome Engineering, University of Minnesota, Minneapolis, MN, USA*
TIAN-ZUO WANG • *State Key Laboratory of Vegetation and Environmental Change, Institute of Botany, The Chinese Academy of Sciences, Beijing, China*
GEORG WEILLER • *Plant Science Division, Research School of Biology, College of Medicine, Biology, and Environment, The Australian National University, Canberra, ACT, Australia*
JAMES L. WELLER • *School of Plant Science, University of Tasmania, Hobart, TAS, Australia*
GUIJUN YAN • *Faculty of Science, School of Plant Biology, The University of Western Australia, Crawley, WA, Australia; The UWA Institute of Agriculture, The University of Western Australia, Crawley, WA, Australia*
HUAAN YANG • *Department of Agriculture and Food Western Australia, South Perth, WA, Australia*
JUAN CARLOS SERRANI YARCE • *Plant Biology Division, The Samuel Roberts Noble Foundation, Ardmore, OK, USA*
WEN-HAO ZHANG • *State Key Laboratory of Vegetation and Environmental Change, Institute of Botany, The Chinese Academy of Sciences, Beijing, China*

Chapter 1

The Model Legume Genomes

Steven B. Cannon

Abstract

The primary model legumes to date have been *Medicago truncatula* and *Lotus japonicus*. Both species are tractable both genetically and in the greenhouse, and for both, substantial sets of tools and resources for molecular genetic research have been assembled. As sequencing costs have declined, however, additional legume genomes have been sequenced, and the funding available to crops such as soybean has enabled these to be developed to the status of genetic models in their own right. This chapter, therefore, describes a broader set of model species in the legumes, and discusses similarities and differences between the genomes sequenced to date, as well as computational resources available for various legume species. Genome structural characteristics in, for example, *Medicago truncatula* and *Glycine max*, can have large impacts on the kinds of functional genomic research that may be carried out in these species. Both of these genomes have substantial redundancy for many gene families, but the nature of the redundancy is different in the two genomes—with the redundancy typically being in the form of local gene duplications in *Medicago*, and in whole-genome-duplication-derived duplications in *Glycine*. Similar considerations (about gene environments and genome structure) will likely need to be taken into account for any model or crop species.

Key words *Medicago truncatula*, *Lotus japonicus*, *Glycine max*, *Phaseolus vulgaris*, Legume genomes, Gene families

1 Introduction

Through the period of the late 1990s and early 2000s the term "model legumes" typically referred to *Medicago truncatula* and *Lotus japonicus*, as both of these species had substantial, active research communities that were working to develop genomics resources for the species (ESTs, BAC libraries, genome sequences, mutant collections, etc.). The legume research community also developed an extensive set of protocols for model biology using these plants (such as transformation methods and phenotyping methods). While both *Medicago* and *Lotus* remain crucial models for the legumes, the genetic and genomic resources available for other legume species have expanded the set of species that might be considered "models" for this plant family. As of late 2012,

Ray J. Rose (ed.), *Legume Genomics: Methods and Protocols*, Methods in Molecular Biology, vol. 1069, DOI 10.1007/978-1-62703-613-9_1, © Springer Science+Business Media, LLC 2013

nearly complete genome sequences have been generated for six legume species (*Medicago truncatula* [*Mt*], *Lotus japonicus* [*Lj*], *Glycine max* [*Gm*], *Phaseolus vulgaris* [*Pv*], *Cicer arietinum* [*Ca*], and *Cajanus cajan* [*Cc*]); and several other genome sequencing projects are under way (including *Arachis hypogaea* [*Ah*], *Medicago sativa* [*Ms*], *Lupinus angustifolius* [*La*]). Additionally, the various species have certain strengths that the others may lack. For example, numerous classical traits have been characterized in pea (*Pisum sativum* [*Ps*]). *Gm*, *Pv*, and *Mt* have particularly high-quality genome sequences (arguably more complete than for the other species). *Gm* and *Pv* have received extensive QTL mapping and phenotyping over several decades. *Gm* has several facile, reliable transformation systems.

Numerous studies have established that extensive synteny exists between many legumes, and particularly among species within the "warm-season legume" group and the "cool-season legume" group (both described below). However, genome sequencing has also revealed limitations in the ability to infer functional similarities on the basis of corresponding genomic position. In particular, some classes of genes evolve rapidly; and some patterns of genomic evolution can obscure or complicate comparisons and inferences.

This review presents an overview of some relevant legume taxonomy, and then describes some features of legume genomes sequenced to date, focusing particularly on evolutionary and structural differences between the *Mt* and *Gm* genomes.

2 Some Legume Taxonomy and Genomic Conservation and Change

The legumes underwent a burst of speciation shortly after the origin of the family. Many lineages are therefore distantly separated from one another, and this has implications for the ability to translate information between species.

The legume family has traditionally been classified into three subfamilies: Papilionoideae, Mimosoideae, and Caesalpinioideae. Molecular systematic work has since shown that only the Papilionoideae and Mimosoideae are monophyletic, and that the Caesalpinioideae is composed of multiple early-diverging lineages (Fig. 1). The family originated at around 60 Mya, based on estimates from fossil-calibrated phylogenies [1]. Radiation into the several caesalpinioid and the mimosoid and papilionoid clades occurred within about 10 My after the origin of the family [1].

Most agronomically important species fall within the Papilionoideae. The papilionoid clade itself underwent early radiation into numerous clades, including genistoids (with agronomically important lupins, e.g., *Lupinus angustifolius* or *L. albus*) and dalbergioids (withagronomically important peanuts, *Arachis hypogaea*).

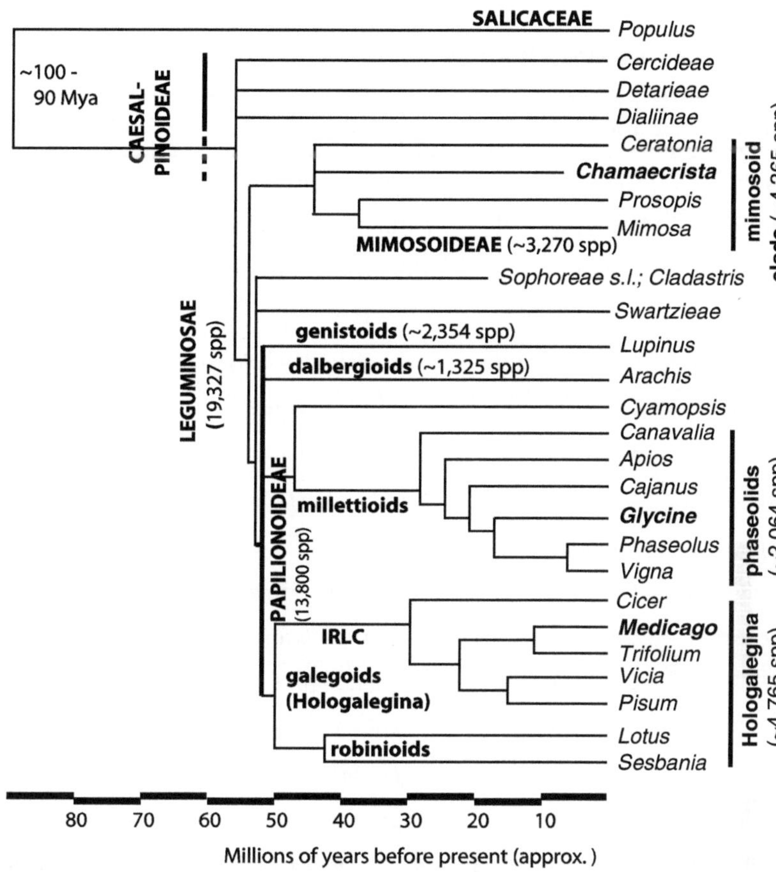

Fig. 1 Taxonomic relationships among selected legume genera. Each genus contains one or more food crop, or a genomic model (*Medicago*, *Lotus*, *Chamaecrista*). *IRLC* inverted-repeat-loss clade. Approximate inferred speciation dates follow the timings in Lavin et al. [1]. The phylogeny is after Lavin et al. [1] and Lewis et al. [20]. Common names and uses are in Table 1. Figure from Cannon et al. [21], with permission

However, the majority of agronomic species fall into one of the two other papilionoid clades: the Hologalegina clade (often referred to as the "galegoid" or "cool-season legumes": chickpea, pea, faba bean, lentil, clovers, medics; Table 1), or the sister phaseolid clade (often referred to as the "warm-season legumes": soybean, cowpea, common bean, various other bean species; Table 1). The cool-season legume clade is also referred to as the "inverted repeat loss clade" (IRLC in Fig. 1)—describing the configuration of chloroplast genes in this clade.

Lotus japonicus, a forage species and a significant model in the legumes, is within the Hologalegina clade, but is sister to the IRLC and diverged shortly after separation from the phaseolid clade. *Lotus* and *Medicago* share a common ancestor at around 50 Mya,

Table 1
Selected food and model legumes

Clade	Tribe	Binomial	Common name	Uses	Note
Cercicdae	Cercicdeae	*Tylosema esculentum*	Morama bean	s,t	D,P
Detarieae	Detarieae	*Detarium senegalense*	Sweet detar	s	P
Detarieae	Detarieae	*Tamarindus indica*	Tamarind	p	P
Umtiza	Caesalpinieae	*Ceratonia siliqua*	Carob	s,p	P
Caesalpinieae	Caesalpinieae	***Chamaecrista fasciculata***	Partridge pea	m	D,P
Caesalpinieae	Caesalpinieae	*Cordeauxia edulis*	Yeheb-nut	s	D,P
Mimosoid	Mimoseae	*Parkia speciosa*	Petai	s,p,f	D,P
Mimosoid	Mimoseae	*Prosopis glandulosa*	Honey mesquite	s,p,f	D,P
Mimosoid	Mimoseae	*Inga edulis*	Ice-cream bean	p	P
Indigoferoid	Indigofereae	*Cyamopsis tetragonoloba*	Guar/cluster bean	s,p,f	
Genistoid	Genisteae	*Aspalathus linearis*	Rooibos tea	l	D,P
Genistoid	Genisteae	*Lupinus albus*	**White lupin**	s	[a]
Genistoid	Genisteae	*Lupinus angustifolius*	Narrow-leaved lupin	s	[a]
Genistoid	Genisteae	*Lupinus luteus*	**Yellow lupin**	s	[a]
Genistoid	Genisteae	*Lupinus mutabilis*	Andean lupin; tarwi	s	C[a]
Dalbergioid	Aeschynomeneae	*Arachis hypogaea*	**Peanut/groundnut**	s	
Galegoid	Galegeae	*Glycyrrhiza glabra*	Licorice	t	P
Galegoid	Cicereae	*Cicer arietinum*	**Chickpea**	s	
Galegoid	Trifolieae	***Medicago truncatula***	Barrel medic	f,m	
Galegoid	Vicieae/Fabeae	*Lathyrus sativus*	Grass pea	s,f	D[a]
Galegoid	Vicieae/Fabeae	*Lens culinaris*	**Lentil**	s	
Galegoid	Vicieae/Fabeae	Pisum sativum	**Pea**	s,p,f,m	
Galegoid	Vicieae/Fabeae	*Vicia faba*	**Fava bean**	s	
Robinioid	Loteae	*Lotus tetragonolobus*	Asparagus pea	p	
Robinioid	Loteae	***Lotus japonicus***	Birdsfoot trefoil	f,m	P
Robinioid	Sesbanieae	*Sesbania spp.*	Agati	f,l,s,p	F,P
Millettioid	Phaseoleae	*Pediomelum esculentum*	Breadroot, prairie turnip	t	D,P
Millettioid	Phaseoleae	*Apios americana*	Potato bean; groundnut	t	P
Millettioid	Phaseoleae	*Cajanus cajan*	**Pigeonpea**	s,p	D,P
Millettioid	Phaseoleae	*Canavalia ensiformis*	Jack bean/velvet bean	s,p,f	[a]
Millettioid	Phaseoleae	*Lablab purpureus*	Hyacinth bean	s,p,f	
Millettioid	Phaseoleae	***Glycine max***	**Soybean**	s,m	

(continued)

Table 1 (continued)

Clade	Tribe	Binomial	Common name	Uses	Note
Millettioid	Phaseoleae	*Pachyrhizus erosus*	**Jicama/yam bean**	t	
Millettioid	Phaseoleae	*Phaseolus coccineus*	Scarlet runner bean	s,p	
Millettioid	Phaseoleae	*Phaseolus lunatus*	Lima bean	s	[a]
Millettioid	Phaseoleae	***Phaseolus vulgaris***	**Common bean**	s,p	
Millettioid	Phaseoleae	*Phaseolus acutifolius*	Tepary bean	s,p	D
Millettioid	Phaseoleae	*Macrotyloma geocarpum*	Hausa groundnut	s	D
Millettioid	Phaseoleae	*Psophocarpus spp.*	Winged bean	p,t	
Millettioid	Phaseoleae	*Vigna angularis*	**Adzuki bean**	s	
Millettioid	Phaseoleae	*Vigna aconitifolia*	Moth bean	s	
Millettioid	Phaseoleae	*Vigna mungo* and *radiata*	**Mung bean**	s	
Millettioid	Phaseoleae	*Vigna subterranea*	Bambara groundnut	s	D
Millettioid	Phaseoleae	*Vigna unguiculata*	**Cowpea**	S,p	

Sequenced legume genomes are underlined. Model or "major" crop legumes are in bold text. Primary uses: s = seed; t = tuber or root; p = pod or pod wall; l = leaf; f = forage; m = model. Comments: D = drought-, C = cold-, or F = flooding-tolerant; P = perennial

[a]Varieties may contain toxins (alkaloids or cyanogenic glycosides) removable in preparation. Table from Cannon et al. [22], with permission

and *Lotus* and *Medicago* each has a common ancestor with phaseolid species at around 55 Mya [1].

Outside the papilionoids are many species that are important species in tropical forests and savannas, as well as some species used for food, including the caesalpinioid species *Tylosema esculenta* (morama bean) and *Cordeauxia edulis* (yeheb nut), *Ceratonia siliqua* (carob), and *Tamarindus indica* tamarind). See other agronomic and model species in Table 1.

Around the time of the origin of the papilionoid clade, a polyploidy event (or whole-genome duplication, WGD) occurred that affects all papilionoids evaluated to date, including all phaseolid and galegoid species [2]. Although several basal genera have not yet been evaluated (e.g., Sophoreae, Cladastris, and Swartzieae), we may at least approximately refer to this polyploidy event as the "papilionoid WGD." This early WGD is important for legume researchers, as it means that many genes in the model and agronomic papilionoid legumes exist as paralogous duplicates (albeit now having evolved independently for ~55 Mya), and perhaps have acquired distinct functions. In fact, an older genome triplication (or paleohexaploidy) event means that many genes may have numerous paralogs. The paleohexaploidy is estimated to have occurred between 100 and 150 Mya [3, 4], prior to the separation of the order Vitales (containing grape) from the eurosids (containing the Malvidae or "eurosids II," containing Arabidopsis; and the Fabidae I or "eurosids I," containing the legumes).

Considering that some legume genera have undergone yet another genome duplication, the number of expected paralogous genes in a given gene family can be substantial. For example, in soybean, one would expect (in the absence of other gene gains or losses) $3 \times 2 \times 2 = 12$ paralogous genes—counting up from an ancestral gene prior to the pre-eurosid paleohexaploidy, through the papilionoid WGD, and then through the *Glycine* WGD. This pattern is in fact seen in both syntenic regions in the soybean genome compared with itself [5] and in selected gene families (Fig. 2). This pattern of expansion of gene copy number provides opportunity for gene sub- and neofunctionalization—though also may make biologists' jobs more challenging, as functional redundancy (or at least partial redundancy) may be rampant in some

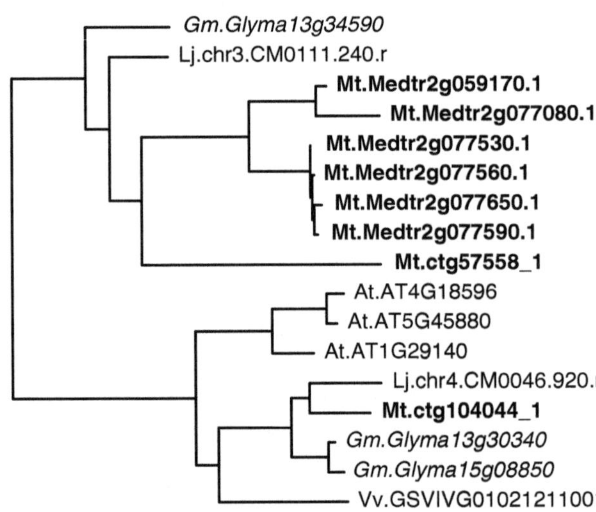

Fig. 2 A gene family (of major pollen allergen proteins), showing whole-genome-duplication-derived paralogs in soybean, and tandem gene duplications in *Medicago*. Vv = *Vitis vinifera*; AT = *Arabidopsis thaliana*; Lj = *Lotus japonicus*; Mt = *Medicago truncatula*; Gm = *Glycine max*. Chromosome numbers are indicated as, e.g., Medtr2 or AT4 or Glyma13. Note that in *At*, *Lj*, and *Gm*, all of the more-recent paralogs come from different chromosomes (and presumably derive from whole-genome duplications), whereas most of the *Mt* genes come from near one another on one chromosome, and presumably derive from tandem or local gene duplications. Also note that there are two old clades in this gene family: one containing representatives from *Gm*, *Lj*, and *Mt*, and another containing representatives from *Gm*, *Lj*, and *Mt* as well as *At* and *Vv*. This family may therefore contain two of the three expected gene lineages deriving from the dicot paleohexaploidy. The gene phylogeny is from the Legume Information System (LIS)

gene families. However, as is discussed below, genetic redundancy can vary dramatically both by gene family and by plant lineage.

Some genomics work has also been done on Caesalpinoid species such as *Chamaecrista fasciculata* (*Cf*, partridge pea) [2, 6], and understanding of some of these early-diverging lineages will be important for a full understanding of evolution of the family and of traits and features within the family. In particular, the non-papilionoid lineages are of interest both because many do not nodulate (though most species in the Mimosoideae do), and most may lack the polyploidy that is shared by at least the large majority of Papilionoid species [2]. The capacity of mimosoid species to nodulate, despite lack of the polyploidy shared by the papilionoid legumes, indicates that the evolution of nodulation did not strictly depend on a prior (recent) polyploidy. One model is that primitive forms of plant-rhizobial symbiosis existed in some of the early legumes, and that elaboration of function and structure then evolved independently in at least the mimosoid and papilionoid lineages [7]. In this model, the polyploidy near the origin of the papilionoids may have been important for evolution of some aspects of the papilionoid nodule—while evolution of similar structures and functions in *Chamaecrista* and other mimosoid legumes would have arisen through other subfunctionalization events in critical genes. The nod-factor receptor, NFP, has a paralog, LYR1, which derived from the papilionoid WGD [6]. Similarly, a transcription factor important in nodule development, ERN1, has a WGD-derived paralog, ERN2. LYR1 and ERN2 are most active during mycorrhizal colonization, supporting a model in which signaling pathways in the nodule have been repurposed from older pathways that originated in the mycorrhizal symbiosis [6].

3 Translating Information About Gene Function Between Species

A key premise behind biological models is that information obtained in the model will be applicable in many other organisms. The extent to which this premise holds with regard to gene function depends on various genomic and other factors. Gene redundancy, evolutionary conservation, regulation, and functional constraints may differ for genes in different plant lineages. For example, genes involved in pathogen defense may evolve rapidly, under shifting pathogen pressures—evolving the capacity to recognize new effector molecules over thousands of years. In contrast, core metabolic, structural, developmental, or regulatory genes may be conserved over very long evolutionary time frames—maintaining their essential sequence and protein structures over tens or even hundreds of millions of years.

Several factors determine the likelihood that a given gene will have retained its function in orthologs from two legume species.

- Is a gene under pressure for diversification within a population or over time? For example, two sets of paralogs within the chlorophyll a–b-binding protein family evolve under two regimes: one set changing slowly over time, and apparently constrained by their roles in a multimeric protein complex in Photosystem II, and one set changing rapidly over time, consistent with their roles in tuning the light harvesting complex for different light conditions [8].
- What is the typical copy number of the gene? Does a gene typically occur as a singleton, or within genomic clusters (which can cause expansions or contractions in copy number)? For example, while the proteasome 20S subunits (discussed next) tend to be genomically isolated, members of the nucleotide-binding site, leucine-rich repeat (NBS-LRR) disease resistance (R) genes, often occur in clusters, where they are subject to copy number expansion or contraction [9]. In *Mt*, 764 NBS-LRR genes have been identified, and approximately 90 % of these are in clusters [6]. In *Gm*, 370 NBS-LRR genes have been identified, and approximately 50 % of these are in clusters [10].
- In Arabidopsis (At), Gm, and Mt, the counts of proteasome subunits are 23, 33, and 18, respectively, differing by history of polyploidy and by apparently infrequent gene losses. In each case, the genes are spread throughout the genome, with paralogous genes clearly deriving from genome duplications ([8] and this report). The *At* and *Gm* genomes have each undergone two rounds of polyploidy since their divergence from one another, while the *Mt* genome has undergone one round of polyploidy. After each polyploidy, at least one gene in each of the conserved proteasome subunit subfamilies has been retained, with post-genome-duplication gene losses in each lineage apparently under functional constraints to retain approximate stoichiometry in the proteasome complex. The pattern of gene evolution in this family is far more conservative than in the NBS-LRRs, for example.
- What is the regulatory and structural context around a gene? Genes may be located in gene-dense euchromatic regions, with frequent recombination, or in gene-poor, recombination-suppressed, transposon-dense pericentromeric regions. These effects of genomic location may affect the evolutionary trajectory of a gene. This is exemplified by in-depth studies of two paralogous clusters of NBS-LRR disease resistance (R) genes in soybean [11–13]. One cluster is in a euchromatic region of soybean chromosome 13, and the other is in the pericentromeric region of soybean chromosome 15. The pericentromeric cluster appears to have been shrinking over time (with four NBS-LRR genes appearing to be functional), and the euchromatic cluster to have both expanded in copy

number and in gene diversity (with 13 NBS-LRR genes appearing to be functional) [11]. The genomic context of these two "matched clusters" has led one to grow and the other to atrophy.

4 Genome Differences in the Legumes

Comparisons of the *Gm* and *Mt* genomes show some striking differences, with implications for functional and applied genomics work in these and other legume species. The first difference, already mentioned, is the paleopolyploidy (or WGD) in the *Glycine* lineage, estimated to have occurred between ~5 and 13 Mya [10]. Correspondingly, the *Gm* genome is about twice the size of the *Mt* genome (1,100 Mb vs. ~500 Mbp). This might be expected, considering WGD in *Glycine*; however, difference is not as a straightforward doubling of the ancestral genetic material to produce one genome twice the size of the other. First, the *Gm* pericentromeres are enormous, comprising about 610 Mbp, in contrast to the ~125 Mbp pericentromeres in *Mt* [6]. Second, the proportions of genes located in the euchromatin and pericentromeres are different: about 78 % of the *Gm* genes are in about 400 Mbp of euchromatin, while about 95 % of genes are in about 375 Mbp of euchromatin in *Mt* [6, 10].

Despite these differences, the numbers of genes in the two genomes are similar. The counts of "high-confidence" genes are 46,430 in *Gm* and 47,846 in *Mt*. How can this be? In a comparison of large syntenic regions between the euchromatic regions of the two genomes, between Mt5 and Gm01, 16.3 Mbp of *Mt* corresponds with 12.3 Mbp of *Gm* [6]. This means that the *Mt* region takes up about 1.33 times more space. So soybean is a little more "space efficient" in terms of gene packing—at least in the euchromatic regions. However, there are between 1.54 and 2.07 times more *Mt* genes in this space than in the corresponding *Gm* region (depending on the level of evidence required for the genes). The similar count of genes in the two genomes is due to a high rate of local (tandem) gene duplications in *Mt*—which has brought the number of genes in *Mt* to approximately the gene count in *Gm*. A schematic of the gene distributions is shown in Fig. 3. In summary, *Mt* has a somewhat higher gene density than *Gm*, but the *Mt* regions take up more space than *Gm*, as more genes have duplicated locally than in *Gm*. However, the WGD in *Gm* makes up for what might otherwise be a gene deficit with respect to *Mt*.

Practical implications for functional genomics are that both *Mt* and *Gm* may cause researchers trouble due to gene redundancy. In either genome, a gene of interest may have no or one or several paralogs. In *Mt*, the papilionoid WGD-derived paralogs should be

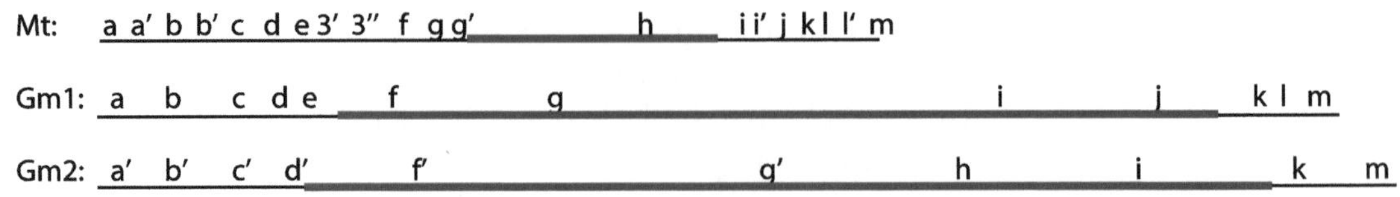

Fig. 3 Schematic of gene organization in *Medicago truncatula* and *Glycine max*. Gene density is higher in *Mt* than in *Gm*, and syntenic regions are spread out (occupy more genomic space) than in *Gm*. Many genes in *Mt* occur as tandemly duplicated clusters of paralogs. *Gm* is duplicated, and also stores more genes in the big pericentromeres

on different chromosomes (if both have been retained), while additional paralogs may have arisen within local clusters, and be of varying ages and perhaps including remnant pseudogenes. In *Gm*, there are likely to be paralogs for most genes (four, assuming no losses after either WGD), with one highly similar paralog from the *Glycine* WGD; but clusters of tandem-duplicate-derived genes are less common than in *Mt*.

Another noteworthy difference between the genomes is that the evolutionary rate has been elevated in the *Medicago* lineage relative to in *Glycine*. The average Ks rate in *Mt* with respect to the papilionoid WGD is about 1.5 times faster than the Ks rate in *Gm* with respect to the papilionoid WGD. Also, nucleotide diversity within the *Mt* population is around three times higher than it is in *Gm*, with *Mt* being $\theta_w = 0.0043$ bp^{-1} and *Gm* being $\theta_W = 0.0017–0.0023$ bp^{-1} (for cultivated and wild collections, respectively) [14, 15]. The rate of linkage disequilibrium (LD) decay is also much higher in *Mt* than in *Gm*, with average pairwise r^2 values falling to values of ~0.2 over a span of ~5–10 kb in *Mt* and over 100–500 kb in *Gm* (depending in both species on genomic region and on collection of genotypes considered) [14, 15]. All of these factors have implications for breeding and comparative and functional genomics in these species. For example, high rates of genetic variety in *Mt* may make interesting mutations more common, but smaller LD may make it more difficult to use linked markers in association studies.

5 Other Legume Genomes

In general, strong synteny is seen across species within the cool-season legumes, and across species within the warm-season legumes. However, there may still be major differences between two species in either of these groups. For example, although pea and *Mt* share chromosome-scale synteny across most chromosomes [16], the pea genome is, at 4,768 Mbp [17], nearly ten times larger than the *Mt* genome. This seems to be due to a much larger number of transposons in pea than in *Mt* [18].

Similarly, among the warm-season legumes, most genera within the Phaseoleae appear to have retained generally chromosome-scale synteny, and most genera within this group have 1N=11 chromosomes (with the prominent exception of *Glycine*, with 1N=20 or 1N=40, depending on ploidy). Yet there are clearly major differences in local gene organization between, for example, *Gm* and *Phaseolus vulgaris* (*Pv*, common bean). Ashfield et al. [11] have shown that corresponding clusters of disease resistance genes in *Gm* and *Pv* have expanded dramatically in *Pv*, and (as mentioned previously), expanded moderately in one homoeolog of *Gm*, and degraded (lost gene copies) in the other homoeolog of *Gm*.

An additional factor to consider when working with sequenced legume genomes is that not all of the genomes have been sequenced to similar depth or quality, or by the same sequencing methods. The soybean genome was sequenced using a whole-genome shotgun (WGS) approach, to about 8× coverage, with most of the sequence contributed by Sanger-generated BAC- and fosmid-end pairs. This generated substantial contiguity and large scaffold sizes. In contrast, the *Medicago* and *Lotus* genomes (at least as of the *Mt* 3.5 and *Lj* 2.5 versions) used primarily a clone-by-clone sequencing strategy, with each bacterial artificial chromosome (BAC) clone spanning ~100–150 kb (though a new WGD-derived *Mt* genome assembly version is anticipated in early 2013). Multi-clone scaffolds for these projects, therefore, were only as large as could be assembled by clone walking. The L50 is 1.27 Mb for *Mt* 3.5 scaffolds, 10 Mb for *Gm* 1.0 scaffolds [6, 10]. The *Mt* v3.5 genome assembly also excludes most pericentromeric DNA, as the *Mt* genome sequencing consortium focused on sequencing in euchromatic regions. Genome assemblies assembled using different WGS strategies can also differ dramatically. The pigeonpea (*Cajanus cajan*, *Cc*) genome (1.0 version), sequenced with a WGS strategy but with short-read sequencing of small- and medium-insert fragment libraries [19], has scaffold L50 of 0.52 Mb—about 19th the length of the soybean scaffold L50.

Similarly, the quality and completeness of gene modeling may differ dramatically between genomes. A fragmented genome sequence makes gene predictions difficult if gaps are present across any portions of the gene sequence. Different transposon compositions and distributions, different gene modeling software, and different repeat-prediction protocols may all contribute to differences in gene models for various legume genomes.

6 Tools for Genomic Exploration in the Legumes

A number of Web sites are useful for legume biologists. These can be generally categorized as "broad" or "specialized," with the former covering species across a clade and the latter focusing on a single species. Broad or clade-wide resources include Phytozome and PLAZA (both endeavoring to span all sequenced genomes),

Table 2
Selected online researchers for legume researchers

Broad resources (spanning a clade or all plants)		
Legume Information System (LIS)	http://comparative-legumes.org	Genetic maps, QTLs, genome browsers, sequence search, gene families
LegumeIP	http://plantgrn.noble.org/LegumeIP/	*Medicago*, soybean, and lotus gene families and gene expression networks
Phytozome	http://www.phytozome.net	Genome browsers, gene families, sequence, and keyword search
PLAZA	http://bioinformatics.psb.ugent.be/plaza	Genome synteny, gene families, sequence and keyword search
SyMAP synteny browser	http://www.symapdb.org	Genome synteny
Specific resources (focused on single species)		
SoyBase	http://soybase.org	Soybean genome browser, QTL data, genetic maps, mutants, gene expression, metabolic pathways
SoyDB	http://casp.rnet.missouri.edu/soydb/	Soybean gene expression data, metabolic pathways, transcription factors
Medicago truncatula HapMap project	http://www.medicagohapmap.org	*Medicago* genome browser, haplotype data for many *Medicago* accessions
Kazusa Lotus genome project	http://www.kazusa.or.jp/lotus/	*Lotus* genome browser, genetic map
Clover Garden	http://clovergarden.jp	Red and white clover genetic maps

Resources are divided approximately by taxonomic breadth: those spanning many plant species or families first, and those focused on single species below

the Legume Information System (spanning the legume family), and the SyMAP synteny browser (covering the grasses, legumes, and Rosaceae and Cucurbitaceae). These and selected specialized resources are shown in Table 2, along with brief descriptions of some key strengths of each resource. As these Web resources and tools change rather quickly and are all rather extensive, they are not described in detail here.

7 Summary

Researchers working in legume molecular genetics are blessed with a wealth of genomic resources, including the genome sequences for around half a dozen diverse legume species, from models and crops. Information gained about gene functions and plant traits, gained through lab and field studies in any one of these species, should be translatable to other species in the family. Researchers should bear in mind, however, evolutionary, structural, and technical differences between the various sequenced genomes. The behavior of gene families may differ markedly between gene families within a genome, and between species. As sequencing and gene modeling strategies continue to change, it seems likely that every new genome will show significant technical differences with respect to genomes previously sequenced using different strategies or technologies. Legume researchers, hunting for gems in a rich bazaar of genomic data and tools, should keep in mind *caveat emptor* as an important precautionary phrase.

References

1. Lavin M, Herendeen PS, Wojciechowski MF (2005) Evolutionary rates analysis of Leguminosae implicates a rapid diversification of lineages during the tertiary. Syst Biol 54: 575–594
2. Cannon SB, Ilut D, Farmer AD, Maki SL, May GD, Singer SR et al (2010) Polyploidy did not predate the evolution of nodulation in all legumes. PLoS One 5:e11630
3. Vierstra RD (2003) The ubiquitin/26S proteasome pathway, the complex last chapter in the life of many plant proteins. Trends Plant Sci 8:135–142
4. Tang H, Wang X, Bowers JE, Ming R, Alam M, Paterson AH (2008) Unraveling ancient hexaploidy through multiply-aligned angiosperm gene maps. Genome Res 18:1944–1954
5. Severin AJ, Cannon SB, Graham MM, Grant D, Shoemaker RC (2011) Changes in twelve homoeologous genomic regions in soybean following three rounds of polyploidy. Plant Cell 23:3129–3136
6. Young ND, Debelle F, Oldroyd GE, Geurts R, Cannon SB, Udvardi MK et al (2011) The Medicago genome provides insight into the evolution of rhizobial symbioses. Nature 480:520–524
7. Madsen LH, Tirichine L, Jurkiewicz A, Sullivan JT, Heckmann AB, Bek AS et al (2010) The molecular network governing nodule organogenesis and infection in the model legume *Lotus japonicus*. Nat Commun 1:10
8. Cannon SB, Mitra A, Baumgarten AM, Young ND, May G (2004) The fates of segmental and tandem gene duplications in large *Arabidopsis thaliana* gene families. BMC Plant Biol 4:10
9. McHale L, Tan X, Koehl P, Michelmore RW (2006) Plant NBS-LRR proteins: adaptable guards. Genome Biol 7:212
10. Schmutz J, Cannon SB, Schlueter J, Ma J, Mitros T, Nelson W et al (2010) Genome sequence of the paleopolyploid soybean. Nature 463:178–183
11. Ashfield T, Egan AN, Pfeil BE, Chen NW, Podicheti R, Ratnaparkhe MB (2012) Evolution of a complex disease resistance gene cluster in diploid Phaseolus and tetraploid Glycine. Plant Physiol 159:336–354
12. Innes RW, Ameline-Torregrosa C, Ashfield T, Cannon E, Cannon SB, Chacko B et al (2008) Differential accumulation of retroelements and diversification of NB-LRR disease resistance genes in duplicated regions following polyploidy in the ancestor of soybean. Plant Physiol 148:1740–1759
13. Wawrzynski A, Ashfield T, Chen NW, Mammadov J, Nguyen A, Podicheti R et al (2008) Replication of nonautonomous retroelements in soybean appears to be both recent and common. Plant Physiol 148:1760–1771

14. Branca A, Paape TD, Zhou P, Briskine R, Farmer AD, Mudge J et al (2011) Whole-genome nucleotide diversity, recombination, and linkage disequilibrium in the model legume Medicago truncatula. Proc Natl Acad Sci USA 108:E864–E870
15. Hyten DL, Choi IY, Song Q, Shoemaker RC, Nelson RL, Costa JM et al (2007) Highly variable patterns of linkage disequilibrium in multiple soybean populations. Genetics 175: 1937–1944
16. Bordat A, Savois V, Nicolas M, Salse J, Chauveau A, Bourgeois M et al (2011) Translational genomics in Legumes allowed placing in silico 5460 unigenes on the Pea functional map and identified candidate genes in *Pisum sativum* L. G3 (Bethesda) 1:93–103
17. Bennett MD, Smith JB (1976) Nuclear DNA amounts in angiosperms. Philos Trans R Soc Lond B 274:227–274
18. Jing R, Knox MR, Lee JM, Vershinin AV, Ambrose M, Ellis TH et al (2005) Insertional polymorphism and antiquity of PDR1 retrotransposon insertions in pisum species. Genetics 171:741–752
19. Varshney RK, Chen W, Li Y, Bharti AK, Saxena RK, Schlueter JA et al (2012) Draft genome sequence of pigeonpea (*Cajanus cajan*), an orphan legume crop of resource-poor farmers. Nat Biotechnol 30:83–89
20. Lewis G, Schrire B, Mackind B, Lock M (2005) Legumes of the world. Royal Botanic Gardens, Kew
21. Cannon SB, Sato S, Tabata S, Young ND, May GD (2011) Legumes as a model plant family. In: Pratap A, Kumar J (eds) Biology and breeding of food legumes. CABI Press, Cambridge MA, pp 348–361
22. Cannon SB, May GD, Jackson SA (2009) Three sequenced legume genomes and many crop species: rich opportunities for translational genomics. Plant Physiol 151: 970–977

Chapter 2

Fluorescent In Situ Hybridization (FISH) on Pachytene Chromosomes as a Tool for Genome Characterization

René Geurts and Hans de Jong

Abstract

A growing number of international genome consortia have initiated large-scale sequencing projects for most of the major crop species. This huge amount of information not only boosted genetic and physical mapping research, but it also enabled novel applications on the level of chromosome biology including molecular cytogenetics that supports plant genetics, genomics, and breeding programs. The simultaneous detection of a large number of BAC-based probes by multicolor fluorescent in situ hybridization (FISH) can provide a rapid overview of super-contig and gap distribution on euchromatin chromosome areas and will display directly and precisely the positions of chromosome rearrangements. Furthermore, hybridizations of BACs on the chromosomes of related species can confirm genomic colinearity, or the occurrence of inversions and translocations events. This cross-species FISH together with meiotic pairing studies is a powerful source of information that elucidates the nature of genome rearrangements, and the consequences of such rearrangements for introgressive hybridizations. In this chapter we describe a general-purpose protocol for FISH on pachytene chromosomes.

Key words Fluorescent in situ hybridization, FISH, Pachytene, BAC-FISH painting

1 Introduction

Whole-genome sequencing has revolutionized biological sciences. While large consortia laboriously sequenced most model organisms, more recent sequencing efforts made genome information available for almost every important eukaryotic organism. In the field of legumes (*Fabaceae*) the genomes of soybean (*Glycine max*), *Medicago truncatula*, *Lotus japonicus*, and pigeon pea (*Cajanus cajan*) are already made available and there are possibly many more to come [1–4]. However, while sequencing itself has become very effective and relatively easy, sequence assembly has not become easier. One of the tools that are helpful in the assembly is the chromosomal mapping of single copy and repetitive sequences by fluorescent in situ hybridization (FISH) [5–8].

Ray J. Rose (ed.), *Legume Genomics: Methods and Protocols*, Methods in Molecular Biology, vol. 1069, DOI 10.1007/978-1-62703-613-9_2,

FISH provides the possibility to determine the physical position of a certain sequence on a chromosome. In FISH, DNA from a clone is fluorescently labeled and used as probe to identify the location of the specific sequence on a set of metaphase chromosomes or pachytene bivalents, the latter giving on average a 20-fold higher spatial resolution [9]. Thus an approximate genomic position can be determined for each individual clone. The probe DNA can vary widely in size. Most often labeled fragments of 50–150 kb genomic DNA (e.g., a bacterial artificial chromosome (BAC) clone) are used as probes for hybridization on chromosomal targets. BAC libraries are generally available for model species and for most of the important crops. Besides such insert clones, small DNA fragments of even 500 bp can be used as a probe instead. In such cases the detection will be indirect through application of a signal amplification step using specific detection amplification systems. A disadvantage of using large probes, like BAC clones, can be the presence of tandem or dispersed repeats in the sequence, which will hybridize to multiple loci, and so can cause erratic mapping of the unique sequences. To block repeats from hybridization, an excess of repeat sequences is added to the hybridization mix. The most accepted method in plant science is Cot DNA, a pool of highly and middle repetitive DNA obtained from sheared single-strand genomic DNA and reannealed according to the principles of DNA reassociation kinetics [10]. Cot DNA is used to study genome structure and organization and has also been used to simplify the sequencing of genomes that contain large amounts of repetitive sequence [11]. In this way repeats can be effectively suppressed in the labeled probe, allowing the localization of almost any genomic clone in a medium-sized plant genome.

Here, we describe a general protocol for FISH on pachytene chromosome bivalents using young flower buds as starting material. The advantage of this procedure is that besides chromosome complements at meiotic pachytene that can be observed, also chromosome sets at mitotic (pro-)metaphase are abundantly present, which can be used simultaneously for analyses. Additionally, a protocol is included for masking repeat sequences with Cot-DNA pre-hybridization.

2 Materials

2.1 Flower Buds for Chromosome Preparations

1. Anthers from young flower buds.
2. Fixative: Glacial acetic acid:ethanol 98 % (1:3).
3. 70 % ethanol.
4. Grease-free microscope slides, 24 × 24 mm coverslips, clean slide boxes.
5. 45 and 60 % acetic acid.

6. Counterstaining solution: 1 % carmine in 45 % acetic acid.
7. For the sodium citrate buffer we add 10 mM citric acid to 10 mM sodium citrate ($Na_3C_6H_5O_7$, which acts as the base) until a pH of 4.5 was obtained.
8. Proteolytic enzymes: Cytohelicase, pectolyase Y-23, and cellulase RS (Sigma). An enzyme mixture is made (1 % w/v each) in the sodium citrate buffer.
9. Microscope with high-resolution (N.A. > 0.5) phase contrast optics.

2.2 DNA Isolation and Labeling

1. Plasmids containing DNA repeat sequences: For example, 5S rDNA (e.g., clone pCT4.2 [12]), 45S rDNA (e.g., clone pTA71 [13]), and/or telomere repeats (e.g., clone pAtT4 [14]).
2. Plasmids containing DNA that will be used as probe (e.g., BAC clones).
3. Plasmid DNA is isolated with a plasmid DNA isolation kit (mini and midiprep) (e.g., Qiagen).
4. Biotin-dUTP and digoxigenin (Dig)-dUTP nick translation kits for labeling of the DNA clones (Roche).
5. For direct labeling of DNA Cy3.5-dCTP (GEHealthcare, gelifesciences.com), Fluorescein-12-dUTP (PerkinElmer), or CyTM3-dUTP (Amersham Biosciences, gelifesciences.com).

2.3 Cot-100 Isolation and Labeling

1. Genomic DNA isolated, e.g., by the CTAB method [15].
2. Sonicator (e.g., Virbra cell sonicator, www.ecomcat.co.uk).
3. Formamide.
4. A rotating oven.
5. DNA S1 nuclease.
6. Chloroform–isoamyl alcohol (24:1).
7. 100 % ethanol.
8. HB50 (50 % deionized formamide, 2× SSC, 50 mM sodium phosphate, pH 7).

2.4 FISH

1. Hybridization mixture: 20 ng labeled probe (plus 1 μg Cot-100 DNA if required), 10 μL 20 % dextran sulfate in HB50 (50 % deionized formamide, 2× SSC, 50 mM sodium phosphate pH 7, 20 % dextran sulfate).
2. Hot plate.
3. 50 % formamide/2× SSC three times, 5 min each.
4. Avidin-Texas Red to detect biotin-labeled DNA (Roche).
5. Biotin-conjugated goat-anti-Avidin to amplify biotin signal (Roche).

6. Sheep-antidigoxigenin-fluorescein (FITC) to detect Dig-labeled DNA probes (Roche).
7. Rabbit-anti-sheep-FITC to amplify Dig signal (Roche).
8. 5 μg/mL DAPI (4′, 6-diamidino-2-phenylindole) in Vectashield antifade solution (Vector Laboratories).

2.5 Microscopy and Image Capturing

1. Microscope equipped with high N.A. plan apochromatics and epifluorescence illumination with filter sets for DAPI, FITC, and Cy3 fluorescence.
2. High-resolution camera, e.g., Photometrics Sensys 1,305 × 1,024 pixel CCD camera, which is controlled by Genus Image Analysis Workstation software (Applied Imaging Corporation).
3. Adobe Photoshop.
4. ImageJ processing software (http://rsb.info.nih.gov/ij).

3 Methods

3.1 Flower Buds for Chromosome Preparations

Chromosome slides can be made with cell spreads from different tissue types containing cell complements at different mitotic or meiotic stages. In most cases anthers are used containing pollen mother cells at pachytene, the stage in meiotic prophase I in which homologous chromosomes are fully paired. Pachytene chromosomes are always much longer than mitotic metaphase counterparts resulting in a significant higher resolution when compared to metaphase chromosomes. Also, pachytene chromosomes display well-differentiated patterns of heterochromatin and euchromatin, which is most helpful for identifying specific chromosome regions [9].

1. Plants should be grown under such conditions that they abundantly flower. To find anthers in the appropriate stage of development, young flower buds are collected in the late morning and directly fixed in freshly prepared glacial acetic acid:ethanol 98 % (1:3) and left in the fixative for a few hours. It is recommended to use at least 20× the amount of fixative than plant material. Refresh the fixative if the solution turns yellowish green. The plant material can be stored in the fixative at −20 °C for several days. If storage for a longer time is needed, the fixative should be replaced by ethanol 70 %. Now the samples can be stored at 4 °C. In this condition buds can be kept for months.
2. To stage the pollen mother cells containing meiocytes, young anthers are selected that are still transparent or light green. These anthers are squashed in a drop of 45 % acetic acid on a microscopic slide covered with a 24 × 24 mm coverslip.

Eventually a solution of 1 % carmine in 45 % acetic acid can be used to counterstain.

3. The developmental stage of the anthers is checked using a phase contrast microscope. Pollen mother cells at meiosis are recognized by a large cytoplasm and are surrounded by a clear callose wall. Anthers that are too early in development have no differentiated cells, whereas anthers that are too late in development contain pollen grains. As an example, for *Medicago truncatula* we found that anthers of 2 mm in length contain meiotic cells [16].
4. To prepare microscope slides containing chromosome spreads suited for hybridization flower buds in the appropriate stage of development are rinsed twice in Milli-Q water and 10 mM sodium citrate buffer (pH 4.5), and subsequently digested in an enzyme mixture of 1 % cytohelicase, 1 % pectolyase Y-23, and 1 % cellulase RS in the citrate buffer at 37 °C for 1–3 h [16]. Enzyme activity may decrease in time, so it is worth monitoring the progress of the enzymatic digestion of the material. Anthers that are sufficiently digested do not show remnants of cell walls and thick cytoplasm of the cells in the chromosome slides, whereas too long exposure to the pectolytic enzyme mixture will result in broken cells and affected chromatin structure.
5. After enzyme treatment, the soft flower buds are carefully rinsed with Milli-Q water and kept on ice for further treatment. Flower buds are transferred individually to clean grease-free microscopic slides. Supporting tissues are removed with dissecting needles as much as possible. Immediately thereafter 2 μL Milli-Q of water is added and the anthers carefully dissected and homogenized with fine needles.
6. The cells are macerated by adding 80 μL of 60 % acetic acid, mixed well, and spread over a small circle of 1–2 cm across the slide while carefully heating the slide on a hot plate for 2 min at 50 °C. Freshly prepared ice-cold acetic acid:ethanol (1:3) is added in a circle around the cells and the slides are left to dry.
7. Every slide (without coverslip) is carefully screened under the microscope with high-resolution (N.A. > 0.5) phase contrast optics to determine the developmental stage of the anther pollen mother cells, the quality of chromosome spreading, the absence of cytoplasm, and the presence of well-differentiated heterochromatin. Carefully selecting high-quality slides is most crucial for a successful FISH experiment. If needed, duration and temperature of the maceration and cell spreading should be adapted until satisfactory results are obtained. Only the best slides should be selected and can be stored in a dust-free microscopic box at 4 °C.

3.2 DNA Isolation and Labeling

In the start-up phase of FISH experiments a number of test experiments can be carried out with the ubiquitous 5S, 45S rDNA, and telomere repeats as probes. We generally use clone pTA71, which contains a 9.1 kb fragment of 45S rDNA of wheat [13], clone pCT4.2, which contains a 5S rDNA of *Arabidopsis thaliana* [12], and clone pAtT4, which contains a telomeric repeat 5′-CCCTAAA-3′ of *A. thaliana* [14].

1. Plasmid DNA is isolated with a standard miniprep kit and labeled with either biotin-dUTP or Dig-dUTP by nick translation according to the manufacturer's manual.
2. For DNA isolation of BACs the alkaline lysate method is used [17] or any commercial plasmid DNA purification kit, e.g., for Midi preparation.
3. For indirect labeling, the isolated BAC DNA is labeled with either biotin-nick translation mix or Dig-nick translation mix. The biotin-labeled BAC probes are detected by Avidin-Texas Red and amplified by biotin-conjugated goat-anti-Avidin and Avidin-Texas Red. The Dig-labeled BAC probes are detected by sheep-anti-Dig-fluorescein (FITC) and amplified by rabbit-anti-sheep-FITC. For direct labeling, the isolated plasmid DNA is labeled with Cy3.5-dCTP, fluorescein-12-dUTP, or CyTM3-dUTP, without further detection and amplification steps.

3.3 Cot-100 Isolation and Labeling

The Cot-fraction of genomic DNA consists of highly repetitive tandem arrays (such as 45S rDNA, satellite repeats, and telomeres) and dispersed repeats (mostly LTR retrotransposons). In most plant species with small-to-moderate genome size such repetitive sequences will be enriched in the DNA Cot-100 fraction. Here we use a modified protocol described by Zwick et al. for preparing Cot-100 DNA [18]. The calculations should first be done for every new species. The Cot-DNAs thus obtained are first tested as a probe in FISH before using it as blocking agent.

1. Total genomic DNA is isolated using the cetyl trimethyl ammonium bromide (CTAB) method [15] and sonicated to fragment size of about 500 bp.
2. To produce Cot-100, the re-association temperature (Tre) has to be determined, which is: $Tre = Tm - 25$ (Tm: melting temperature). The Tm can be calculated by the following formula: $Tm = 81.5 + 16.6 \quad \log M + 41 \times (\%G + C) - 500/L - 0.62F$ in which M = molar concentration of monovalent cations (=salt concentration), $\%G + C = G + C$ fraction of the genomic DNA, L = length of sheared DNA, and F = molar formamide concentration. The re-annealing time (Ts) should be calculated with the following equation: $Ts = 100/\text{molar (M) DNA concentration}$. The molar concentration of the DNA can be calculated using the average molecular weight for a deoxynucleotide monophosphate: 339 g/mol (*see* **Note 1**).

3. The sheared genomic DNA is denatured at 95 °C for 10 min, and re-annealed at 65 °C for the calculated time period in a rotating oven. The remaining single-strand DNA is removed using 1 U/μg DNA S1 nuclease at 37 °C for 90 min.
4. The reaction is stopped and extracted by adding equal volumes of chloroform–isoamyl alcohol (24:1), mixed well, and centrifuged at 2,250 × *g* for 10 min. The upper layer is transferred into a new tube. DNA was precipitated with 2.5 volumes of ice-cold 100 % ethanol at −20 °C, overnight, and then centrifuged at 4 °C with 18,000 × *g* for 30 min.
5. DNA is air-dried and dissolved in 20 μL HB50 (50 % deionized formamide, 2× SSC, 50 mM sodium phosphate, pH 7).

3.4 FISH

The FISH protocol as described here is largely based on the protocols previously described by Zhong et al. and Kuliková et al. with few adaptations [16, 19].

1. For FISH with the 5S rDNA, 45S rDNA, telomere, and other repetitive sequences, the DNA is labeled with dUTPs conjugated with a fluorophore (FITC, Cy3, etc.), which requires no further probe detection and signal amplification procedures.
2. In case of BACs used as probe, addition of Cot-100 DNA as a competitor for blocking repeats that are present in the BACs may be required. Cot-100 DNA is added in a 50:1 ratio compared to labeled probe.
3. A hybridization mixture contains 20 ng labeled probe (plus 1 μg Cot-100 DNA if required) and 10 μL 20 % dextran sulfate in HB50 (50 % deionized formamide, 2× SSC, 50 mM 1 M sodium phosphate pH 7, 20 % dextran sulfate), and is diluted with HB50 to a total volume of 20–40 μL.
4. The hybridization mix containing the probe is transferred onto the chromosome slide, denatured on a hot plate at 80 °C for 3 min, and incubated in a moist chamber at 37 °C for at least 20 h.
5. After hybridization the slides are rinsed in 50 % formamide/2× SSC three times, 5 min each.
6. Probes labeled with Dig-dUTP are detected with sheep-anti-Dig-fluorescein (FITC) and amplified with rabbit-anti-sheep-FITC, resulting in green signal. Probes labeled with biotin-dUTP are detected with streptavidin CY3 and amplified with streptavidin biotin, resulting in red signal. For the directly labeled probes, the detection and amplification steps are omitted.
7. The slides are washed three times in 70 %, 90 %, and 100 % ethanol, respectively, and air-dried. Chromosomes are counterstained with 5 μg/mL 4′, 6-diamidino-2-phenylindole (DAPI) in Vectashield antifade solution (*see* **Note 2**).

3.5 Microscopy and Image Capturing

1. Slides are studied using a microscope equipped with high N.A. plan apochromatics and epifluorescence illumination with filter sets for DAPI, FITC, and Cy3 fluorescence.
2. Images are captured by a CCD specially designed for fluorescent images.
3. Images are slightly improved by changing the thresholds for removing background and enhancing contrast. DAPI images can be subtly sharpened with a 7 × 7 Hi-Gauss high-pass spatial filter or by de-convolution to accentuate minor details and heterochromatin differentiation of the chromosomes.
4. The different FISH signals are merged in a multichannel mode (Fig. 1). Fluorescent images are displayed in grey or light blue

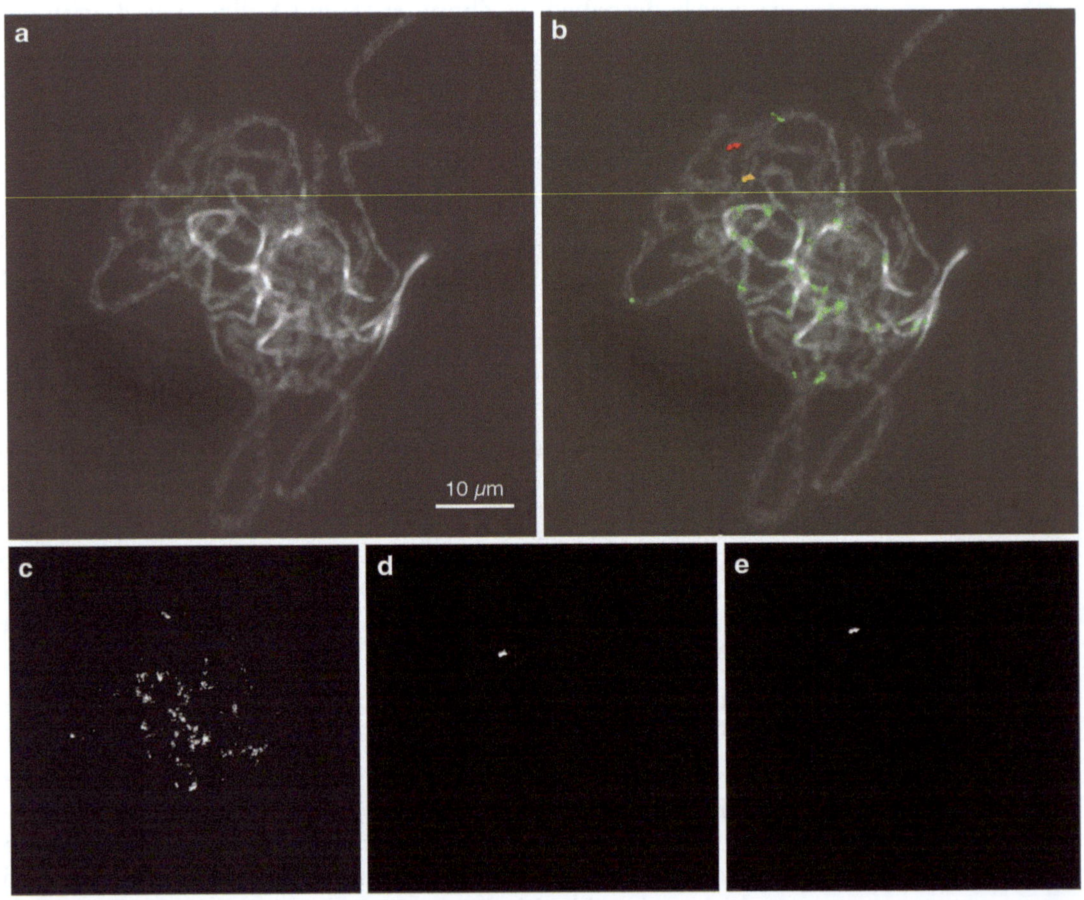

Fig. 1 Example of BAC-FISH painting on a pollen mother cell at pachytene of *Medicago truncatula*. (**a**) Well-spread pachytene chromosomes stained with DAPI; (**b**) three-color FISH with BACs labeled with the green fluorescing FITC (**c**), the orange fluorescing Cy3 (**d**), and the red fluorescing Cy3.5 (**e**), respectively. The individual BAC images were separately captured with a black/white camera and their images stacked in pseudo-colored spot channels using Adobe Photoshop software. Note that the green fluorescing BAC has multiple foci due to repetitive sequences in the BAC DNA

for DAPI and pseudo-colored for the other signals. The images can be processed in specialized image capturing software or in Adobe Photoshop CS. Open-source ImageJ image processing software can be sued for measurements and for straightening the chromosomes using the plug-in of Kocsis et al. [20].

5. For chromosome lengths and gap sizes the images of several different pachytene spreads have to be measured. The comparison is also important to establish variation in the heterochromatin patterns.

4 Notes

1. The choice of Cot-DNA for blocking requires some experience. For small-genome plants (like *Arabidopsis*) with low amounts of repetitive sequences in the euchromatin it is in general not needed to prepare Cot-DNA for blocking, but larger genomes, especially with BACs containing retrotransposons regions need Cot-DNA. It is recommended first to isolate different Cot fractions (Cot-1 Cot10, Cot100, Cot500) and use their DNA as probe in a FISH experiment to see where most of the repetitive sequences reside. This information can then be used to decide which Cot fraction is the best for blocking.
2. The same basic protocol can be used for cross-species chromosome painting, using BAC probes on the chromosomes of related species. In such cases hybridization and washing stringencies may be lowered if donor DNA in the probe and recipient DNA in the chromosome slide are not highly homologous. To do so, the formamide concentration is lowered from 50 to 35 % or 20 % [21].

References

1. Schmutz J, Cannon SB, Schlueter J, Ma J, Mitros T, Nelson W et al (2010) Genome sequence of the palaeopolyploid soybean. Nature 463:178–183
2. Young ND, Debellé F, Oldroyd GED, Geurts R, Cannon SB, Udvardi MK et al (2011) The Medicago genome provides insight into the evolution of rhizobial symbioses. Nature 480: 520–524
3. Sato S, Nakamura Y, Kaneko T, Asamizu E, Kato T, Nakao M et al (2008) Genome structure of the legume, *Lotus japonicus*. DNA Res 15:227–239
4. Varshney RK, Chen W, Li Y, Bharti AK, Saxena RK, Schlueter JA et al (2012) Draft genome sequence of pigeonpea (*Cajanus cajan*), an orphan legume crop of resource-poor farmers. Nat Biotechnol 30:83–89
5. Schubert I, Fransz PF, Fuchs J, de Jong JH (2001) Chromosome painting in plants. Methods Cell Sci 23:57–69
6. Sharma AK, Sharma A (2001) Chromosome painting—principles, strategies and scope. Methods Cell Sci 23:1–5
7. Kato A, Vega JM, Han F, Lamb JC, Birchler JA (2005) Advances in plant chromosome identification and cytogenetic techniques. Curr Opin Plant Biol 8:148–154
8. Jiang J, Gill BS (2006) Current status and the future of fluorescence in situ hybridization (FISH) in plant genome research. Genome 49:1057–1068

9. De Jong JH, Fransz PF, Zabel P (1999) High resolution FISH in plants—techniques and applications. Trends Plant Sci 4:258–263
10. Peterson DG, Schulze SR, Sciara EB, Lee SA, Bowers JE, Nagel A et al (2002) Integration of Cot analysis, DNA cloning, and high-throughput sequencing facilitates genome characterization and gene discovery. Genome Res 12:795–807
11. Yuan Y, San Miguel PJ, Bennetzen JL (2003) High-Cot sequence analysis of the maize genome. Plant J 34:249–255
12. Campell BR, Song YG, Posch TE, Cullis CA, Town CD (1992) Sequence and organization of 5s ribosomal RNA-encoding genes of *Arabidopsis thaliana*. Gene 112:225–228
13. Gerlach WL, Bedbrook JR (1979) Cloning and characterization of ribosomal-Rna genes from wheat and barley. Nucleic Acids Res 7: 1869–1885
14. Richards EJ, Ausubel FM (1988) Isolation of a higher eukaryotic telomere from *Arabidopsis thaliana*. Cell 53:127–136
15. Porebski S, Bailey LG, Baum BR (1997) Modification of a CTAB DNA extraction protocol for plants containing high polysaccharide and polyphenol components. Plant Mol Biol Rep 15:8–15
16. Kulikova O, Gualtieri G, Geurts R, Kim D-J, Cook D, Huguet T, de Jong JH, Fransz PF, Bisseling T (2001) Integration of the FISH pachytene and genetic maps of *Medicago truncatula*. Plant J 27:49–58
17. Woo SS, Jiang J, Gill BS, Paterson AH, Wing RA (1994) Construction and characterization *of bacterial artificial chromosome library of Sorghum bicolor*. Nucleic Acids Res 22: 4922–4931
18. Zwick MS, Hanson RE, Islam-Faridi MN, Stelly DM, Wing RA, Price HJ, McKnight TD (1997) A rapid procedure for the isolation of Cot-1 DNA from plants. Genome 40:138–142
19. Zhong XB, de Jong JH, Zabel P (1996) Preparation of tomato meiotic pachytene and mitotic metaphase chromosomes suitable for fluorescence in situ hybridization (FISH). Chromosome Res 4:24–28
20. Kocsis E, Trus BL, Steer CJ, Bisher ME, Steven AC (1991) Image averaging of flexible fibrous macromolecules: the clathrin triskelion has an elastic proximal segment. J Struct Biol 107: 6–14
21. Mandáková T, Lysak MA (2008) Chromosomal phylogeny and karyotype evolution in x=7 crucifer species (Brassicaceae). Plant Cell 20: 2559–2570

Chapter 3

Targeted Mutagenesis for Functional Analysis of Gene Duplication in Legumes

Shaun J. Curtin, Justin E. Anderson, Colby G. Starker, Nicholas J. Baltes, Dhananjay Mani, Daniel F. Voytas, and Robert M. Stupar

Abstract

Assessment of gene function oftentimes requires mutant populations that can be screened by forward or reverse genetic analysis. The situation becomes more complicated in polyploidy or paleopolyploid genomes that have two or more copies for most genes. Here we describe a method for engineering zinc-finger nucleases (ZFNs) for the purpose of creating targeted mutations in the paleopolyploid soybean genome. ZFNs are recombinant proteins composed of an engineered zinc-finger array fused to a nonspecific cleavage domain. When engineered to recognize a specific nucleotide sequence, the cleavage domain will generate highly mutagenic DNA double-strand breaks frequently resulting in insertions and deletions at the target locus. Depending on the number of target sites present within the genome, this method has the capacity to target either single- or multi-copy gene families. In this chapter, we describe an inexpensive, rapid, and user-friendly approach for ZFN assembly and application in soybean based on the previously described context-dependent assembly method.

Key words Legume, Soybean, ZFN, TALEN, Mutagenesis, Duplication, Polyploidy

1 Introduction

Assessing gene function in a plant species typically requires specific genetic resources, such as stable mutants or transgenic stocks, or the transient knockdown of gene expression using techniques such as virus-induced gene silencing (VIGS). Unfortunately, these stocks and methodological approaches have not been well developed for many crop species, including most legumes. Furthermore, assessing gene function in polyploid or paleopolyploid species presents an even greater challenge, as multi-mutant knockouts (double-, triple-, etc.) may be required to confer a phenotypic change. Targeted mutagenesis approaches may be particularly useful for polyploid or paleopolyploid species because (assuming that the DNA sequences of duplicated genes are reasonably conserved) it is possible to simultaneously target multiple members of homoeologous or otherwise

Ray J. Rose (ed.), *Legume Genomics: Methods and Protocols*, Methods in Molecular Biology, vol. 1069,
DOI 10.1007/978-1-62703-613-9_3, © Springer Science+Business Media, LLC 2013

paralogous gene families, resulting in a combination of single- and multi-gene mutations.

Targeted mutagenesis can be accomplished by using endonuclease-based genome engineering technologies, such as zinc-finger nucleases (ZFNs), transcription activator-like effector nucleases (TALENs), and meganucleases (MNs). These customizable endonucleases can be used to generate DNA sequence modifications at specific targeted loci. Modifications may include knockout mutations, edited genes, site-specific transgene insertions, and targeted structural changes (deletions, inversions, and translocations) [1]. When delivered to plant cell nuclei, nucleases generate double-strand breaks (DSBs) at a targeted locus, which can be repaired by the error-prone nonhomologous end joining (NHEJ) pathway, or by the more precise homologous recombination (HR) pathway. Due to a strong preference for DSBs to be repaired by the NHEJ pathway, nucleases can direct mutations within a plant genome. The current procedure for generating plants (including soybean) with a nuclease-induced DNA sequence modification requires four steps: (1) development of an engineered nuclease construct, (2) delivery of the construct into the plant's genomic DNA, (3) induction of nuclease expression, and (4) identification of plants carrying the modified gene(s) [1].

Genome engineering technologies have been developed and refined in several model organisms, and a long series of landmark developments have been published in recent years [2–6]. Furthermore, there have been recent efforts to utilize genome engineering technologies in crop species [5, 7–13], with the promise that they may serve as a new crop improvement strategy. To date, soybean (*Glycine max*) is the only legume species that has experienced targeted DNA sequence modifications using a genome engineering approach. Specific genes have been successfully targeted for mutagenesis in somatic [6] and germline [12] cells of soybean using engineered ZFNs.

This chapter describes established protocols for ZFN-based mutagenesis in soybean for either single-copy or duplicated genes. We focus the method on construct design and hairy root transformation. We present the current state of the art in ZFN construct design, which is based on a string of technological advances that have been previously described [5, 6, 12, 14–16]. The hairy root protocol is presented as a rapid means to screen for ZFN-induced mutations in somatic cells. The procedure is highly adaptable for whole-plant mutagenesis; however, we do not describe the whole-plant transformation protocol here. Soybean whole-plant transformation protocols can be found in previous publications [12, 17].

The approaches presented can be translated to any plant species capable of genetic transformation that has a sequenced (or partially sequenced) genome. At the time of this submission, the legume genomes that have been sequenced include *Lotus japonicus* [18], soybean [19], *Glycine soja* [20], *Medicago truncatula* [21], pigeon pea (*Cajanus cajan*) [22], and common bean (*Phaseolus vulgaris*) (available online at http://mazorka.langebio.cinvestav.mx/phaseolus/ and www.phytozome.org). The methods presented here will certainly advance beyond the current state by the time additional sequenced legumes and other polyploidy plant genomes emerge. We anticipate that major advancements will occur within the realm of enzyme design for targeted double-stranded breaks, such as the recent developments that have been made in TALEN technologies [23, 24]. Furthermore, it is likely that significant advances will be made in modulating double-stranded break repair, resulting in higher frequencies of mutagenesis and/or gene editing. This chapter presents a "user guide" to the current methods used in soybean-targeted mutagenesis while assuming that this type of approach will soon become more efficient, flexible, and accessible to a wider range of taxa.

2 Materials

1. Restriction enzymes XbaI, ApaI, BamHI-HF, NheI, NheI-HF, BglII, XhoI, SpeI, EcoRI-HF, KpnI-HF, BSA, and their buffers (New England Biolabs).
2. Plasmid DNA for the FokI expression vector (4R aa linker), the FokI expression vector (8 aa linker), middle finger units or finger 2 (F2) templates in pGem-T Easy, and the inducible 218 binary vector (for estrogen-inducible transgene expression) can be ordered through Addgene (http://www.addgene.org/) [12].
3. 1.0, 1.5, and 2.0 % agarose solution in 500 mL of 0.5× TBE with 25 μL of 10 mg/mL ethidium bromide for various gels.
4. QIAquick PCR Purification Kit (Qiagen).
5. QIAquick gel extraction kit (Qiagen).
6. DNeasy Plant Mini Kit (Qiagen).
7. QIAprep Spin Miniprep Kit (Qiagen).
8. T4 ligase with 2× ligation buffer (Promega).
9. DH5α competent cells.
10. Fast Cycling PCR kit (Qiagen).
11. Hot Start Taq*Plus* kit (Qiagen).
12. KOD Hot Start DNA Polymerase (EMD Chemicals Inc.).
13. All-purpose bleach (5.25 % hypochlorite).
14. 12 N Hydrochloric acid.
15. *Agrobacterium rhizogenes* strain (K599).

16. Spectinomycin stock: Dissolve 1 g of spectinomycin in 20 mL of dH_2O. Filter sterilize, aliquot out into micro-centrifuge tubes, and store at −20 °C.
17. Kanamycin stock: Dissolve 1 g of kanamycin in 20 mL of dH_2O. Filter sterilize, aliquot out into micro-centrifuge tubes, and store at −20 °C.
18. Carbenicillin stock: Dissolve 10 g of carbenicillin in 20 mL of dH_2O. Filter sterilize, aliquot out into micro-centrifuge tubes, and store at −20 °C.
19. LB spectinomycin or kanamycin plates: 25 g of LB broth (10 g/L tryptone, 5 g/L yeast extract, 10 g/L NaCl) in 1 L dH_2O and 15 g/L granulated agar. Autoclave the medium, cool to 65 °C, and add 1 mL of spectinomycin or kanamycin stock.
20. 50 mM stock solution of 17β-estradiol: 134 mg (Sigma-Aldrich) in 10 mL of 100 % ethanol. Aliquot into tubes and store at −20 °C. Caution: A chemical fume hood should be used when handling 17β-estradiol.
21. LB liquid medium with spectinomycin or kanamycin: 25 g of LB broth in 1 L dH_2O. Autoclave the medium, cool to at least 65 °C, and add 1 mL of spectinomycin or kanamycin stock.
22. ¼ MS solid media plates for soybean germination: 1.1 g Murashige and Skoog basal salt mixture and 750 mg/L $MgCl_2$ in dH_2O. Adjust pH to 5.8 with 1 M KOH solution and add 8 g/L granulated agar. Autoclave the medium.
23. ¼ MS liquid media for *Agrobacterium* resuspension: 1.1 g Murashige and Skoog basal salt mixture and 750 mg/L $MgCl_2$ in dH_2O. Adjust pH to 5.8 with 1 M KOH solution. Autoclave liquid medium.
24. ¼ MS solid media plates for transformed cotyledons: 1.1 g Murashige and Skoog basal salt mixture and 750 mg/L $MgCl_2$ in dH_2O. Adjust pH to 5.8 with 1 M KOH solution and add 8 g/L granulated agar. Autoclave the medium, cool to 65 °C, and add carbenicillin 500 mg/L and 10 μM 17β-estradiol. Store at 4 °C for no more than 4 days. Caution: A chemical fume hood should be used when handling 17β-estradiol.

3 Methods

3.1 Using the Web-Based Tool ZiFit to Identify Potential ZFN Targets

1. Select a gene of interest (GOI) using the "BROWSE Genome" or "BLAST Genome" function of Phytozome (http://www.phytozome.net/soybean). Highlight and copy genomic DNA sequence.
2. Open a tab at http://zifit.partners.org/. Select "ZiFiT" and then "Proceed to ZiFiT". Under context-dependent assembly (CoDA) click "Design Zinc Finger Nucleases" and uncheck

"Exon/Intron Case Sensitivity". Paste genomic DNA sequence of GOI and "Submit" (*see* **Note 1**).

3. On the main page under "Sort By" will be a list of ZFN target sites. The colored DNA triplets indicate the individual DNA-binding sites for each of the three fingers in the zinc-finger array (ZFA). The lower strand of colored bases is where the left array will bind and the upper strand is where the right array will bind (*see* **Note 2**).
4. When selecting a ZFN it is worth considering the future screening assay. For the PCR enrichment assay, a restriction enzyme that encompasses the part of the ZFN target site is required. Copy the 25–27-bp target and paste into a DNA sequence analysis tool that finds restriction enzyme sites such as NEB cutter. Verify potential off-targeting sites of the ZFN by querying the soybean genome with the ZFN target sequence.
5. Return to ZiFiT and the list of ZFN candidates. Click the "+" next to a ZFN candidate to expand the design parameters. Collect and collate finger and recognition helix data into a spreadsheet for future analysis and record keeping.
6. At this point, the user has two options (Fig. 1):

   Option A: The ZFAs can be used according to the ZiFit output. Click on the blue-colored "ZF DNA sequence" link to obtain a pop-up window containing the full DNA sequence encoding both left and right arrays. If this option is selected, then skip to **step 11**.

   Option B: ZFAs can be modified to utilize the preferred codon usage of soybean (*see* **steps 7–10**).

3.2 Optional: ZFA Modification for the Preferred Codon Usage of Soybean

7. ZFAs can be modified to utilize the preferred codon usage of soybean. This method uses a Perl script designed to generate the complete sequence of individual ZFAs.

   For PC: Download Perl Programming Language. Go to http://www.perl.org/ and click "Download Perl". Click "ActiveState Perl" link "Download ActivePerl 5.14.2 for Windows".

   For Mac: Confirm Perl Programming language is loaded on your device. Navigate to Applications folder, and then Utilities, and click on "Terminal". In the window that opens type "perl –v" and then press return. The window should reply "This is perl..." and your current version. If not follow PC directions above to "Download ActivePerl 5.14.2 for Mac OS X."

8. Open the Stupar Lab Homepage at http://stuparlab.cfans.umn.edu/. Click on the "Protocols" tab, followed by "ZFN Creator". Click "Perl program for ZFA creator" to download program for soybean preferred codon usage. Copy the program "ZF_DNA_Sequence_SoyCodons.pl" from the "Downloads" folder and paste to the desktop.

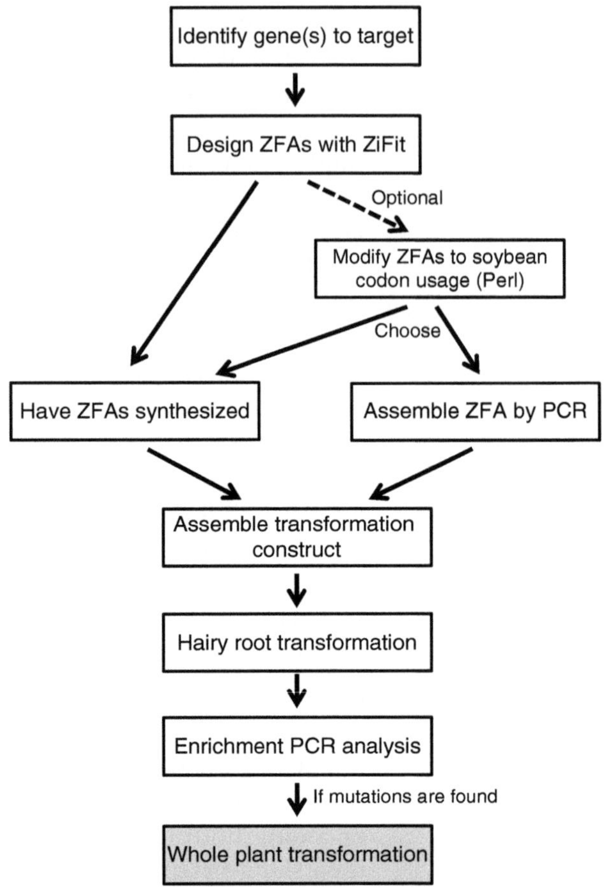

Fig. 1 The stepwise process of design, assembly, and testing of a CoDA ZFN in soybean

9. *For PC*: On the desktop double click on "ZF_DNA_Sequence_SoyCodons.pl". A programming window will appear with "Enter output file name:" Input the name of the intended ZFA and press return.

   For Mac: Navigate to Applications folder, and then Utilities, click on "Terminal" to open the terminal window, and type "perl " (be sure to include a space after perl). Drag the "ZF_DNA_Sequence_SoyCodons.pl" file from your desktop into the Terminal window and press "Enter". Input the name of the intended ZFA data file in the programming window after "Enter output file name:" and press return.

10. Return to the ZiFIT results page from the http://zifit.partners.org/website and input the amino acid recognition helices for each finger into the perl script (e.g., Left F1 Helix (i.e., KRQHLEY), press "Enter," and continue with the remaining recognition helices). Press enter and an output text file, with your title, will appear on the desktop containing the ZFA

sequence and primers information. Users can choose to have the ZFA commercially synthesized (**step 11**) or assemble the ZFA by PCR (**steps 12–14**). Repeat **steps 9** and **10** to obtain sequences for the right ZFA.

3.3 Commercial Synthesis of ZFAs

11. The ZiFit output or the Perl script output can both be used to order synthesized ZFAs (~300 bp each) using a commercial gene synthesis provider. If this option is chosen, then skip to **step 15**.

3.4 PCR Assembly of ZFAs

12. The Perl script output contains F1 forward and F3 reverse primer sequences for each ZFA. Order these primers for PCR reaction in **step 13**.
13. Carry out PCR using Qiagen Fast Cycling PCR kit. Combine the following reagents into a 200 μL tube: 10 μL Qiagen Fast Cycling PCR mix, 0.5 μL F1 and F3 of combined forward and reverse primer pairs (5 μM), 0.5 μL (25-50 ng) F2 template, and 9.5 μL dH_2O (Fig. 2) (*see* **Note 3**).

Cycling conditions	
1 cycle	95 °C for 5 min
35 cycles	96 °C for 5 s 60 °C for 5 s 68 °C for 6 s
1 cycle	72 °C for 1 min

*Spike PCR reaction after ten cycles with 1.5*μL *of 10*μM *primer pair* (*see below*) *to extend the ZFA fragment with the correct restriction sites* (Fig. 2) (*see* **Note 4**).

For the left-ZFA and right-ZFA—4R aa linker (5- and 6-bp spacer)

XbaI.ZFA F1-AGTGGTTGG TCTAGACCCG GGGAGCGCC CCT TCCAGTGTCGCATTTGCATGCGGAACTTT

BamHI.ZFA R1-TTCAGATTTCACTAGCTGGGATCCCCTCAGG TGGGTTTTTAGGTG

14. Run 5 μL of the PCR aliquot on a 1.5 % gel to confirm successful PCR. PCR purify the remaining 15 μL sample using a Qiagen PCR Purification Kit. Elute the PCR using 23 μL of the EB buffer.

3.5 Assembly of the Transformation Construct

15. Digest both the left and right ZFAs (purified PCR-assembled product or commercially synthesized product) using XbaI and BamHI-HF restriction endonucleases. Digest 23 μL of PCR-purified product by combing 3 μL buffer, 3 μL 10× BSA, 0.5 μL XbaI, and 0.5 μL BamHI-HF (for a final reaction volume of 30 μL); incubate for 3 h at 37 °C (*see* **Note 5**).

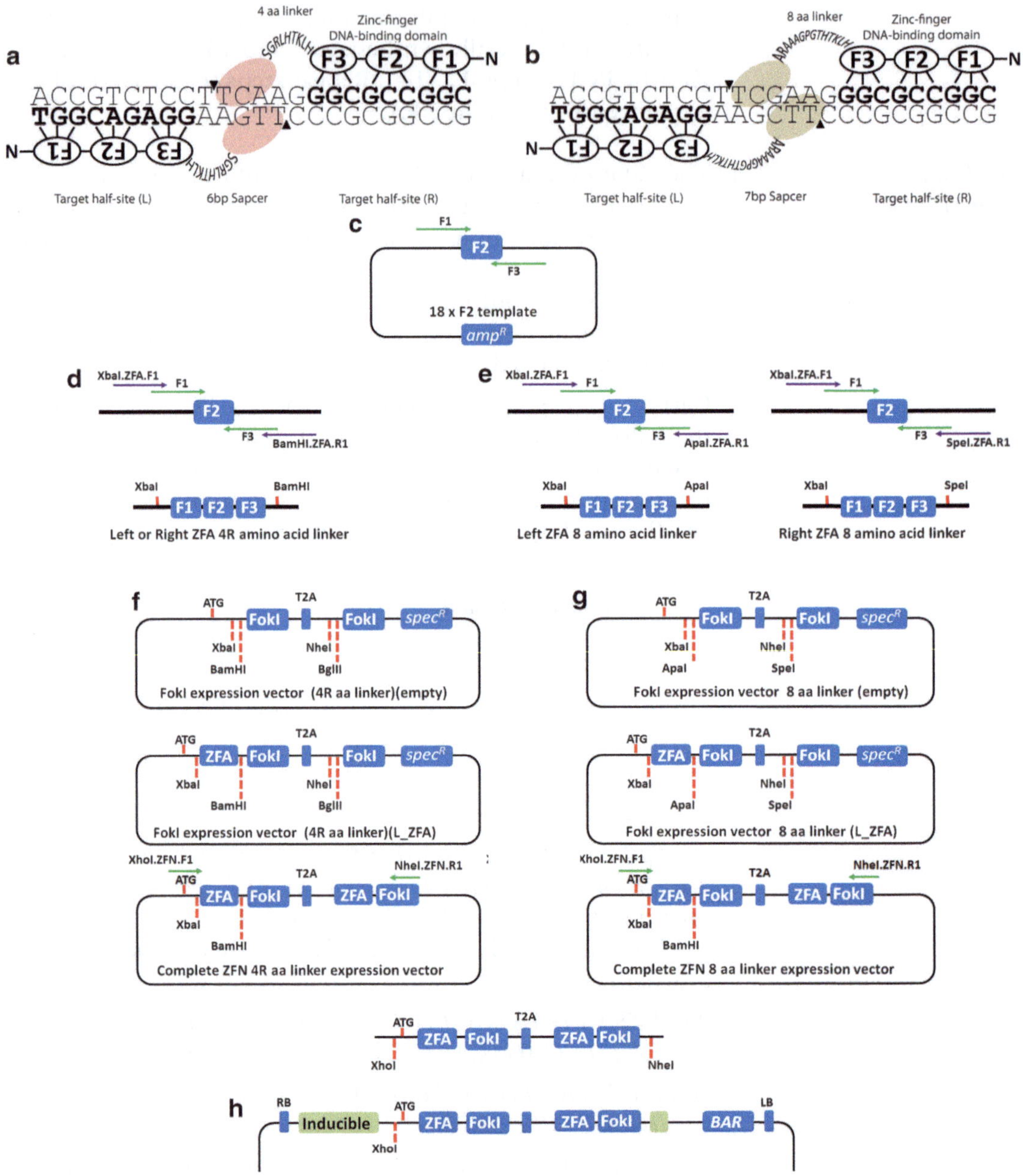

Fig. 2 The assembly of the CoDA ZFN. (**a**) A representation of a ZFN bound to DNA with the 4R aa inter-domain linker used for target sites with a 5 or a 6 bp spacer. (**b**) A ZFN bound to DNA with the 8 aa inter-domain linker used for target sites with a 7 bp spacer. (**c**) The CoDA platform has 18 distinct common middle fingers (F2) which are used as templates to build ZFAs by PCR. (**d**) A schematic representation of the PCR assay used to build both left and right ZFAs for the 4R aa linker FokI expression vector. The forward and reverse primers each encodes respective F1 unit and F3 unit (indicated by *green arrows*). The CoDA platform has potentially 319 N-terminal F1 and 344 C-terminal F3 fingers. After ten cycles of PCR, the reaction is spiked with a primer pair (indicated by the *purple arrows*) to extend to the array with XbaI and BamHI restriction sites for downstream cloning. (**e**) The spiking reaction is different for the 8 aa linker left and right arrays and uses the primers with XbaI/ApaI and XbaI/SpeI, respectively. (**f**) The ZFN is constructed by first ligating the left array into the FokI expression (4R aa linker) vector followed by the right array. (**g**) The same strategy is carried out for the FokI expression vector (8 aa linker). (**h**) The complete ZFN is then cloned into a binary vector of choice and transformed into *Agrobacterium*

16. In a separate reaction linearize the FokI expression vector by digestion with XbaI and BamHI-HF. Combine the following reagents into a 200 μL tube: 1 μg of FokI expression vector, 3 μL buffer, 3 μL 10× BSA, 0.5 μL XbaI, 0.5 μL BamHI-HF, and sterile water up to a final volume of 30 μL. Incubate for 3 h at 37 °C and gel purify the backbone (~4.0 kb) using QIAquick gel extraction kit (*see* **Note 6**).
17. Use a Promega ligation kit to ligate the left ZFA into the XbaI/BamHI sites of the linearized FokI expression vector. Combine the following reagents into a 200 μL tube: 5 μL of 2× ligation buffer, 1.5 μL of XbaI/BamHI linearized FokI expression vector, 3.0 μL left ZFA XbaI/BamHI gel eluted fragment, and 0.5 μL T4 ligase. Incubate ligation overnight at 16 °C or at room temperature for 1–2 h.
18. Transform ligation reaction into *E. coli* (DH5α) by combining ~40 μL chemically competent cells with 5 μL of the ligation reaction into a 1.5 mL micro-centrifuge tube. Incubate on ice for 5 min, and heat shock for 30 s at 42 °C. Return to ice for 5 min and add 300 μL LB. Incubate and shake for 30–60 min at 37 °C. Plate 300 μL of the transformation on LB/spec50 (LB agar plates supplemented with 50 mg/L spectinomycin) and incubate overnight at 37 °C (*see* **Note 7**).
19. Select three single colonies from the transformation and inoculate 4 mL cultures of LB (supplemented with 50 mg/L spectinomycin). Incubate and shake overnight at 220 rpm, 37 °C. On the following day purify plasmids from the LB cultures using a QIAprep spin miniprep kit.
20. Confirm successful ligation of the left ZFA array by a restriction endonuclease diagnostic assay using XbaI and BamHI-HF to identify desired plasmids containing the left ZFA (~300 bp). Combine the following reagents into a 200 μL tube: 10 μL of left ZFA/FokI expression plasmid from **step 19**, 3 μL buffer, 3 μL 10× BSA, 0.25 μL XbaI, and 0.25 μL BamHI-HF in sterile water to final volume of 30 μL. Incubate for 1 h at 37 °C and run samples on a 1.0 % gel to confirm L-ZFA insertion.
21. The recombinant plasmid encoding the left ZFA-FokI expression vector is digested with NheI and BglII to produce XbaI/BamHI-compatible DNA ends for the ligation of the right ZFA from **step 15**. Combine the following reagents into a 200 μL tube: 10 μL of L-ZFA/FokI expression plasmid, 3 μL buffer, 3 μL 10× BSA, 0.5 μL NheI, and 0.5 μL BglII in sterile water to final volume of 30 μL. Incubate for 3 h at 37 °C and gel isolate the left ZFA/FokI expression vector (~4.4 kb) using a QIAquick gel extraction kit.
22. Use a Promega ligation kit to ligate the right ZFA XbaI/BamHI fragment into sites of the NheI/BglII linearized left ZFA/FokI expression vector. Combine the following reagents into a 200 μL tube: 5 μL of 2× ligation buffer, 1.5 μL of

NheI/BglII linearized left ZFA/FokI expression vector, 3.0 μL right ZFA XbaI/BamHI gel eluted fragment, and 0.5 μL T4 ligase. Incubate ligation overnight at 16 °C or at room temperature for 1–2 h (*see* **Note 8**).

23. Transform the ligation into *E. coli* (DH5α) and spread on LB plates (supplemented with 50 mg/L spectinomycin).
24. Select three single colonies from the transformation and inoculate 4 mL cultures of LB (supplemented with 50 mg/L spectinomycin). Incubate and shake overnight at 220 rpm, 37 °C. On the following day purify plasmids from the LB cultures using a QIAprep spin miniprep kit.
25. Confirm successful ligation of the right ZFA array by a restriction endonuclease diagnostic assay using KpnI-HF and EcoRI-HF to visualize a (~624-bp) fragment confirming the incorporation of the right ZFA. Combine the following reagents into a 200 μL tube: 10 μL of completed ZFN cassette from **step 24**, 3 μL buffer, 3 μL 10× BSA, 0.25 μL EcoRI-HF, and 0.25 μL KpnI-HF in sterile water to final volume of 30 μL. Incubate for 1 h at 37 °C and run samples on a 1.0 % gel.
26. Verify the DNA sequence of both left and right ZFAs in the completed ZFN cassette by a sequence reaction with the following primers:

    ZFN R1 5′-CACATATTTCAATTTATGACG-3′

    ZFN R2 5′-TTGAGCTTGTGGCGCAGCTCG-3′

27. Upon sequence confirmation of the completed ZFN cassette, PCR amplify the ZFN cassette with the primers below to obtain an ~2,020 bp PCR amplicon encompassing the complete ZFN open reading frame. Combine the following reagents into a 200 μL tube (*see* **Note 9**).

    32.5 μL dH_2O 5 μL 10× Buffer for KOD hot start DNA polymerase.

    3 μL 25 mM $MgSO_4$.

    5 μL dNTPs (2 mM each).

    1.5 μL *XhoI.ZFN.F1 5′-GATCACTCGAGATGGATTATAAG GATCACGATGGCGACTACAAGG-3′* (10 μM).

    1.5 μL *NheI.ZFN.R1 5′-GATCGCTAGCTCAATAAAAG TTTATCTCGCCGTTATTAAATTTCC-3′* (10 μM).

    0.5 μL template completed ZFN cassette from **step 26**.

    1 μL KOD hot start DNA polymerase.

Cycling conditions	
1 cycle	95 °C for 2 min
30 cycles	95 °C for 20 s 60 °C for 10 s 70 °C for 40 s

28. Run 25 μL of the PCR product from **step 27** on a 1.0 % gel to confirm successful PCR. Gel purify the ~2-kb fragment using a QIAquick gel extraction kit.

29. Digest the gel isolated PCR product with XhoI and NheI-HF to produce XhoI- and SpeI-compatible DNA ends for ligation into the inducible binary vector. Combine the following reagents into a 200 μL tube: 23 μL of gel-isolated PCR product from **step 28**, 3 μL buffer, 3 μL 10× BSA, 0.5 μL XhoI, and 0.5 μL NheI-HF in sterile water to final volume of 30 μL. Incubate for 3 h at 37 °C.

30. Linearize the inducible binary vector with XhoI and SpeI. Combine the following reagents into a 200 μL tube: 10 μL of inducible binary vector, 3 μL buffer, 3 μL 10× BSA, 0.5 μL XhoI, and 0.5 μL SpeI in sterile water to final volume of 30 μL. Incubate for 3 h at 37 °C.

31. Use a Promega ligation kit to ligate the ZFN cassette amplicon (~2,020 bp) with XhoI/NheI-HF ends to the compatible XhoI/SpeI sites of the inducible binary vector. Combine the following reagents into a 200 μL tube: 5 μL of 2× ligation buffer, 1.5 μL of XhoI/SpeI linearized inducible binary vector, 3.0 μL gel-isolated XhoI/NheI-HF-digested ZFN cassette from **step 29**, and 0.5 μL T4 ligase. Incubate ligation overnight at 16 °C or at room temperature for 1–2 h.

32. Transform the ligation into *E. coli* (DH5α) and spread on LB plates (supplemented with 50 mg/L kanamycin).

33. Select three single colonies from the transformation and inoculate 10 mL cultures of LB (supplemented with 50 mg/L kanamycin). Incubate and shake overnight at 220 rpm, 37 °C. On the following day purify plasmids from the LB cultures using a QIAprep spin miniprep kit (*see* **Note 10**).

34. Perform diagnostic digestion of purified plasmids with EcoRI-HF (*see* **Note 11**). Combine the following reagents into a 200 μL tube: 15 μL of inducible binary vector, 3 μL buffer, and 0.5 μL EcoRI-HF in sterile water to final volume of 30 μL. Incubate for 1 h at 37 °C and run samples on a 1.0 % gel.

35. Verify the DNA sequence of the inducible ZFN binary vector using the following sequencing primers and confirm correct open reading frame:

    LexA F1 5′-TGGATATGTATATGGTGGTAATGC-3′

    pea3A R1 5′-ATGCATTGAACTTGACGAACG-3′

    ZFN R1 5′-CACATATTTCAATTTATGACG-3′

36. Transform into *Agrobacterium rhizogenes* K599 strain using standard electroporation techniques. Transformed *Agrobacterium* cells are spread onto LB plates with kanamycin (50 mg/L) and incubated at 28 °C for 48 h.

37. Perform colony PCR of three colonies to confirm transformation of the plasmid into *Agrobacterium*. Combine the following reagents into a 200 μL tube: 10 μL Hot Start Taq *Plus* master mix, 1 μL of the combined primers (below) (10 μM), 9 μL sterile water:

BAR_F1 5′-CATCGTCAACCACTACATCGAGACAA-3′

BAR_R1 5′-GAAACTTTATTGCCAAATGTTTGAACG A-3′

Cycling conditions	
1 cycle	95 °C for 5 min
35 cycles	94 °C for 30 s 56 °C for 30 s 72 °C for 60 s
1 cycle	72 °C for 10 min

38. Inoculate three single colonies from the LB/kanamycin (50 mg/L) plates into three 10 mL sterile tubes of liquid LB medium supplemented with kanamycin (50 mg/L). Incubate and shake overnight at 28 °C, 220 rpm. Prepare sample glycerol stocks and store at −80 °C until needed (*see* **Note 12**).

3.6 Preparation of Soybean Cotyledons for Hairy-Root Transformation

39. A soybean cotyledon transformation for hairy-root production is carried using a modified protocol from the Donald Danforth Plant Science Center (Kevin Lutke, Christopher G. Taylor, and Manjula Govindarajulu). Seed of the soybean cv. Bert is surface sterilized with chlorine gas for 16–24 h. Disinfected seeds are placed (six per plate) on ¼ MS solid media and germinated for 7 days at 24 °C under 18:6-h photoperiod (~150 μmol/s/m^2).
40. An *Agrobacterium* culture is prepared (10 mL LB; 50 mg/L kanamycin), inoculated from a glycerol stock harboring an appropriate ZFN binary vector, and incubated at 28 °C for 2 days. The bacterial culture is pelleted at 3000 × *g* for 10 min, resuspended, and made up to a volume to 50 mL with ¼ MS liquid media. A spectrophotometer is used to confirm the OD_{600nm} concentration of the re-suspended cultures. Adjust OD_{600nm} to approximately 0.2–0.3 and transfer solution to a labeled 90 × 25 mm plate.
41. 7-day-old cotyledons are harvested and prepared for transformation by cutting between the hypocotyl and the half-way point of the cotyledon and transferred to labeled plates containing the liquid culture. The seed coat is removed if necessary. The plates are placed in a vacuum chamber and a vacuum is drawn for 5 min and held under vacuum for a further 20 min.

Release the vacuum and place cotyledons flat side up on sterile 70 mm filter paper inserted into a 90 × 25 mm plate (the filter paper should be wet from the liquid culture). Plates are wrapped with Parafilm (Pechiney Plastic Packaging, IL) and incubated in the dark at 28 °C for 3 days (*see* **Note 13**).

42. Cotyledons are removed from the filter paper, placed in a fresh 90 × 25 mm plate with 50 mL of ¼ MS liquid media with carbenicillin (500 mg/L), and gently agitated for 30 min.
43. After the wash, the cotyledons are inserted aseptically into ¼ MS solid media (carbenicillin 500 mg/L; 10 μM 17β-estradiol). The estradiol in the media is used to induce ZFN transgene expression. The estradiol is labile and breaks down in tissue culture after 2–3 days limiting ZFN expression and associated toxic side effects. The cut surface of the cotyledon faces out of the media. Six cotyledons are arranged per plate and wrapped with Parafilm. Plates are incubated for 14 days at 24 °C under 18:6-h photoperiod (~150 μmol/s/m^2) until hairy-roots appear from the cotyledons.

3.7 PCR Enrichment Assay for Verification of Targeted Mutations

44. Hairy-roots are harvested from cotyledons approximately 14 days after transformation and flash frozen with liquid nitrogen. DNA is extracted from the hairy-roots using a DNeasy Kit according to the manufacturer's instruction (*see* **Note 14**).
45. Digest 100 ng of genomic DNA from individual or pooled hairy-root samples with the restriction enzyme that recognizes the DNA sequence in the ZFN spacer (*see* **step 4**). This step digests the wild-type DNA sequences, thereby enriching the DNA template with (newly) mutated sequences (Fig. 3). Combine the following reagents into a 200 μL tube: 100 ng genomic DNA, 3 μL buffer, 3 μL 10× BSA, and 1.0 μL the appropriate restriction enzyme in sterile water to final volume of 30 μL. Incubate for 3 h at 37 °C.
46. PCR primers are carefully designed to amplify the target locus ensuring that they are close enough to obtain reliable DNA sequence for downstream cloning and analysis. Primers are typically positioned approximately 200 bp from the target site (Fig. 3a) (*see* **Note 15**).
47. Perform PCR on the digested genomic DNA with ZFN target-specific primers.

    10 μL Hot Start Taq *Plus* master mix.

    1 μL Primer pair (10 μM).

    5 μL Digested hairy-root genomic DNA.

    4 μL dH_2O.

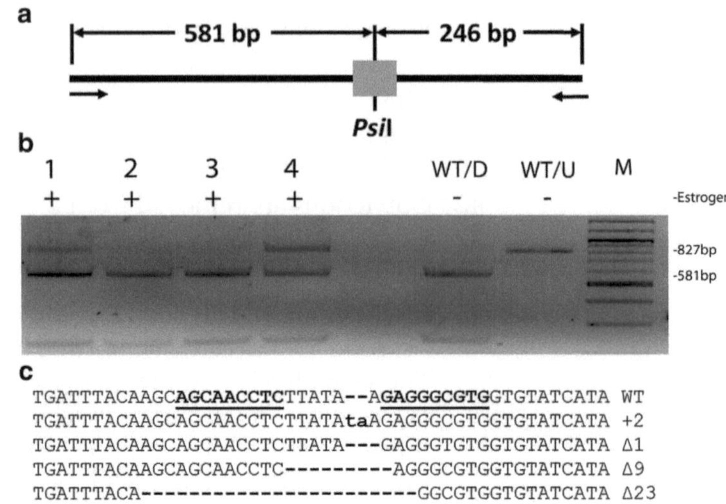

Fig. 3 Detection of ZFN-induced mutations in soybean hairy-root tissue. (**a**) A schematic strategy highlighting an example of a restriction endonuclease PCR assays used to enrich mutated DNA sequences from soybean hairy-root tissue. A ZFN was designed to target a soybean locus and was screened with a primer set. Mutated sequences that no longer have the PsiI restriction site will be resistant to the PCR product digestion. (**b**) The PCR products from four root samples, including an undigested (WT/U) and digested (WT/D) wild-type control, were separated on a 2 % agarose gel. *Lanes* showing undigested products (*lanes 1* and *4*) indicate that a portion of the hairy-root cells have novel mutations at the restriction sites. (**c**) Sequences of ZFN-induced mutations in soybean hairy-root tissues

Cycling conditions	
1 cycle	95 °C for 5 min
35 cycles	94 °C for 30 s 56 °C for 30 s 72 °C for 60 s
1 cycle	72 °C for 10 min

48. Run 5 μL of the PCR aliquot on a 1.0 % gel to confirm successful PCR. PCR purify the remaining 15 μL sample using a QiagenPCR Purification Kit. Elute the purified PCR product using 23 μL of the EB buffer.
49. Post PCR, digest the purified PCR product using the same restriction enzyme used for enrichment (**step 45**). Combine the following reagents into a 200 μL tube: 23 μL of PCR-purified product, 3 μL buffer, 3 μL 10× BSA, and 1.0 μL the appropriate restriction enzyme in sterile water to final volume of 30 μL. Incubate for 3 h at 37 °C.
50. Separate digested PCR product on a 2 % gel to visualize potential restriction enzyme-resistant PCR fragments. The presence

of digestion-resistant PCR fragments indicates the potential disruption of the enzyme's recognition site by ZFN-induced NHEJ mutations (Fig. 3b) (*see* **Note 16**).

51. Confirm ZFN mutations by cloning the digested PCR reaction into a PCR cloning vector of choice. Perform colony PCR on individual colonies and sequence the PCR product. Analyze for mutations at the target site to confirm ZFN functionality (Fig. 3c).

4 Notes

1. A ZiFiT Graphic Summary window will appear with small color bars above a large red one. The graphic summary pop-up represents targets along the gene (red bar). ZFN sites in the graphic are color-coded based on ZFN target spacer size (5 bp = blue; 6 bp = green; 7 bp = gold). The 5, 6, and 7 bp refer to the size in nucleotides of the spacer region between the left and right ZFA-binding sites. Targets with 5 or 6 bp spacer sites require a ZFA with an inter-domain linker of 4 amino acids (aa). The ZFA inter-domain linker is defined as the number of aa between the last conserved histidine of the third zinc-finger and the first residue of the FokI cleavage domain [25]. We have developed FokI expression vectors with both the 4 and 8 aa inter-domain linkers suitable for 5–6 bp and 7 bp spacer target sites, respectively. However to date we only have data confirming the function of the 4 aa linker (4R aa linker; *see* [25]). ZFNs using the 4R linker are preferred until future confirmation of the 8 aa linker in legumes.
2. Quality check ZFN target sites by verifying whether the ZFN target site hits an exon region of the GOI. Copy the 25–27 bp target from the upper strand. Return to your gene's genomic sequence at http://www.phytozome.net/soybean. Click Edit → Find and paste your 25–27-bp target to find the target region. Determine if your ZFN target site falls in an exon region. Confirm ZFN target consists primarily of GNN codons and a limited number of "T"s since they appear to have weaker binding affinities. An array with less than four "T"s is recommended.
3. A Qiagen Fast Cycling PCR was used to generate the ZFAs since the reaction takes approximately ~35 min. The extension of the ZFA with XbaI and BamHi restriction sites is carried out by spiking the reaction with the appropriate primer pair 10 min into the reaction. This reaction has been successfully carried with various polymerases including Pfu Turbo (Strategene).
4. Engineering ZFAs with an 8 aa inter-domain linker for ZFN target sites with 7-bp spacers requires a specific set of primer

pairs for the extension of the ZFAs with the appropriate restriction sites.

For the left ZFA—8 aa linker (7-bp spacer):

XbaI.ZFA F1 5′-AGTGGTTGGTCTAGACCCGGGGAGCG CC CCTTCCAGTGTCGCATTTGCATGCGGAA CTTT-3′

ZFA.ApaI R1 5′-ATACGAGGGCCCTAGCAGCAGCT CCTGGTCCAGTGTGGGTTTTTAGGTG-3′

For the right ZFA—8 aa linker (7-bp spacer):

XbaI.ZFA F1 5′-AGTGGTTGGTCTAGACCCGGGGAGCG CCCCTTCCAGTGTCGCATTTGCATGCGGAACT TT-3′

ZFA.SpeI R1 5′-TTTTGACT AGTGCCCTAGCAGCAGCTC CTGGTCCAGTGTGGGTTTTTAGGTG-3′

5. When engineering ZFNs with the 8 aa inter-domain linker, the PCR of the left and right ZFAs will be carried out with specific primer pairs utilizing different restriction sites. Therefore the digestion of the left ZFA PCR product will be digested with XbaI and ApaI and the right ZFA PCR product will be digested with XbaI and SpeI. Follow the digestion parameters suggested in **step 13**.
6. For the 8 aa linker (7-bp spacer) use a specific plasmid for FokI expression vector (8 aa linker) and linearize with XbaI and ApaI. Combine the following reagents into a 200 μL tube: 1 μg of FokI (7-bp) expression vector, 3 μL buffer, 3 μL 10× BSA, 0.5 μL XbaI, and 0.5 μL ApaI in sterile water up to final volume of 30 μL. Incubate for 1 h at 25 °C, followed by 3 h at 37 °C, and gel isolate the backbone (~4.0 kb) using QIAquick gel extraction kit.
7. An excellent protocol for preparation of highly competent bacterial cells for chemical transformation has been described [26].
8. Digestion of the FokI expression 8 aa linker (L-ZFA) requires the restriction enzymes XbaI and SpeI to facilitate ligation of the right ZFA. However it will be necessary to remove 5′ phosphate groups from these DNA ends using Antarctic phosphatase to prevent self-ligation due to the compatible cohesive DNA ends created by these enzymes. The ligation of the right ZFA will need to be appropriately screened to select for correct orientation since the right ZFA XbaI/SpeI fragment can potentially ligate in the other direction.
9. It is important to use a proofreading polymerase such as KOD Hot Start or Pfu Turbo (Stratagene) and/or sequence the product.
10. The inducible vector used in our transformations is a low-copy plasmid and requires 10 mL mini-prep volumes for acceptable concentrations.

11. Ligation of the ZFN cassette into the inducible binary vector is sometimes incomplete. Therefore a diagnostic restriction enzyme digest using the EcoRI is recommended to highlight the correct clone.
12. Prepare glycerol stocks from each colony by combining 200 μL of the inoculated LB medium with 200 μL of 60% glycerol solution (the final solution will be 30% glycerol). Incubate at room temperature for 1 h and store at −80 °C.
13. Harvesting of the cotyledons including the transfer of the plates into the vacuum chamber is carried out aseptically.
14. Since the enrichment PCR is typically very sensitive, it is therefore possible to pool roots for the DNA extraction step.
15. For a successful enrichment PCR assay, only one restriction site should be present in the PCR amplicon, i.e., at the ZFN target site.
16. Use a restriction enzyme-digested and undigested positive control to provide indication of a successful digestion. Repeat digestion if the positive control has not fully digested.

Acknowledgment

This work was supported by the Minnesota Soybean Research and Promotion Council (project #19-12C) and the United Soybean Board (project #2506).

References

1. Curtin SJ, Voytas DF, Stupar RM (2012) Genome engineering of crops with designer nucleases. Plant Genome 5:42–50
2. Kim YG, Cha J, Chandrasegaran S (1996) Hybrid restriction enzymes: zinc finger fusions to Fok I cleavage domain. Proc Natl Acad Sci U S A 93:1156–1160
3. Bibikova M, Beumer K, Trautman JK, Carroll D (2003) Enhancing gene targeting with designed zinc finger nucleases. Science 300:764
4. Porteus MH, Baltimore D (2003) Chimeric nucleases stimulate gene targeting in human cells. Science 300:763
5. Maeder ML, Thibodeau-Beganny S, Osiak A, Wright DA, Anthony RM, Eichtinger M et al (2008) Rapid "open-source" engineering of customized zinc-finger nucleases for highly efficient gene modification. Mol Cell 31:294–301
6. Sander JD, Dahlborg EJ, Goodwin MJ, Cade L, Zhang F, Cifuentes D et al (2011) Selection-free zinc-finger-nuclease engineering by context-dependent assembly (CoDA). Nat Methods 8:67–69
7. Lloyd A, Plaisier CL, Carroll D, Drews GN (2005) Targeted mutagenesis using zinc-finger nucleases in Arabidopsis. Proc Natl Acad Sci U S A 102:2232–2237
8. Wright DA, Townsend JA, Winfrey RJ, Irwin PA, Rajagopal J, Lonosky PM et al (2005) High-frequency homologous recombination in plants mediated by zinc-finger nucleases. Plant J 44:693–705
9. Townsend JA, Wright DA, Winfrey RJ, Fu F, Maeder ML, Joung JK, Voytas DF (2009) High-frequency modification of plant genes using engineered zinc-finger nucleases. Nature 459:442–445
10. Shukla VK, Doyon Y, Miller JC, DeKelver RC, Moehle EA, Worden SE et al (2009) Precise genome modification in the crop species *Zea mays* using zinc-finger nucleases. Nature 459:437–441
11. Gao H, Smith J, Yang M, Jones S, Djukanovic V, Nicholson MG et al (2010) Heritable targeted mutagenesis in maize using a designed endonuclease. Plant J 61:176–187

12. Curtin SJ, Zhang F, Sander JD, Haun WJ, Starker C, Baltes NJ et al (2011) Targeted mutagenesis of duplicated genes in soybean with zinc-finger nucleases. Plant Physiol 156: 466–473
13. Li T, Liu B, Spalding MH, Weeks DP, Yang B (2012) High-efficiency TALEN-based gene editing produces disease-resistant rice. Nat Biotechnol 30:390–392
14. Wright DA, Thibodeau-Beganny S, Sander JD, Winfrey RJ, Hirsh AS, Eichtinger M et al (2006) Standardized reagents and protocols for engineering zinc finger nucleases by modular assembly. Nat Protoc 1:1637–1652
15. Ramirez CL, Foley JE, Wright DA, Muller-Lerch F, Rahman SH, Cornu TI et al (2008) Unexpected failure rates for modular assembly of engineered zinc fingers. Nat Methods 5:374–375
16. Zhang F, Maeder ML, Unger-Wallace E, Hoshaw JP, Reyon D, Christian M et al (2010) High frequency targeted mutagenesis in Arabidopsis thaliana using zinc finger nucleases. Proc Natl Acad Sci U S A 107:12028–12033
17. Paz MM, Martinez JC, Kalvig AB, Fonger TM, Wang K (2006) Improved cotyledonary node method using an alternative explant derived from mature seed for efficient Agrobacterium-mediated soybean transformation. Plant Cell Rep 25:206–213
18. Sato S, Nakamura Y, Kaneko T, Asamizu E, Kato T, Nakao M et al (2008) Genome structure of the legume, *Lotus japonicus*. DNA Res 15:227–239
19. Schmutz J, Cannon SB, Schlueter J, Ma J, Mitros T, Nelson W et al (2010) Genome sequence of the palaeopolyploid soybean. Nature 463:178–183
20. Kim MY, Lee S, Van K, Kim TH, Jeong SC, Choi IY et al (2010) Whole-genome sequencing and intensive analysis of the undomesticated soybean (*Glycine soja* Sieb. and Zucc.) genome. Proc Natl Acad Sci U S A 107:22032–22037
21. Young ND, Debelle F, Oldroyd GE, Geurts R, Cannon SB, Udvardi MK et al (2011) The Medicago genome provides insight into the evolution of rhizobial symbioses. Nature 480:520–524
22. Varshney RK, Chen W, Li Y, Bharti AK, Saxena RK, Schlueter JA et al (2012) Draft genome sequence of pigeonpea (*Cajanus cajan*), an orphan legume crop of resource-poor farmers. Nat Biotechnol 30:83–89
23. Christian M, Cermak T, Doyle EL, Schmidt C, Zhang F, Hummel A et al (2010) Targeting DNA double-strand breaks with TAL effector nucleases. Genetics 186:757–761
24. Cermak T, Doyle EL, Christian M, Wang L, Zhang Y, Schmidt C et al (2011) Efficient design and assembly of custom TALEN and other TAL effector-based constructs for DNA targeting. Nucleic Acids Res 39:e82
25. Händel EM, Alwin S, Cathomen T (2008) Expanding or restricting the target site repertoire of zinc-finger nucleases: the inter-domain linker as a major determinant of target site selectivity. Mol Ther 17:104–111
26. Maeder ML, Thibodeau-Beganny S, Sander JD, Voytas DF, Joung JK (2009) Oligomerized pool engineering (OPEN): an "open-source" protocol for making customized zinc-finger arrays. Nat Protoc 4:1471–1501

Chapter 4

RNA-Seq for Transcriptome Analysis in Non-model Plants

Rohini Garg and Mukesh Jain

Abstract

Sequencing of mRNA using next-generation sequencing (NGS) technologies (RNA-seq) has the potential to reveal unprecedented complexity of the transcriptomes. The transcriptome sequencing of an organism provides quick insights into the gene space, opportunity to isolate genes of interest, development of functional markers, quantitation of gene expression, and comparative genomic studies. Although becoming cheaper, transcriptome sequencing still remains an expensive endeavor. Further, the assembly of millions and billions of RNA-seq reads to construct the complete transcriptome poses great informatics challenges. Here, first we outline various important issues from experimental design to data analysis, including various strategies of transcriptome assembly, which need substantial consideration for a successful RNA-seq experiment. Further, we describe a method for using RNA-seq to characterize the transcriptome of a plant species, taking the example of a legume crop plant chickpea. Our aim is to provide a quick start guide to the nonexpert researchers for NGS-based transcriptome analysis.

Key words RNA-seq, Transcriptome, Assembly, Next-generation sequencing, Quality control, Expression analysis

1 Introduction

Analysis of the transcriptome is very important to interpret the gene content of an organism where the genome sequence is not available. One of the most important applications of next-generation sequencing (NGS) technologies is the sequencing of an mRNA population (RNA-seq). RNA-seq provides a powerful and cost-effective means to construct and characterize the transcriptome of an organism for gene discovery and measurement of gene expression [1–4]. Although the cost of sequencing is decreasing rapidly, these technologies are still expensive and pose substantial computational challenges, most important being the accurate assembly of millions/billions of sequence reads to construct the complete transcriptome.

Legumes represent the second most important source of human food and animal forage after cereals. Legumes are an important source of nutrition in developing countries and possess the unique

Ray J. Rose (ed.), *Legume Genomics: Methods and Protocols*, Methods in Molecular Biology, vol. 1069,
DOI 10.1007/978-1-62703-613-9_4, © Springer Science+Business Media, LLC 2013

ability to fix atmospheric nitrogen. Despite their importance, legumes have been orphans from the genomic revolution. However, it is only recently genomic resources have been developed for a few important legume species, including soybean, *Lotus*, *Medicago*, and chickpea [5–8]. From the recent studies, it is clear that legume biology is rapidly undergoing transformation due to application of genomic approaches. However, considering the large number of domesticated legume species [9], it is very important to extend the use of genomic approaches to other legume species for their genetic enhancement.

RNA-seq provides a unique opportunity to sequence and characterize the transcriptome of non-model organisms very rapidly [10, 11]. Recently, we sequenced the transcriptome of chickpea, a non-model crop legume plant, using different NGS technologies. The optimized assembly of short-read sequence data generated using Illumina platform resulted in 53,409 transcripts with an average length of 523 bp representing 28 Mb of total transcriptome sequence [12]. However, hybrid assembly of optimized primary assemblies of short-read (Illumina) and long-read (Roche 454) data generated 34,760 transcripts in total with an average length of 1,020 bp representing 35.5 Mb of total transcriptome sequence [7]. The assembly of sequence data involved various quality control steps, filtering of high-quality data, and optimization of assembly parameters using different assembly programs followed by assembly assessment. Here, we describe important considerations and general workflow of an RNA-seq experiment followed by a method for transcriptome assembly and analysis, which can be applied to any of the legumes or other plant species.

Firstly, the important considerations about RNA-seq experiments, which need prior attention, are discussed briefly to ensure that the question of interest is properly addressed. The general workflow of an RNA-seq experiment has also been outlined and illustrated in Fig. 1 to help understand the whole procedure of transcriptome analysis for beginning researchers.

1.1 Experiment Design

The experimental design is very critical to a transcriptome sequencing project. The knowledge of species background and choice of tissue samples, biological replication, sequencing technology, read-type, and sequencing depth are important issues, which need to be considered carefully (*see* **Note 1**).

1.1.1 Species Background Information

Before starting an RNA-seq experiment, knowing some background information about the plant species is very important. The knowledge of genome size, ploidy level, and repetitive DNA content can help evaluate the complexity of the transcriptome to be analyzed. The prevalence of polyploidy and genome duplications

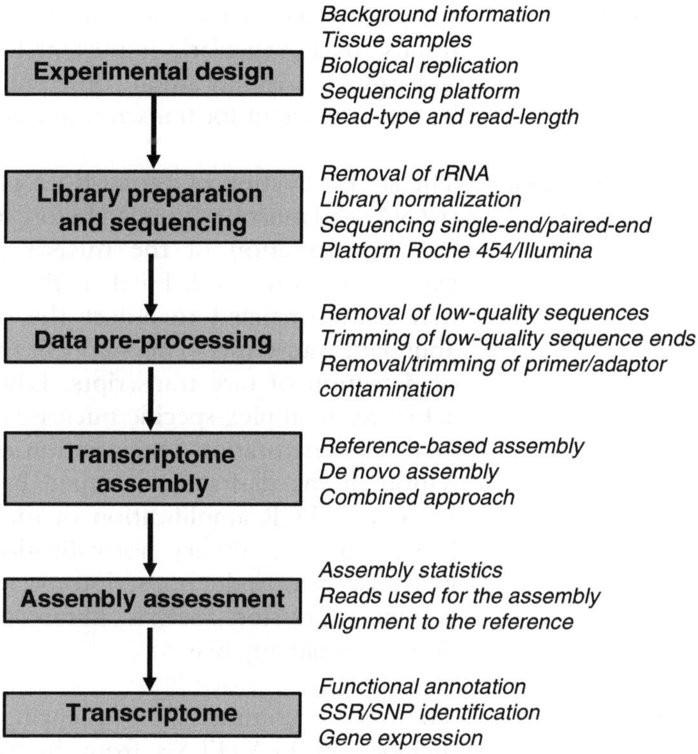

Fig. 1 Steps involved in RNA-seq-based transcriptome analysis

can complicate the transcriptome assembly. Further, knowing the level of similarity and divergence with the most closely related species for which sequence data are available can also provide a clue about the complexity of genome/transcriptome of the species being investigated. In fact, this information may help design the transcriptome assembly and downstream data analysis strategies.

1.1.2 Choice of Tissue Samples

The appropriate choice of tissue samples is also an important aspect to achieve the aim of transcriptome analysis. If the aim is transcriptome construction, choice of a broad range of tissue samples representing the whole life cycle of plant will be most appropriate to obtain the representation of most of the transcripts expressed in the plant. Otherwise, the tissue samples (tissues/organs, developmental stages, treatments, and controls) should be chosen according to the biological question being asked. In a transcriptome construction study, the pooling of RNA from different tissue samples is a good option to reduce the cost of library preparation. However, the sequencing of individual samples may serve the purpose for both transcriptome construction and gene expression studies.

1.1.3 Biological Replication

Biological replication of the data is very important to present accurate results. It is particularly important for gene expression studies to take care of variability among different samples. However, it may not be very important for transcriptome construction studies.

1.1.4 rRNA Removal and Library Normalization

The largest fraction of a total RNA preparation is ribosomal RNA (rRNA) and must be removed before library preparation to increase the representation of the mRNA population, especially those expressed at low level. Further, the sequencing libraries prepared may be normalized to adjust the overrepresentation of highly abundant transcripts. This step will further increase the possibility of detection of rare transcripts. Library normalization could be achieved by duplex-specific nuclease (DSN) normalization, which involves denaturation–reassociation of cDNA followed by degradation of the ds-fraction formed by abundant transcripts using DSN, and PCR amplification of the remaining ss-DNA fraction [13]. However, library normalization depends on the question being asked from the transcriptome analysis experiment. If the aim is gene expression analysis, library normalization should not be done to avoid any bias.

1.1.5 Choice of Sequencing Platform and Read-Type

The major commercial sequencing platforms available today include GS FLX/FLX+ from Roche 454, Genome Analyzer/HiSeq from Illumina, SOLiD from Life Technologies, Heliscope from Helicos, and PacBIO from Pacific Biosciences. These platforms have been well reviewed elsewhere [14–17]. Among these, Roche 454 and Illumina platforms have been used most commonly to construct the transcriptome and study gene expression. Roche 454 sequencing provides the advantage of larger reads (average of 400 bp), but the total number of reads generated per run is much less, thus making it relatively expensive. However, the Illumina platform generates a very large number of reads making it very cost-effective, but the read length is shorter (up to 150 bp). Due to shorter read length, very high sequencing depth is required confounding the computation problems of assembly. The paired-end sequencing (sequencing of cDNA fragments from both ends) on the Illumina platform provides a better alternative to single-end (sequencing of cDNA fragments from only one end) short-read sequencing and can offer the advantage of long-range exon connectivity as obtained using Roche 454 technology (*see* **Note 2**). The choice of sequencing platform depends on the technology access and budget constraints. Although sequence data from a single platform can be used, use of more than one platform may provide a better picture of the transcriptome (*see* **Note 3**).

1.1.6 Sequencing Depth

The sequencing depth is defined as the average number of reads representing a given base in the reconstructed sequence. Considering the cost of an RNA-seq experiment, the depth of sequencing should also be considered carefully so as to present most of the expressed transcript sequences at low cost (*see* **Note 4**). This is also dependent on the availability of a reference genome sequence for assembly. In case of the availability of a reference sequence, a lower sequencing depth (as low as 10×) may be sufficient to represent most of the transcripts. However, when the reference sequence is not available, a very high sequencing depth (more than 30×) is required for de novo construction of the transcriptome.

1.1.7 Computational Requirements

The availability of computational resources may also be an important determinant of experimental design (*see* **Note 4**). For a non-model species without a reference sequence, due to the requirement of a large amount of sequence data, computational requirements will be very high as compared to the species for which a reference sequence is available. In fact, the assembly process can consume hundreds of gigabytes of random access memory (RAM) depending on the assembly strategy and dataset. For example, assembly of the chickpea transcriptome datasets was performed using a server with 48 processors and 128 GB RAM [12]. However, choice of a sequencing platform like Roche 454, which produces less data with longer sequences, might decrease the computational requirements with some limitations, such as non-detection of rare transcripts, incomplete transcripts, and sequencing errors.

1.2 Sequencing

Once an experiment has been properly designed, libraries are prepared and sequencing is performed accordingly. This step is relatively easy as several commercial service providers for all the sequencing platforms are available, who can prepare the library and sequence the RNA samples at relatively lower cost. However, the service provider must be chosen carefully to ensure high quality of sequence data; otherwise it may lead to rejection of a lot of data at the preprocessing step, complicate the transcriptome assembly and other downstream analysis, and generate ambiguous output. The addition of barcodes (multiplexing) during library preparation will allow mixing of samples during the sequencing step and may lead to sequencing of multiple samples more efficiently at lower cost. In general, for RNA-seq library preparation, rRNA is removed from the DNA-free total RNA samples. This can be achieved by selecting poly(A) mRNAs or removing rRNA by hybridization-based depletion methods. RNA is fragmented and reverse transcribed into cDNA followed by ligation of sequencing adaptors and selection of desired fragment size. Finally, one or both ends of the cDNAs are sequenced using NGS technologies. To increase the representation of less abundant transcripts, normalized RNA-seq libraries may be constructed. Another consideration is whether to include a PCR amplification step to prepare the library or to employ an amplification-free approach used for single-molecule sequencing technologies from Helicos and Pacific Biosciences [17]. The use of an amplification-free approach ensures minimal redundancy in the sequence data and higher coverage of the transcripts.

1.3 Data Preprocessing

All the sequencing platforms result in some artifacts, including read errors, contamination of adaptor/primer sequences, and low-quality reads which can impact the downstream data processing greatly. The removal of such sequence contamination and artifacts is crucial for accuracy and computational efficiency of transcriptome assembly (*see* **Note 5**). It has been well demonstrated that the quality of sequence data affects the assembly output [7, 12, 18]. Thus, the sequence data should be preprocessed using such quality control tools to filter high-quality reads for further analysis. Further, removal of redundant clonal reads can also be considered to enhance the computational efficiency and decrease computational power requirement. This step is relatively simple and several computational pipelines have been developed for data preprocessing [18–21].

1.4 Transcriptome Assembly

Assembly refers to the process of aligning and merging of smaller DNA sequences to construct the longer original sequence. Assembly process is quite challenging using NGS data due to small read-length and large volume of data, which is further complicated by duplicated/repetitive sequences. Several assembly programs based on different algorithms have been developed for assembly and implemented successfully using different strategies [22, 23]. To generate the best optimal transcriptome, the assembly strategy should be carefully designed (*see* **Notes 6** and **7**). Depending on the availability of a reference genome, the transcriptome assembly approach may be chosen from the following strategies.

1.4.1 Reference-Based Assembly

If the genome sequence for the target transcriptome is available, the reference-based assembly should be preferred. In the reference-based assembly, firstly, the RNA-seq reads are aligned to the reference genome followed by clustering of overlapping reads to build a graph representing all isoforms. Finally, all the individual transcript isoforms are resolved from the graph [23]. Cufflinks and Scripture are the examples of complete pipelines that provide utility of reference-based assembly [24–26]. Reference-based assembly decreases the requirement of computational power substantially and is very sensitive, even for the lowly expressed genes. Further, it allows the discovery of novel transcripts not currently annotated from the genome, novel/alternative transcript isoforms, and alternative transcriptional start and stop sites. Although it is highly recommended to use the genome sequence of the same plant species as reference, it is possible to use the genome sequence of a closely related species as a reference with the disadvantage of missing diverged genomic/transcript regions. Further, it is also recommended that the genome sequence being used as a reference should be of very high quality to avoid carryover of genome misassemblies and deletions in the transcriptome assembly.

1.4.2 De Novo Assembly

De novo assembly means construction of the transcriptome using RNA-seq reads without the requirement of a reference genome/transcriptome. De novo assembly using NGS data is a major challenge as it demands large datasets and high computational resources. Several de novo assembly programs (such as Velvet, Oases, ABySS, SOAPdenovo, and Trinity) have been developed for this purpose and a few transcriptomes have also been constructed successfully [7, 12, 22, 27–31]. Although most of these programs have been developed for short reads, long reads can also be used for the assembly. De novo assembly is very important for the organisms where a reference genome is not available and can provide an initial set of transcripts. Further, it can detect novel transcripts from the genomic regions, which have not been sequenced and trans-spliced transcripts.

1.4.3 Combined Approach

Reference-based and de novo approaches can be combined to generate a better transcriptome. In this combined approach, first the reads are assembled using the reference genome as discussed above followed by de novo assembly of the remaining reads that fail to assemble in a reference-based approach. Alternatively, the transcripts generated from the reference assembly can be combined with the remaining non-assembled reads for de novo assembly in the second step. This approach leads to the assembly of most of the reads and provides advantages of both the reference and de novo assembly approaches. No automated pipeline/program implementing this combined approach has been developed as of now. The overlap-layout-consensus (OLC)-based assembly program such as TIGR Gene Indices Clustering tool (TGICL) can be chosen for this assembly. Recently, we reported the assembly of the wild chickpea transcriptome by this approach using the transcriptome of cultivated chickpea as a reference in the reference-based assembly followed by de novo assembly of the contigs generated along with the remaining unassembled reads [32].

1.5 Assembly Evaluation

Although no standard criteria have been established to assess the quality of assembly, a few metrics, including accuracy, completeness, contiguity, chimerism, and variant resolution based on a set of well-established reference transcripts, have been proposed to compare the output of different assemblers and assembly strategies [23]. However, the well-established reference transcripts are not available for most of the non-model plant species for assembly assessment. Therefore, a few criteria, including assembly statistics, total number of reads assembled, and alignment to the reference transcriptome/proteome, may be used to obtain an idea of the quality of transcriptome assembly. The first criteria is the assembly statistics, including total number of transcripts generated, total assembly size, average length, and N50 length (N50 length is the length of the smallest contig in the set comprising the largest contigs whose combined length represents 50 % of the total assembly size) of the transcripts. In addition, the average number of reads per assembled transcript, number of reads per transcript as a function of transcript length, and average depth per transcript (average number of reads per bp along a transcript) may also be considered for the assembly assessment. Another criterion is to look for the number of sequence reads, which have been utilized for the transcriptome assembly. Ideally, all the sequence reads should be assembled to produce assembled transcripts. However, practically it does not happen and a significant number of reads remain unassembled. Thus, the measurement of the number of reads assembled and how many of them are mapped uniquely will provide clues about the quality of assembly. A large number of non-assembled reads indicate nonoptimal assembly and large number of non-uniquely assembled reads might represent redundant and chimeric transcripts. Further, alignment of transcriptome assembly against a trusted reference transcriptome/proteome from a closely related species is also an important criterion for assembly assessment. The percentage coverage of reference transcripts can give an idea about the completeness and accuracy of the transcriptome assembly. It may be noted that optimization of one or more of these criteria may affect the other criteria negatively. Therefore, in many cases the selection of best assembly will depend upon the underlying biological question and careful logical judgment of the researcher (*see* **Note 8**). In addition, the accuracy of assembly could be evaluated experimentally by PCR amplification of a few representative transcripts followed by traditional Sanger sequencing.

1.6 Downstream Analysis

1.6.1 Characterization and Annotation

To achieve the aim of transcriptome characterization, putative function is assigned to the transcripts generated. To predict the putative function, the most common method is BLAST search of transcript sequences in various databases or proteins from a closely related species. Another important aspect is to assign the gene ontology (GO) terms to the transcripts, which represent their broad functional category. GO terms may be assigned on the basis of homology to proteins of closely related species or using specific databases like Blast2GO (http://blast2go.com/b2ghome). In addition, annotation may be integrated with other information such as assignment of metabolic pathways to the transcripts from the KEGG database (http://www.genome.jp/kegg/). Specific gene families involved in pathways of interest may be characterized. Another important aspect of functional characterization is identification of transcription factor content of the transcriptome, which can provide insight into the regulatory protein repertoire of the plant species.

1.6.2 Detection of Microsatellites and Sequence Variations

The identification of microsatellites (simple sequence repeats) is another important application of transcriptome analysis. Simple perl scripts such as MISA and SSRIT can be used on the transcriptome sequence to detect microsatellites [33, 34], which can be developed into functional markers for breeding and mapping purposes. In addition, if the transcriptome sequence is generated from two or more different genotypes/varieties, single-nucleotide polymorphisms (SNPs) can also be identified, which are important for the generation of linkage/genetic maps and breeding purposes. Many RNA-seq studies have focused specifically on the detection of microsatellites and SNPs only.

1.6.3 Gene Expression Analysis

RNA-seq is a powerful tool to measure gene expression and its use has been well demonstrated and reviewed in many articles [1, 4, 7, 35]. A very good correlation has been observed among the gene expression studies based on RNA-seq data and other approaches. In fact, RNA-seq provides several advantages over other approaches to quantitate gene expression [4]. If the transcriptome assembly has been generated from different tissue samples, gene expression can be determined by mapping the reads from individual tissue samples on the transcripts. Several programs and pipelines are available to quantitate gene expression using RNA-seq data [3, 4, 36].

2 Materials

2.1 Sequencing Platforms

Roche 454 GS FLX/FLX+ and Illumina Genome Analyzer II/HiSeq sequencers or well-experienced commercial service providers.

2.2 Quality Control Tool

NGS QC Toolkit [18] Sequence quality control tool: http://www.nipgr.res.in/ngsqctoolkit.html.

2.3 Assembly Tools

1. Velvet [37] sequence assembly tool: http://www.ebi.ac.uk/~zerbino/velvet.
2. Oases [38] sequence assembly tool: http://www.ebi.ac.uk/~zerbino/oases.
3. ABySS [39] sequence assembly tool: http://www.bcgsc.ca/platform/bioinfo/software/abyss.
4. SOAPdenovo sequence assembly tool: http://soap.genomics.org.cn/soapdenovo.html.
5. CLC Genomics Workbench sequence assembly tool: http://www.clcbio.com/index.php?id=1240.
6. GS denovo assembler (Newbler) sequence assembly tool: http://www.454.com/products/analysis-software/index.asp.
7. MIRA [40] sequence assembly tool: http://www.chevreux.org/projects_mira.html.

8. TGICL [41] sequence assembly tool: http://sourceforge.net/projects/tgicl/files/latest/download.
9. CAP3 [42] sequence assembly tool: http://seq.cs.iastate.edu/cap3.html.

3 Methods

3.1 Sequencing

To start the transcriptome analysis using RNA-seq, the experiment should be carefully designed taking into consideration all the factors discussed above. The biological question should also be well defined to make the appropriate choices for sequencing and data analysis strategies (*see* **Notes 1** and **2**). Once the experiment is designed, sequencing is performed accordingly on the selected sequencing platform(s). We recommend sequencing on both Roche 454 and Illumina platforms to obtain longer reads and high sequence depth, respectively, for construction of a de novo transcriptome from a non-model plant species (*see* **Note 3**). Further, for the Illumina platform, we recommend paired-end sequencing with at least 75 bp read length to generate better assembly. The sequencing in one complete run on the Roche 454 platform (more than one million reads) and two lanes on the Illumina IIA platform (more than 60 million paired-end reads) should provide a comprehensive picture of the transcriptome (*see* **Note 4**). However, depth of sequencing may be increased depending on the availability of funding. We generated about 135 million short reads using Illumina (72 bp paired-end and 51 bp single-end) and about 2.6 million reads using Roche 454 platforms from different tissue samples of chickpea for de novo transcriptome construction [7, 12].

3.2 Data Preprocessing

Prior to any downstream processing, the raw sequenced data needs to be cleaned for low-quality sequences and contamination of adaptors and primers. Many programs/tools are available, which can be used for quality control of sequence data generated [18–21]. Recently, we have developed an NGSQC toolkit for quality control and filtering of high-quality data from Roche 454 and Illumina platforms [18]. Various parameters can be provided for optimal quality control and filtering according to the available dataset in this toolkit. We outline important steps for quality control of sequencing reads using the NGSQC toolkit.

1. Download the NGSQC toolkit and install it on the server or the desktop computer along with all the dependencies according to the user manual.
2. For Illumina data (using IlluQC.pl or IlluQC_PRLL.pl), we recommend filtering out of reads having contamination of primers/adaptor sequences and low-quality reads with more than 30 % bases having a Phred quality score of <20 (default –l

and –s parameters). After filtering, various statistics files in text and graph formats along with a file for high-quality filtered reads are generated.

3. A few bases at the ends can also be trimmed after looking at their average quality in all the reads using TrimmingReads.pl script. The trimming of a few low-quality bases from the end(s) may generate better assembly results.
4. For Roche 454 data also (using 454QC.pl or 454QC_PRLL.pl), the low-quality reads (with more than 30 % of bases having a Phred quality score of <20) should be filtered and reads containing primer/adaptor sequences should be trimmed (default –l and –s parameters).
5. Another important issue in the quality of Roche 454 data is due to the presence of homopolymers. With the increase in length of homopolymer, the signal intensity distribution broadens, which results in ambiguous base call [43]. Such data may lead to frame-shift errors in the transcript sequences. Although some assembly programs take care of the homopolymer problem, the utility is not available in all the programs. Therefore, trimming of reads from 3′-end of the reads after the encounter of first homopolymer may be considered (–n parameter, default is off). However, this may lead to loss of a substantial amount of important sequence data. Therefore, we recommend using this option, only if a very high depth of sequencing data is available.
6. Minimum length cutoff (–m parameter) may also be specified to filter out the smaller sequences for Roche 454 data. We recommend setting it at 100 (default) to discard sequences less than 100 bp long.

In our study, the quality filtering of ~135 million reads obtained from the Illumina platform resulted in 107 million (79 %) high-quality reads [12]. We trimmed a few bases from the 3′-end of Illumina reads also to reveal the effect of trimming low-quality bases on the assembly output. Likewise, filtering of more than 2.5 million reads from Roche 454 platform resulted in about 1.9 million (75 %) high-quality reads of 372 bp average length [7].

3.3 Transcriptome Assembly

Considering that sequence data has been generated on both Roche 454 and Illumina platforms, the transcriptome assembly can be performed using the following steps (*see* **Note 6**).

3.3.1 Optimization of Illumina Short-Read Assembly

1. Download all the available assemblers (for example, Velvet, Oases, ABySS, SOAPdenovo, Trinity, and CLC) developed for de novo assembly of short-read data and install them as per given instructions (*see* **Note 7**).
2. The beginners may choose a small dataset (~one million reads) to start the analysis and get familiar with functioning of the programs, various parameters, and output.

3. Once familiar with the tools, the whole dataset can be used for the assembly using different parameters. Most of the de novo assembly programs for short reads are based on the *de Bruijn* graph algorithm. For the *de Bruijn*-based algorithms, *k*-mer value is the most important parameter, which affects the assembly output considerably. Therefore, it is recommended to use different *k*-mer values for the assembly along with other parameters.
4. The assembly output at different parameters should be assessed based on several criteria as discussed in Subheading 1.5 to select the best assembly from each individual tool.
5. The best assemblies obtained from an individual tool should be compared and evaluated further to select the optimum assembly.

We optimized de novo assembly of RNA-seq data (107 million high-quality reads) of chickpea obtained from the Illumina platform using different assembly programs, including Velvet, Oases, ABySS, SOAPdenovo, and CLC. Based on the assembly assessment criteria described above (Subheading 1.5), the best assembly generated 53,409 transcripts with an average length of 523 bp using the Oases program [12]. It has been suggested that using the multiple-*k* method, in which multiple *k*-mer values are used for the de novo transcriptome assembly, gives better assembly in terms of contiguity and gene identification [29, 44]. However, we did not find this strategy useful in our unpublished studies on transcriptome assemblies of some non-model plant species. This suggests that the assembly results may be highly context dependent and need to be optimized for each dataset (*see* **Note 8**).

3.3.2 Optimization of Roche 454 Long-Read Assembly

1. Download all the available assemblers (for example, Newbler, MIRA, TGICL, CAP3, and CLC) developed for the assembly of Roche 454 data and install them as per given instructions (*see* **Note** 7). Although TGICL and CAP3 programs have originally been developed for the assembly of expressed sequence tags (ESTs) derived from Sanger sequencing, they work well with Roche 454 data also. Some of the programs have been developed to assemble both short-read and long-read data (such as MIRA and CLC). Many of the short-read assembly programs can also be used for the assembly of Roche 454 data; however, they may not give optimal results.
2. The beginners may choose a small dataset (~0.1 million reads) to start the analysis and get familiar with the functioning of the programs, various parameters, and output.
3. Now the whole data can be used for the assembly. Various assembly parameters can be used to obtain the best assembly. Likewise, the best assembly is to be decided based on several assessment criteria as discussed above (Subheading 1.5).

A comparison of performance of different assemblers on 454 transcriptome data suggested that merged assembly of primary de novo assemblies obtained from different programs results in better transcriptome assembly [45]. However, in our study, we found that merged assembly generated redundant and chimeric transcripts [7]. Therefore, different assembly strategies should be used carefully and their output may be greatly context dependent. The optimization of de novo assembly of chickpea transcriptome using long-read Roche 454 data resulted in generation of variable statistics on different assessment criteria [7]. We selected three best assemblies to proceed further for hybrid assembly.

3.3.3 Hybrid Assembly

1. For the hybrid assembly two approaches can be used. First, all the Illumina and Roche 454 data can be used together for the assembly. Although several programs have the capability of handling both the data together, computational requirements are very high and their performance can also vary. For example, ABySS can handle both the datasets together and produces a reasonably good assembly. However, although MIRA can also handle both the datasets, assembly of Illumina data is very slow and may take several days depending on the volume of data.
2. Therefore, we recommend hybrid assembly of the best primary assemblies obtained individually from both the datasets as a second approach.
3. An OLC-based assembly program can be used for this purpose. We found TGICL to be very efficient for this purpose.

The hybrid assembly of optimized assemblies of short-read Illumina and long-read Roche 454 data generated 34,760 transcripts with average length of more than 1 kb for chickpea [7]. The hybrid assembly was 1.3 times larger in total size with about two times increase in average transcript length as compared to the de novo assembly of only short-read data [7, 12].

3.4 Transcriptome Characterization

Once the transcriptome assembly has been finalized, we can proceed for characterization of the transcriptome. The analysis may include analysis of GC content, similarity/divergence with other species, functional annotation, gene ontology categorization, identification of transcription factor encoding genes, genes/pathways of interest, and identification of microsatellites. In addition, if the transcriptome sequence of two or more species is available, polymorphic microsatellite and sequence variations such as SNPs can be identified, which can be developed as functional markers for association/genetic mapping studies. Gene expression can also be analyzed by mapping reads from individual tissue samples on the transcriptome assembly and assigning read count to each transcript in all the tissue samples followed by statistical analysis.

3.5 Conclusions

Although transcriptome assembly is quite challenging, it is easier with proper experimental design taking into consideration all the factors. Several studies suggest that RNA-seq can be used for excellent transcriptome assembly and characterization to answer diverse biological questions with a good experimental design and accurate choice of assembly programs and strategy. Since computational data analysis is highly context dependent, the above recommendations, discussions, and method are meant to serve as guidelines to initiate an RNA-seq experiment for transcriptome analysis of an organism.

4 Notes

1. Experimental design is very critical to transcriptome analysis using RNA-seq and all the factors, including species background, choice of tissue samples, and biological replicates, should be considered very carefully.
2. The right choice of sequencing platform, read-length, and read-type will produce better transcriptome assembly.
3. Using sequence data from Roche 454 and Illumina platforms can provide a better picture of the transcriptome of a plant species.
4. Sequencing depth and computation requirements and their availability should also be considered carefully to finalize the RNA-seq experiment.
5. Data preprocessing is very crucial for transcriptome assembly and downstream analysis. All the low-quality sequences and sequences containing primer/adapter contamination should be removed or trimmed.
6. Assembly strategy should be carefully chosen. In fact, use of all the possible strategies followed by assessment of output to select the best assembly is recommended.
7. Various assemblers with different parameters should be used for construction of the transcriptome to select the optimal assembly.
8. The assessment of assemblies should be done very carefully considering various aspects based on the question to be answered from the study.

Acknowledgments

This work was supported by the Department of Biotechnology, Government of India, under the Next Generation Challenge Programme on Chickpea Genomics (research grant no. BT/PR12919/AGR/02/676/2009). R.G. acknowledges the award of

INSPIRE Faculty Award from the Department of Science & Technology, Government of India. M.J. acknowledges the award of a fellowship under the Cutting-edge Research Enhancement and Scientific Training scheme from the Department of Biotechnology, Government of India.

References

1. Morozova O, Hirst M, Marra MA (2009) Applications of new sequencing technologies for transcriptome analysis. Annu Rev Genomics Hum Genet 10:135–151
2. Costa V, Angelini C, De Feis I et al (2010) Uncovering the complexity of transcriptomes with RNA-Seq. J Biomed Biotechnol 2010: 853916
3. Ozsolak F, Milos PM (2011) RNA sequencing: advances, challenges and opportunities. Nat Rev Genet 12:87–98
4. Jain M (2012) Next-generation sequencing technologies for gene expression profiling in plants. Brief Funct Genomics 11:63–70
5. Cannon SB, May GD, Jackson SA (2009) Three sequenced legume genomes and many crop species: rich opportunities for translational genomics. Plant Physiol 151:970–977
6. Varshney RK, Close TJ, Singh NK et al (2009) Orphan legume crops enter the genomics era! Curr Opin Plant Biol 12:202–210
7. Garg R, Patel RK, Jhanwar S et al (2011) Gene discovery and tissue-specific transcriptome analysis in chickpea with massively parallel pyrosequencing and web resource development. Plant Physiol 156:1661–1678
8. Young ND, Bharti AK (2012) Genome-enabled insights into legume biology. Annu Rev Plant Biol 63:283–305
9. Graham PH, Vance CP (2003) Legumes: importance and constraints to greater use. Plant Physiol 131:872–877
10. Jain M (2011) A next generation approach to the characterization of a non-model plant transcriptome. Curr Sci 101:1435–1439
11. Strickler SR, Bombarely A, Mueller LA (2012) Designing a transcriptome next-generation sequencing project for a nonmodel plant species. Am J Bot 99:257–266
12. Garg R, Patel RK, Tyagi AK et al (2011) *De novo* assembly of chickpea transcriptome using short reads for gene discovery and marker identification. DNA Res 18:53–63
13. Bogdanova EA, Barsova EV, Shagina IA et al (2011) Normalization of full-length-enriched cDNA. Methods Mol Biol 729:85–98
14. Mardis ER (2008) The impact of next-generation sequencing technology on genetics. Trends Genet 24:133–141
15. Morozova O, Marra MA (2008) Applications of next-generation sequencing technologies in functional genomics. Genomics 92: 255–264
16. Simon SA, Zhai J, Nandety RS et al (2009) Short-read sequencing technologies for transcriptional analyses. Annu Rev Plant Biol 60: 305–333
17. Schadt EE, Turner S, Kasarskis A (2010) A window into third-generation sequencing. Hum Mol Genet 19:R227–R240
18. Patel RK, Jain M (2012) NGS QC Toolkit: a toolkit for quality control of next generation sequencing data. PLoS One 7:e30619
19. Blankenberg D, Gordon A, Von Kuster G et al (2010) Manipulation of FASTQ data with Galaxy. Bioinformatics 26:1783–1785
20. Cox MP, Peterson DA, Biggs PJ (2010) SolexaQA: at-a-glance quality assessment of illumina second-generation sequencing data. BMC Bioinformatics 11:485
21. Schmieder R, Edwards R (2011) Quality control and preprocessing of metagenomic datasets. Bioinformatics 27:863–864
22. Miller JR, Koren S, Sutton G (2010) Assembly algorithms for next-generation sequencing data. Genomics 95:315–327
23. Martin JA, Wang Z (2011) Next-generation transcriptome assembly. Nat Rev Genet 12: 671–682
24. Guttman M, Garbe RM, Levin JZ et al (2010) Ab initio reconstruction of cell type-specific transcriptomes in mouse reveals the conserved multi-exonic structure of lincRNAs. Nat Biotechnol 28:503–510
25. Trapnell C, Williams BA, Pertea G et al (2010) Transcript assembly and quantification by RNA-Seq reveals unannotated transcripts and isoform switching during cell differentiation. Nat Biotechnol 28:511–515
26. Trapnell C, Roberts A, Goff L et al (2012) Differential gene and transcript expression analysis of RNA-seq experiments with TopHat and Cufflinks. Nat Protoc 7:562–578
27. Iorizzo M, Senalik DA, Grzebelus D et al (2011) De novo assembly and characterization of the carrot transcriptome reveals novel genes, new markers, and genetic diversity. BMC Genomics 12:389

28. Xia Z, Xu H, Zhai J et al (2011) RNA-Seq analysis and de novo transcriptome assembly of *Hevea brasiliensis*. Plant Mol Biol 77: 299–308
29. Zhao QY, Wang Y, Kong YM et al (2011) Optimizing de novo transcriptome assembly from short-read RNA-Seq data: a comparative study. BMC Bioinformatics 12(Suppl 14):S2
30. Tao X, Gu YH, Wang HY et al (2012) Digital gene expression analysis based on integrated de novo transcriptome assembly of sweet potato [*Ipomoea batatas* (L.) Lam]. PLoS One 7: e36234
31. Wang Y, Zeng X, Iyer NJ et al (2012) Exploring the switchgrass transcriptome using second-generation sequencing technology. PLoS One 7:e34225
32. Jhanwar S, Priya P, Garg R et al (2012) Transcriptome sequencing of wild chickpea as a rich resource for marker development. Plant Biotechnol J 10:690–702
33. Temnykh S, DeClerck G, Lukashova A et al (2001) Computational and experimental analysis of microsatellites in rice (*Oryza sativa* L.): frequency, length variation, transposon associations, and genetic marker potential. Genome Res 11:1441–1452
34. Thiel T, Michalek W, Varshney RK et al (2003) Exploiting EST databases for the development and characterization of gene-derived SSR-markers in barley (*Hordeum vulgare* L.). Theor Appl Genet 106:411–422
35. Weber AP, Weber KL, Carr K et al (2007) Sampling the Arabidopsis transcriptome with massively parallel pyrosequencing. Plant Physiol 144:32–42
36. Wang Z, Gerstein M, Snyder M (2009) RNA-Seq: a revolutionary tool for transcriptomics. Nat Rev Genet 10:57–63
37. Zerbino DR, Birney E (2008) Velvet: algorithms for de novo short read assembly using de Bruijn graphs. Genome Res 18:821–829
38. Schulz MH, Zerbino DR, Vingron M et al (2012) Oases: robust de novo RNA-seq assembly across the dynamic range of expression levels. Bioinformatics 28:1086–1092
39. Simpson JT, Wong K, Jackman SD et al (2009) ABySS: a parallel assembler for short read sequence data. Genome Res 19:1117–1123
40. Chevreux B, Pfisterer T, Drescher B et al (2004) Using the miraEST assembler for reliable and automated mRNA transcript assembly and SNP detection in sequenced ESTs. Genome Res 14:1147–1159
41. Pertea G, Huang X, Liang F et al (2003) TIGR Gene Indices clustering tools (TGICL): a software system for fast clustering of large EST datasets. Bioinformatics 19:651–652
42. Huang X, Madan A (1999) CAP3: a DNA sequence assembly program. Genome Res 9:868–877
43. Margulies M, Egholm M, Altman WE et al (2005) Genome sequencing in microfabricated high-density picolitre reactors. Nature 437:376–380
44. Surget-Groba Y, Montoya-Burgos JI (2010) Optimization of de novo transcriptome assembly from next-generation sequencing data. Genome Res 20:1432–1440
45. Kumar S, Blaxter ML (2010) Comparing de novo assemblers for 454 transcriptome data. BMC Genomics 11:571

Chapter 5

Functional Analysis of Legume Genome Arrays

Nicolas Goffard and Georg Weiller

Abstract

Functional analysis of post-genomics data is essential to identify the biological processes involved in a given investigation. Although most of the ontological tools available are limited to organisms with well-annotated genomes, this chapter provides an overview of two complementary tools—MapMan and GeneBins/PathExpress—that are used to perform a functional analysis of legume gene expression data. MapMan is a stand-alone tool that displays large datasets onto diagrams of metabolic pathways or other processes. Although initially developed for *Arabidopsis thaliana*, MapMan can be extended to other plants by assigning new sequences to their orthologs in the current classification. GeneBins and PathExpress have been developed to perform enrichment analysis of functional groups and metabolic networks, respectively. Based on the KEGG database, these tools can be used with any organism, including the main reference legumes.

Key words GeneBins, PathExpress, MapMan, Gene expression analysis

1 Introduction

High-throughput technologies such as expression microarrays or next-generation sequencing allow the simultaneous measurement of the changes and regulation of genome-wide genes in a given tissue or under certain conditions. The analysis of such data usually occurs in two steps: first, the statistical analysis to determine the list of genes of interest based on differential expression, and second the biological interpretation of those genes to identify the most relevant pathways or processes affected.

Based on the biological knowledgebase available in public databases such as Gene Ontology (GO) [1] or the Kyoto Encyclopedia of Genes and Genomes (KEGG) [2], a number of ontological tools have been developed to identify significantly enriched biological processes in a group of genes [3, 4]. These tools systematically map a large number of genes of interest to the associated biological annotation terms (e.g., GO terms or KEGG canonical pathways) and then evaluate, based on statistical methods such as Chi-square, Fisher's exact test, or hypergeometric distribution, the enrichment

Ray J. Rose (ed.), *Legume Genomics: Methods and Protocols*, Methods in Molecular Biology, vol. 1069,
DOI 10.1007/978-1-62703-613-9_5, © Springer Science+Business Media, LLC 2013

of gene members for each annotation term by comparing the outcome to the reference background (e.g., genome or array). In addition, several methods for in-depth analysis of networks have been developed to identify the relevant regulatory pathways and physical interactions between the genes [5]. Unfortunately, most of these tools are only available for well-annotated genomes.

We present here two complementary solutions, MapMan and GeneBins/PathExpress, which have been developed to overcome this limitation and to support the functional interpretation of legume genomics. MapMan [6, 7] is a software tool that displays large datasets (e.g., gene expression data) onto diagrams of metabolic pathways or other processes and thus enables easier interpretation of results. It is specifically designed for plant-specific pathways and processes. It combines three types of information: (1) a hierarchical classification of genes ("BINs"), (2) images representing a functional context of these genes (e.g., metabolic pathways), and (3) experimental expression data. The transcriptional activities of the binned genes are then displayed on the images using various statistical representations. Although it has initially been tailored to *Arabidopsis thaliana*, MapMan has been extended to other plants such as maize [8], *Medicago truncatula* [9], *Lotus japonicus* [10], tomato [11], potato [12], grapevine [13], and citrus [14], by assigning new sequences to their orthologs in the current classification. It can also be extended to any organism without being restricted to sequence similarity with *Arabidopsis* by using hierarchical functional classification modelled on KEGG ontology [15].

GeneBins [16] is a Web interface for classifying gene expression data, with application to plant genome arrays. It has been developed to search the functional hierarchical classification derived from the KEGG database of a given probe set or to perform an enrichment analysis to interpret a list of identifiers according to their assignment in the ontology. This system is based on the assignment of probe set sequences of any organism to a generic hierarchical classification (i.e., not species specific) based on the similarity to sequences in reference databases such as KEGG [2] or Swiss-Prot/UniProt [17]. To complement this type of analysis, PathExpress [18, 19] has been developed to interpret gene expression data obtained from microarray experiments by identifying the most relevant metabolic pathways and subnetworks associated with a subset of genes. PathExpress is based on a directed graph to model enzymatic reactions in the context of biological pathways. To link gene expression to pathways, PathExpress uses precomputed assignments of the probe sets of supported genome arrays to enzyme commission (EC) numbers, which identifies enzyme entries based on sequence similarities with proteins retrieved from the Swiss-Prot/UniProt database [17]. A graphical pathway representation permits the visualization of the expressed genes in a functional context. GeneBins and PathExpress can be adapted to any organism, and have already been used to interpret transcriptomic and proteomic data in *Medicago truncatula* [20–23], *Lotus japonicus* [24, 25], soybean (*Glycine max*) [26], rice (*Oryza sativa*) [27], bean [28], citrus [14], mandarin [29], coral [30], and human [31].

Recently, several studies have combined the use of MapMan and GeneBins/PathExpress to interpret gene expression results [10, 14, 28].

2 Materials

2.1 MapMan

MapMan is a user-driven tool that displays large datasets (e.g., gene expression data from Arabidopsis Affymetrix arrays) onto diagrams of metabolic pathways or other processes. Downloadable and Web-based versions are available at http://mapman.gabipd.org/web/guest/mapman. By default, MapMan supports *Arabidopsis thaliana*, *Medicago truncatula*, tomato, maize, and barley.

In addition, the Mercator Web application (http://mapman.gabipd.org/web/guest/mercator) can assign MapMan "Bins" to arbitrary DNA or protein input sequences.

2.2 GeneBins

GeneBins is a Web interface for classifying gene expression data, with application to plant genome arrays. It is available at http://bioinfoserver.rsbs.anu.edu.au/utils/GeneBins. As of November 2012, GeneBins supports data for Affymetrix arrays of four plants: *Arabidopsis thaliana*, *Medicago truncatula*, soybean, and rice.

2.3 PathExpress

PathExpress is a Web-based tool to interpret gene expression data in the context of metabolic pathways and networks, based on a directed graph modelling enzymatic reactions from the KEGG database. It is available at http://bioinfoserver.rsbs.anu.edu.au/utils/PathExpress. As of November 2012, PathExpress supports 28 Affymetrix genome arrays, including *Medicago truncatula*, *Lotus japonicus*, rice, maize, tomato, grapevine, cotton, poplar, citrus, wheat, and sugarcane. The full list of supported arrays is available at http://bioinfoserver.rsbs.anu.edu.au/utils/PathExpress/stats.php.

3 Methods

3.1 Input Data

1. The experimental data file for MapMan should contain an identifier (e.g., Affymetrix probe set identifier) in the first column and can contain log (fold change) values or statistical values, which can be used to filter the results (*see* **Note 1**). This file can be prepared as a Microsoft Excel file or tab-delimited text file.
2. The input data for GeneBins and PathExpress consist of the list of genes of interest (e.g., differentially expressed genes under

two or more conditions, defined prior to the functional analysis and using different statistical methods [32]). This list of probe set identifiers (Affymetrix probe set identifiers and/or GenBank accession numbers), belonging to a given genome array, can be pasted into a text box or uploaded from a tab-delimited text file (*see* **Note 1**).

3.2 Statistical Analysis

The principal objective is to determine if the genes associated with a biological process are activated (or repressed) in a given study. In this case, the genes involved in this function should have a higher chance of being selected as a significantly differentially expressed group. The tools statistically examine the enrichment of gene members of each of the annotation terms by comparing the outcome to the reference background (*see* **Note 2**). In addition, because multiple tests are performed, the *p*-values may need to be adjusted (*see* **Note 3**).

1. MapMan includes a statistical module to perform a Wilcoxon rank sum test (similar to a Student's *t*-test but independent of a normal distribution) to identify pathways and processes that change. This test provides a statistical measure to reveal whether the average response of a functional group (BIN) is different from the response of all other BINs. The results are dynamically calculated when an experimental data file is loaded, and are displayed for the selected BIN on the screen.
2. In GeneBins and PathExpress, to detect if a certain functional category or pathway is statistically overrepresented in the selected group of genes, compared to the rest of the genome array, a *p*-value is calculated for all functional entities throughout the classification using the hypergeometric distribution. This *p*-value represents the probability that the intersection of the submitted list of genes with the group of genes belonging to the given functional group occurs by chance.

3.3 Visualization of Results

Functional results are usually represented by a list of pathways or processes ranked by increasing enrichment *p*-values. However, a graphical representation visualizes the expressed genes in a functional context and thus aids the interpretation and understanding of the results.

1. MapMan has been developed to display genomic datasets onto diagrams of metabolic processes and other biological processes. An IMAGEANNOTATOR module is used to link the functional classification of genes with experimental datasets and images of pathways. The image can be custom-built or downloaded from other files. The mapping between the BIN structure and the image is stored in an XML file associated with the image. Genes are represented by color-coded small boxes;

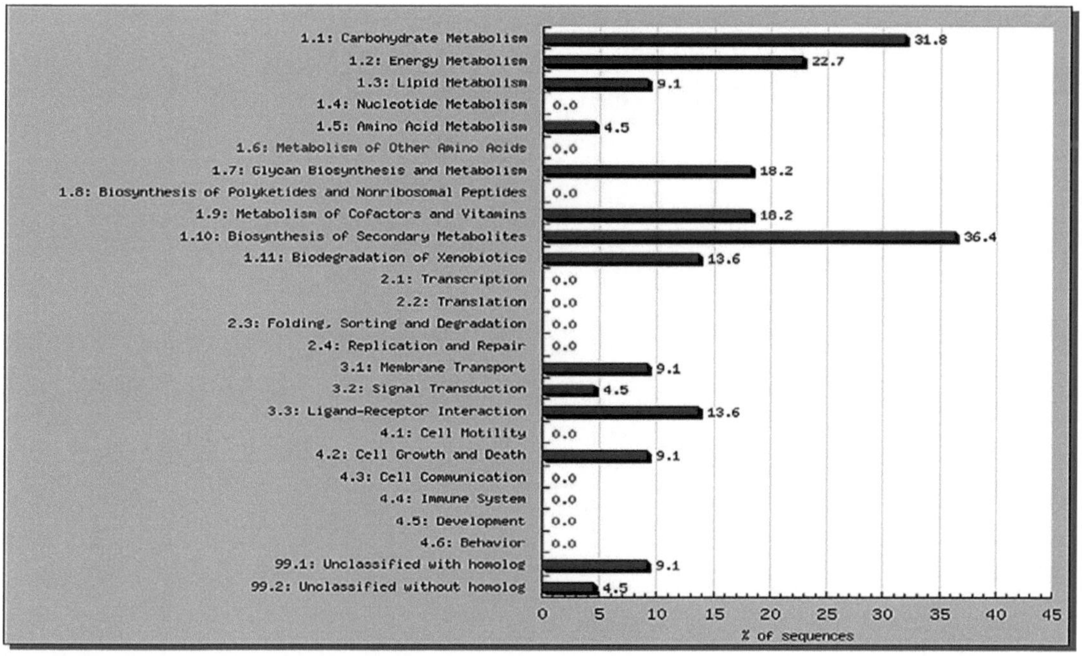

Fig. 1 GeneBins output screen example. The figure shows the functional distribution in the second level of the GeneBins ontology

genes whose expression does not change are colored in white, and an increasingly large increase or decrease is shown as an increasingly intense blue or red color, respectively. Further information on each pathway is available by mouse clicking or mousing over the image.

2. To provide an overview of the functions affected, GeneBins displays a bar plot representing the distribution of the submitted identifiers in the second level of the classification (*see* **Note 4**) (Fig. 1). In addition, to display the gene expression data on images representing a functional context of these genes using MapMan, the complete probe set classification of each organism can be downloaded in the appropriate MapMan format (*see* **Note 5**).

3. In PathExpress, each pathway can be displayed as an automatically generated graphical representation. Enzymatic reactions are highlighted differently if the corresponding enzyme has been identified in the genome array and in the submitted list of identifiers (Fig. 2). Extra information (name of the compounds, definition of the reaction) is displayed as a tool-tip when the mouse is over any of the nodes in the graph. To enhance the visualization of the expression of individual probe sets, all resources (EC assignments and graphical representations) are available to be imported into MapMan.

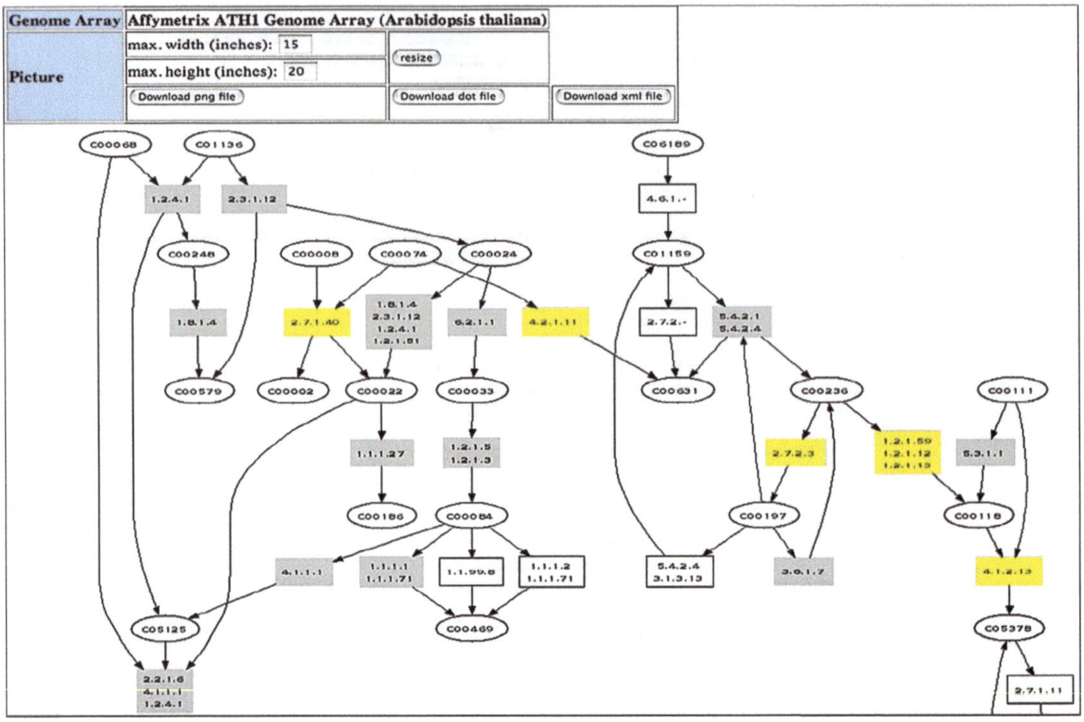

Fig. 2 PathExpress output screen example. The figure shows part of the glycolysis pathway. Metabolites are represented by *ellipses* and reactions by *rectangles*. Reactions mediated by genes included in the reference set (e.g., gene chip) are represented in *grey* if they are not part of the selected list (e.g., differentially expressed) or in *yellow* if they are in the selection list

4 Notes

1. Some bioinformatics tools, such as DAVID [33], are available to convert gene identifiers.
2. It is important to carefully specify the reference set (e.g., genes in an Affymetrix array or genome) from which a subset of selected genes is chosen, because the statistical approach assumes that each gene in the reference set could, in principle, be present in the selected list. The different gene reference background settings may greatly impact the enrichment *p-values*, even when using the same statistical method and annotation content. For example, if the gene list is selected from an Affymetrix array, then the reference list should represent the genes on this array rather than all the genes of the whole genome.
3. When, like here, multiple tests are performed, the resulting *p*-value may need to be corrected because the larger the number of groups tested, the greater is the chance of encountering false positives. Several methods have been developed, such as the conservative Bonferroni correction [34], in which the

p-values are multiplied by the number of comparisons, or the less stringent false discovery rate approach [35], which determines the expected proportion of false-positive results among all rejected hypotheses.

4. As a gene can be assigned to more than one functional group, the sum of percentages can be higher than 100%.
5. The GeneBins classification and annotation have been defined to extend MapMan software [15], with the automatic conversion of pathway pictures from the KEGG database.

References

1. Ashburner M, Ball CA, Blake JA, Botstein D, Butler H, Cherry JM et al (2000) Gene ontology: tool for the unification of biology. The Gene Ontology Consortium. Nat Genet 25:25–29. doi:10.1038/75556
2. Kanehisa M, Goto S, Sato Y, Furumichi M, Tanabe M (2012) KEGG for integration and interpretation of large-scale molecular data sets. Nucleic Acids Res 40:D109–D114. doi:10.1093/nar/gkr988, Database issue
3. da Huang W, Sherman BT, Lempicki RA (2009) Bioinformatics enrichment tools: paths toward the comprehensive functional analysis of large gene lists. Nucleic Acids Res 37:1–13. doi:10.1093/nar/gkn923
4. Khatri P, Sirota M, Butte AJ (2012) Ten years of pathway analysis: current approaches and outstanding challenges. PLoS Comput Biol 8:e1002375. doi:10.1371/journal.pcbi.1002375
5. Bebek G, Koyuturk M, Price ND, Chance MR (2012) Network biology methods integrating biological data for translational science. Brief Bioinform 13:446–459. doi:10.1093/bib/bbr075
6. Usadel B, Nagel A, Thimm O, Redestig H, Blaesing OE, Palacios-Rojas N et al (2005) Extension of the visualization tool MapMan to allow statistical analysis of arrays, display of corresponding genes, and comparison with known responses. Plant Physiol 138:1195–1204. doi:10.1104/pp. 105.060459
7. Thimm O, Blasing O, Gibon Y, Nagel A, Meyer S, Kruger P et al (2004) MAPMAN: a user-driven tool to display genomics data sets onto diagrams of metabolic pathways and other biological processes. Plant J 37:914–939
8. Doehlemann G, Wahl R, Horst RJ, Voll LM, Usadel B, Poree F et al (2008) Reprogramming a maize plant: transcriptional and metabolic changes induced by the fungal biotroph *Ustilago maydis*. Plant J 56:181–195. doi:10.1111/j.1365-313X.2008.03590.x
9. Tellstrom V, Usadel B, Thimm O, Stitt M, Kuster H, Niehaus K (2007) The lipopolysaccharide of *Sinorhizobium meliloti* suppresses defense-associated gene expression in cell cultures of the host plant *Medicago truncatula*. Plant Physiol 143:825–837. doi:10.1104/pp. 106.090985
10. Betti M, Pérez-Delgado C, García-Calderón M, Díaz P, Monza J, Márquez A (2012) Cellular stress following water deprivation in the model legume *Lotus japonicus*. Cells 1:1089–1106
11. Urbanczyk-Wochniak E, Usadel B, Thimm O, Nunes-Nesi A, Carrari F, Davy M et al (2006) Conversion of MapMan to allow the analysis of transcript data from solanaceous species: effects of genetic and environmental alterations in energy metabolism in the leaf. Plant Mol Biol 60:773–792. doi:10.1007/s11103-005-5772-4
12. Rotter A, Usadel B, Baebler S, Stitt M, Gruden K (2007) Adaptation of the MapMan ontology to biotic stress responses: application in solanaceous species. Plant Methods 3:10. doi:10.1186/1746-4811-3-10
13. Rotter A, Camps C, Lohse M, Kappel C, Pilati S, Hren M et al (2009) Gene expression profiling in susceptible interaction of grapevine with its fungal pathogen *Eutypa lata*: extending MapMan ontology for grapevine. BMC Plant Biol 9:104. doi:10.1186/1471-2229-9-104
14. Martinelli F, Uratsu SL, Albrecht U, Reagan RL, Phu ML, Britton M et al (2012) Transcriptome profiling of citrus fruit response to huanglongbing disease. PLoS One 7:e38039. doi:10.1371/journal.pone.0038039
15. Goffard N, Weiller GF (2006) Extending MapMan: application to legume genome arrays. Bioinformatics 22:2958–2959. doi:10.1093/bioinformatics/btl517
16. Goffard N, Weiller GF (2007) GeneBins: a database for classifying gene expression data, with application to plant genome arrays.

BMC Bioinformatics 8:87. doi:10.1186/1471-2105-8-87

17. UniProt C (2012) Reorganizing the protein space at the Universal Protein Resource (UniProt). Nucleic Acids Res 40:D71–D75. doi:10.1093/nar/gkr981, Database issue
18. Goffard N, Frickey T, Weiller GF (2009) PathExpress update: the enzyme neighbourhood method of associating gene-expression data with metabolic pathways. Nucleic Acids Res 37:W335–W339. doi:10.1093/nar/gkp432, Web server issue
19. Goffard N, Weiller GF (2007) PathExpress: a web-based tool to identify relevant pathways in gene expression data. Nucleic Acids Res 35:W176–W181. doi:10.1093/nar/gkm261, Web server issue
20. He J, Benedito VA, Wang M, Murray JD, Zhao PX, Tang Y, Udvardi MK (2009) The *Medicago truncatula* gene expression atlas web server. BMC Bioinformatics 10:441. doi:10.1186/1471-2105-10-441
21. Li D, Su Z, Dong J, Wang T (2009) An expression database for roots of the model legume *Medicago truncatula* under salt stress. BMC Genomics 10:517. doi:10.1186/1471-2164-10-517
22. Imin N, Goffard N, Nizamidin M, Rolfe BG (2008) Genome-wide transcriptional analysis of super-embryogenic *Medicago truncatula* explant cultures. BMC Plant Biol 8:110. doi:10.1186/1471-2229-8-110
23. van Noorden GE, Kerim T, Goffard N, Wiblin R, Pellerone FI, Rolfe BG, Mathesius U (2007) Overlap of proteome changes in *Medicago truncatula* in response to auxin and *Sinorhizobium meliloti*. Plant Physiol 144:1115–1131. doi:10.1104/pp. 107.099978
24. Hogslund N, Radutoiu S, Krusell L, Voroshilova V, Hannah MA, Goffard N et al (2009) Dissection of symbiosis and organ development by integrated transcriptome analysis of Lotus japonicus mutant and wild-type plants. PLoS One 4:e6556. doi:10.1371/journal.pone.0006556
25. Nautrup-Pedersen G, Dam S, Laursen BS, Siegumfeldt AL, Nielsen K, Goffard N et al (2010) Proteome analysis of pod and seed development in the model legume *Lotus japonicus*. J Proteome Res 9:5715–5726. doi:10.1021/pr100511u
26. Mathesius U, Djordjevic MA, Oakes M, Goffard N, Haerizadeh F, Weiller GF et al (2011) Comparative proteomic profiles of the soybean (*Glycine max*) root apex and differentiated root zone. Proteomics 11:1707–1719. doi:10.1002/pmic.201000619
27. Chen S, Li XQ, Zhao A, Wang L, Li X, Shi Q, Chen M, Guo J et al (2009) Genes and pathways induced in early response to defoliation in rice seedlings. Curr Issues Mol Biol 11:81–100
28. Hernandez G, Valdes-Lopez O, Ramirez M, Goffard N, Weiller GF, Aparicio-Fabre R et al (2009) Global changes in the transcript and metabolic profiles during symbiotic nitrogen fixation in phosphorus-stressed common bean plants. Plant Physiol 151:1221–1238. doi:10.1104/pp. 109.143842
29. Zhu A, Li W, Ye J, Sun X, Ding Y, Cheng Y, Deng X (2011) Microarray expression profiling of postharvest Ponkan mandarin (Citrus reticulata) fruit under cold storage reveals regulatory gene candidates and implications on soluble sugars metabolism. J Integr Plant Biol 53:358–374. doi:10.1111/j.1744-7909.2011.01035.x
30. Bay LK, Ulstrup KE, Nielsen HB, Jarmer H, Goffard N, Willis BL et al (2009) Microarray analysis reveals transcriptional plasticity in the reef building coral *Acropora millepora*. Mol Ecol 18:3062–3075. doi:10.1111/j.1365-294X.2009.04257.x
31. Bal G, Kamhieh-Milz J, Futschik M, Haupl T, Salama A, Moldenhauer A (2012) Transcriptional profiling of the hematopoietic support of interleukin-stimulated human umbilical vein endothelial cells (HUVECs). Cell Transplant 21:251–267. doi:10.3727/096368911X580581
32. Vardhanabhuti S, Blakemore SJ, Clark SM, Ghosh S, Stephens RJ, Rajagopalan D (2006) A comparison of statistical tests for detecting differential expression using Affymetrix oligonucleotide microarrays. OMICS 10:555–566. doi:10.1089/omi.2006.10.555
33. Jiao X, Sherman BT, da Huang W, Stephens R, Baseler MW, Lane HC, Lempicki RA (2012) DAVID-WS: a stateful web service to facilitate gene/protein list analysis. Bioinformatics 28:1805–1806. doi:10.1093/bioinformatics/bts251
34. Bonferroni CE (1936) Teoria statistica delle classi e calcolo delle probabilità. Pubblicazioni del Regio Istituto Superiore di Scienze Economiche e Commerciali di Firenze 8:3–62
35. Benjamini Y, Hochberg Y (1995) Controlling the false discovery rate: a practical and powerful approach to multiple testing. J R Stat Soc Ser B 57:289–300

Chapter 6

Genome-Wide Identification of MicroRNAs in *Medicago truncatula* by High-Throughput Sequencing

Tian-Zuo Wang and Wen-Hao Zhang

Abstract

MicroRNAs (miRNAs) are small, endogenous RNAs that play important regulatory roles in development and stress response in plants by negatively regulating gene expression post-transcriptionally. *Medicago truncatula* has been used as a model plant to study functional genomics of legume plants. It has also been widely used to functionally study miRNAs. Identification of miRNAs at the whole-genome level is essential for functional characterization of miRNAs in plants. High-throughput sequencing is a powerful technology to identify miRNAs. In this chapter, the methods used for construction of a small RNA library and high-throughput sequencing involving total RNA isolation, small RNA purification, adapter ligation, reverse transcription, PCR amplification, and Solexa sequencing are described. Bioinformatics and analysis of differential expression of miRNAs including primary disposal, miRNA identification, target prediction, and expression analysis are also discussed. These methodologies associated with identification and functional characterization of miRNAs may provide useful tools for readers to study miRNAs in plants in general and *Medicago truncatula* in particular.

Key words microRNA, *Medicago truncatula*, High-throughput sequencing, Bioinformatics analysis

1 Introduction

There are two most abundant classes of plant small RNAs (sRNAs), endogenous small interfering RNAs (siRNAs), and microRNAs (miRNAs). Of them, miRNAs are one type of endogenous noncoding sRNAs with the approximate length of 21 nt, initially discovered in *Caenorhabditis elegans* as developmental timing regulators [1]. The existence of miRNAs has been widely recognized in organisms including plants, mammals, and viruses. miRNAs have the ability to down-regulate gene expression by targeting corresponding messenger RNAs (mRNAs) in both plants and animals [2, 3].

In plants, the generation of miRNAs is a multistep enzymatic process. The primary transcripts of miRNAs translated by RNA polymerase II are digested to generate pre-miRNAs by DICER-LIKE1 (DCL1). The pre-miRNAs are further digested by DCL1,

Ray J. Rose (ed.), *Legume Genomics: Methods and Protocols*, Methods in Molecular Biology, vol. 1069,
DOI 10.1007/978-1-62703-613-9_6, © Springer Science+Business Media, LLC 2013

leading to the generation of miRNA/miRNA* duplexes. Mature miRNAs which depart from miRNA/miRNA* are incorporated into the RNA-induced silencing complex (RISC) with argonaute protein. miRNAs direct RISC to bind their target mRNAs for cleaving mRNAs with near-perfect complementarity and/or inhibiting translation of those with lower complementarity [4–8].

It has been widely reported that miRNAs play a regulatory role in diverse biochemical and physiological processes in plants. For instance, miRNAs have been shown to play a role in modulation of those processes associated with growth and development in plants, including seed germination [9], leaf morphogenesis [10, 11], and floral organ and root development [6, 12–15]. In addition, there is increasing evidence demonstrating that miRNAs are closely involved in responses of plants to various abiotic and biotic stresses. These include drought [16–20], cold [21–23], salinity [24], nutrition starvation [25–28], metal ion toxicity [29, 30], oxidative stress [31], submergence [32], UV-B radiation [33, 34], and virus [35, 36].

Identification of miRNAs is an essential step to functionally characterize miRNAs in plants. Identification of miRNAs from plant species was firstly done by the traditional Sanger sequencing of relatively small-size cDNA libraries. Using this method, many conserved miRNAs from different species, including Arabidopsis, rice, and poplar, were identified [37]. However, this method is time-consuming and expensive. Moreover, the expression of the non-conserved miRNAs is often low compared to that of the conserved miRNAs [38]. Therefore, many non-conserved miRNAs cannot be detected using the small-scale sequencing method. The development of high-throughput sequencing technology has overcome these problems, and this technology has successfully been used to identify miRNAs in many plant species such as Arabidopsis [39, 40], rice [41], wheat [42], soybean [43], cotton [44], grapevine [45], and poplar [46].

The application of high-throughput sequencing technology has led to systematical identification of numerous miRNAs involved in regulation of many biological and physiological processes in the legume model plant *Medicago truncatula*. According to the newest miRNA database (miRBase 19, released in Aug 2012), *M. truncatula* has the largest number of precursors and mature miRNAs among the reported plant species [47]. This approach has successfully been used to identify novel miRNAs in *M. truncatula*. For instance, Szittya et al. [48] identified 25 conserved miRNA families and 26 new miRNA candidates from *M. truncatula*. Jagadeeswaran et al. [49] identified eight novel sRNAs, of which four are annotated as legume-specific miRNAs. Moreover, the application of high-throughput sequencing has been further extended to functional characterization of miRNAs in *M. truncatula*. For example, Lelandais-Briere et al. [50] identified many miRNAs in root nodules of *M. truncatula*. Wang et al. [51] found that ten known miRNA families and eight new miRNAs were responsive to drought stress. Chen et al. [51] identified eight ethylene-responsive miRNAs [52]. In addition, the involvement of miRNAs in aluminum and mercury toxicity has also been reported in *M. truncatula* [29, 30].

Two types of sequencers have been used in high-throughput sequencing, Roche 454 sequencer and Illumina Solexa sequencer. The two sequencers differ in terms of read length and quantities of reads measured. For instance, longer reads (approx. 400 nt) can be generated by Roche 454 sequencer, while the read length produced by Illumina Solexa sequencer is usually 35 nt. In addition, Illumina Solexa sequencer can generate more reads than Roche 454 sequencer. For example, 844,110 reads and 97,028 unique sequences were obtained from *M. truncatula* by Lelandais-Briere et al. [50] using the Roche 454 sequencer. In contrast, a total of 13,683,619 reads and 4,878,445 unique sequences from two libraries of *M. truncatula* were achieved by the Illumina Solexa sequencer [51]. Given that the length of mature miRNA sequences is about 21 nt, high-throughput sequencing by Illumina Solexa sequencer is more appropriate. Therefore, in this chapter, we shall mainly discuss the methods employed in the study of plant miRNA by high-throughput sequencing using the Illumina Solexa sequencer. The method used for sequencing miRNA has been improved from the method of Hafner et al. [53] and Allen et al. [54].

2 Materials

2.1 Plant Materials

To search for novel miRNAs, one type of material from *M. truncatula* is sufficient. However, if the study aims to compare miRNA expression patterns among different organs and/or tissues, or under varying treatments, several plant materials have to be used. In this chapter, we discuss the two types of sample preparation, and refer to as Sample 1 and Sample 2.

2.2 Reagents

1. Trizol.
2. Chloroform.
3. Isopropanol.
4. 75 and 100 % Ethanol.
5. RNase-free ddH_2O.
6. RNase-free 0.4 M NaCl.
7. ssRNA ladder marker.
8. 3′ RNA adaptors: AppTC GTA TGC CGT CTT CTG CTT G-L (p: phosphate; L: 3′ OH blocking group).

9. 5′ RNA adaptors: rGrUrU rCrArGr ArGrUr UrCrUr ArCrAr GrUrCr CrGrAr CrGrAr UrC.
10. 0.1 % BSA.
11. T4 RNA ligase.
12. PEG6000.
13. 0.1 M DTT.
14. dNTPs (2 mM each).
15. SuperScript II Reverse Transcriptase (Invitrogen).
16. 150 mM KOH/20 mM Tris base.
17. 150 mM HCl.
18. 2× Phusion HF master mix (Finnzymes).
19. 5′ Primer: AAT GAT ACG GCG ACC ACC GAC AGG TTC AGA GTT CTA CAG TCC GA.
20. 3′ Primer: CAA GCA GAA GAC GGC ATA CGA.
21. Agarose gel and low-melting agarose gel.
22. 25 bp DNA ladder.
23. Water-saturated phenol (pH 7.8).
24. Phenol/chloroform/isoamyl alcohol (25:24:1, pH 8.0).
25. 0.5 M NaOH.

2.3 15 % TBE–Urea Denaturing Polyacrylamide Gel

1. 30 % stock acrylamide solution: 29 g acrylamide, 1 g bis-acrylamide. Make up to 100 mL in distilled water and filter. Stable at 4 °C for months.
2. 10× TBE buffer: Tris base 10.8 g, boric acid 5.5 g, $Na_2EDTA \cdot 2H_2O$ 0.93 g. Adjust pH to 8.3 and make up to 100 mL.
3. To make 50 mL of 15 % TBE–urea denaturing polyacrylamide gel, add 25 mL of 30 % stock acrylamide solution, 5 mL of 10× TBE buffer, and 21 g urea. Make up to 50 mL, and add 400 μL of 10 % ammonium persulfate, 40 μL *N*,*N*,*N*′,*N*′-tetramethylethylenediamine. Mix gently.

2.4 Buffers

1. 2× Loading buffer: 18 mM EDTA, 0.025 % bromophenol blue, 0.025 % xylene cyanol in formamide.
2. 10× T4 RNA ligase buffer: 500 mM Tris–HCl (pH 7.5), 100 mM $MgCl_2$, 100 mM DTT, 10 mM ATP.
3. 5× First-strand buffer: 250 mM Tris–HCl (pH 8.3), 375 mM KCl, 15 mM $MgCl_2$.

2.5 Instruments

1. Tabletop centrifuge.
2. Thermomixer.
3. Electrophoresis system.

4. Agilent 2100 (Agilent).
5. PCR instrument (Eppendorf).
6. UV transilluminator.
7. Solexa sequencer (Illumina).

3 Methods

3.1 Construction of Small RNA Library and High-Throughput Sequencing

The outline of sRNA library construction and high-throughput sequencing are shown in Fig. 1. The description of methods is given in the following section.

3.1.1 Isolation of Total RNAs

Total RNAs were extracted from Sample 1 and Sample 2 using the Trizol reagent (*see* **Notes 1–6**). Before gel purification of sRNAs, we recommend using 50 μg of total RNA per sample. As a rule of thumb, 0.1 g of leaf tissue of *M. truncatula* can yield about 20 μg total RNA.

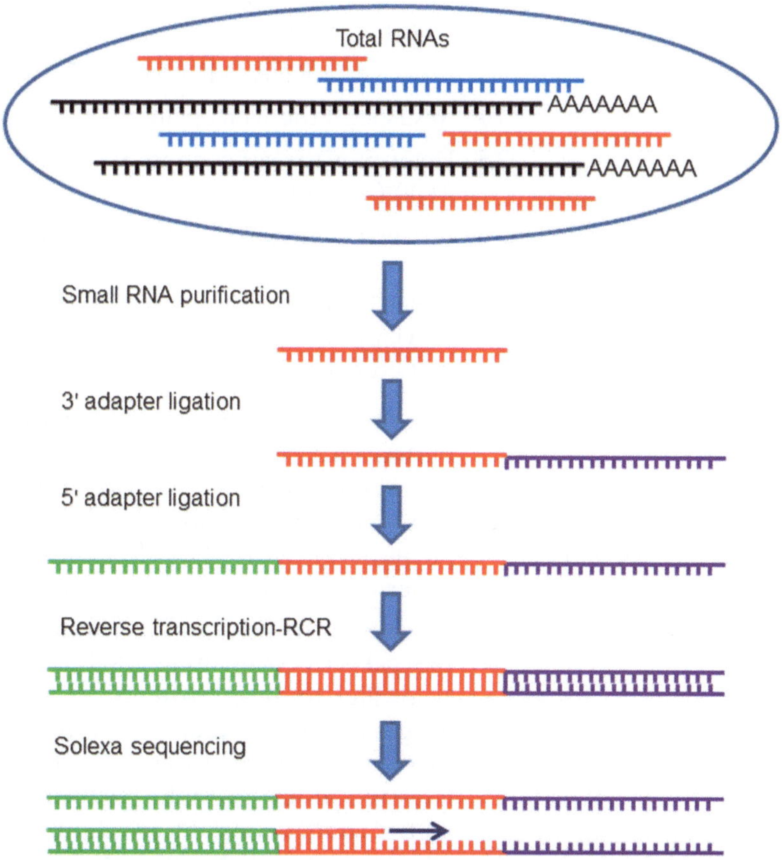

Fig. 1 Outline of small RNA library construction and high-throughput sequencing

1. Add 1 mL Trizol reagent per 0.1 g of sample, which is triturated in a mortar with liquid nitrogen.
2. Incubate the homogenized sample for 5 min at room temperature to permit complete dissociation of the nucleoprotein complex.
3. Add 0.2 mL of chloroform per 1 mL of Trizol reagent used for homogenization. The tube should be capped securely.
4. Shake tube vigorously for 15 s.
5. Incubate for 2–3 min at room temperature.
6. Centrifuge the sample at 12,000 × *g* for 10 min at 4 °C.
7. Place the aqueous phase into a new tube, add 0.5 mL of 100 % isopropanol to the aqueous phase, and mix them gently (*see* **Note 7**).
8. Incubate at room temperature for 10 min.
9. Centrifuge at 12,000 × *g* for 10 min at 4 °C.
10. Remove the supernatant from the tube, and leave the RNA pellet in the tube exclusively.
11. Wash the pellet, with 0.5 mL of 75 % ethanol.
12. Vortex the sample briefly, and then centrifuge the tube at 7,500 × *g* for 5 min at 4 °C. Discard the wash solution.
13. Air-dry the RNA pellet for 5–10 min (*see* **Note 8**).
14. Resuspend the RNA pellet in RNase-free ddH_2O.
15. Determine the concentration and quality of the RNA by Agilent 2100 (*see* **Note 9**).
16. Proceed to downstream application, or store at −70 °C.

3.1.2 Gel Purification of Small RNAs

The total RNA is size-fractionated on a 15 % TBE–urea denaturing polyacrylamide gel and the 18–30 nt sRNAs are then eluted from the excised gel slice.

1. Mix the solution of total RNA with one volume of 2× loading buffer. Denature the RNA by incubation for 30 s at 90 °C, load the RNA and ssRNA ladder marker on a 15 % denaturing polyacrylamide gel, and run the gel at 30 W until the bromophenol blue dye of the gel-loading solution appears within the lower third of the gel (*see* **Notes 10–12**).
2. Dismantle the gel, leaving it mounted on one glass plate. Cut the gel slice containing 18–30 nt sRNAs in smaller pieces according to RNA markers so that they can fit into a preweighed 1.5 mL tube. According to the weight of the gel slices, add 2–3 volumes (v/w) of RNase-free 0.4 M NaCl and elute the sRNAs from the gel by incubating the tube overnight at 4 °C under constant agitation. Collect the supernatant and add 3 volumes of absolute ethanol to precipitate the sRNAs for 2–8 h at 20 °C.

3. Collect the sRNA pellet in a tabletop centrifuge for 15 min at 14,000 ×*g* at 4 °C. Remove the supernatant without perturbing the pellet. Air-dry the RNA pellet to evaporate residual ethanol, as it may inhibit the subsequent enzymatic steps.

3.1.3 Adapter Ligation

Adapters joined to the 3′ end and 5′ end of sRNAs allow for RT-PCR amplification.

1. Ligate the 18–30 nt sRNAs with 3′ RNA adapters by T4 RNA ligase. Prepare a reaction mixture for ligation of the 3′ RNA adapters per 1 μg sRNAs by combining the following components:

sRNAs	1 μg
3′ RNA adapters (100 μM)	1 μL
10× T4 RNA ligase buffer	5 μL
0.1 % BSA	3 μL
T4 RNA ligase	40–50 U
PEG6000	Final 25 %
RNase-free ddH$_2$O	Up to 50 μL

Mix gently and react at 20 °C for 18 h.

2. Gel purification as described in Subheading 3.1.2.
3. Ligate sRNAs with 5′ RNA adapters: follow the way of ligation of the 3′ RNA adapters.
4. Gel purification following the protocols in Subheading 3.1.2.
5. Dissolve the pellet in 5.6 μL RNase-free ddH$_2$O for each sample.

3.1.4 Reverse Transcription

1. Denature the RNA and 3′ primer by incubating the tube for 30 s at 90 °C and transfer the tube to a 50 °C incubator.
2. Add the following reaction mix to each sample:

5× First-strand buffer	3 μL
0.1 M DTT	1.5 μL
10× dNTPs (2 mM each)	4.2 μL

3. Add 0.75 μL of Superscript II reverse transcriptase and mix gently. Incubate for 40 min at 42 °C.
4. Add 40 μL of 150 mM KOH/20 mM Tris base and incubate for 10 min at 90 °C to hydrolyze the RNA.
5. Neutralize the solution by addition of 40 μL of 150 mM HCl to pH 7.0–9.5; thus the subsequent PCR is not inhibited (*see* **Note 13**).

3.1.5 PCR Amplification

1. Perform standard 100 μL PCR reactions for each sample by combining the following components (*see* **Note 14**):

2× Phusion HF master mix	50 μL
cDNA	10 μL
5′ Primer (10 μM)	5 μL
3′ Primer (10 μM)	5 μL
RNase-free ddH_2O	Up to 100 μL

Mix them gently.

2. Use the following cycle conditions for PCR amplification:

Initial denaturation	98 °C	30 s
(15 + *n*) cycles	98 °C 50 °C 72 °C	15 s 85 s 60 s
Final extension	72 °C	10 min
Hold	4 °C	∞

3. Analyze the samples on a 2.5 % agarose gel from the 15th cycle to define the optimal cycle number. The PCR product may appear as a double band with the higher band occurring at the expected length of about 70–80 bp and a lower band corresponding to the 3′ adapter to 5′ adapter ligation products occurring at about 50 bp (*see* **Note 15**).
4. Perform three standard 100 μL PCR reactions, to obtain enough product, with the optimal cycle number (*see* **Note 16**).
5. Purify the PCR product of about 70–80 bp. Run the production of PCR in a 3 % low-melting agarose gel containing a 25 bp DNA ladder for approx. 1 h at 100 V until the marker bands are sufficiently resolved. Excise the band of approximately 70–80 bp on a UV transilluminator. Transfer the gel slice to a 1.5 mL reaction tube and add RNase-free 0.4 M NaCl to a total volume of 500 μL. Melt the agarose at 70 °C for 10 min. Add 1 volume of preheated water-saturated phenol (pH 7.8). Vortex the solutions vigorously and immediately separate the phases by centrifuging at 14,000×*g* for 5 min in a tabletop centrifuge. Collect the solution at the upper phase and extract it in a new tube with 1 volume phenol/chloroform/isoamyl alcohol (25:24:1, pH 8.0), and 1 volume chloroform. Precipitate the DNA by addition of 2 volumes of absolute ethanol and incubation for at least 1 h at 20 °C. Collect the DNA pellet in a tabletop centrifuge at 14,000×*g* at 4 °C for 15 min. Remove the supernatant without disturbing the pellet. Air-dry the DNA pellet to evaporate residual ethanol.

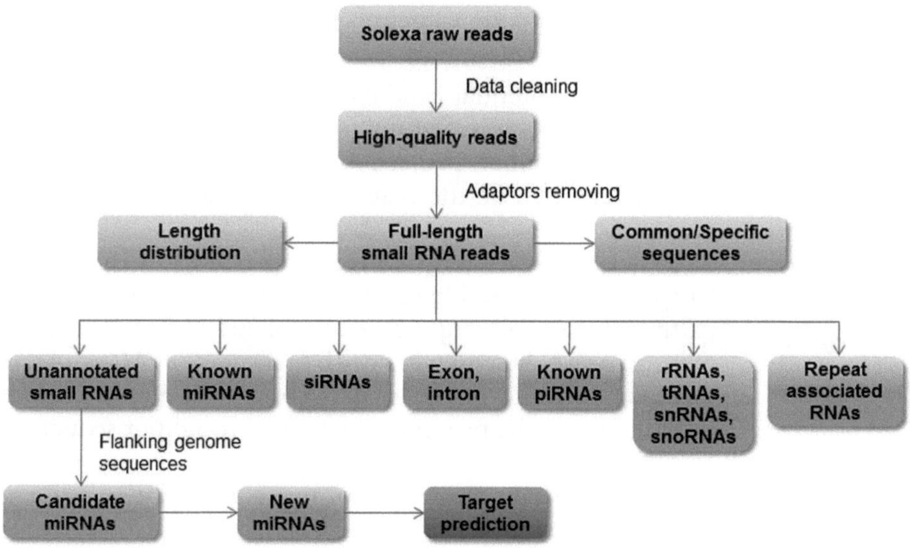

Fig. 2 Flowchart of bioinformatics analysis

3.1.6 Solexa Sequencing

The purified DNA samples are now ready for sequencing by the Illumina Solexa sequencer. The high-throughput sequencing is performed following the manufacturer's instruction.

3.2 Bioinformatics Analysis

Figure 2 shows the procedures sued for bioinformatics analysis. The detailed description of the protocols is given in the following section.

3.2.1 Primary Disposal

1. The raw reads from Solexa sequencing should firstly go through the data cleaning. This includes removal of the low-quality tags and several kinds of contaminants such as tags less than 18 nt, polyA and those tags without adaptors.
2. Remove adaptor sequences of the sRNA reads with exact matches to the adaptor sequences.
3. Analyze the length distribution and common/specific sequences between Sample 1 and Sample 2, including summary of unique tags and total tags.
4. Map high-quality reads to the *M. truncatula* genome sequence (Mt3.5) downloaded from http://medicago.jcvi.org/cgi-bin/medicago/download.cgi/ using SOAP [55] to analyze their expression and distribution on the genome.

3.2.2 Identification of miRNA

1. Remove rRNAs, tRNAs, siRNAs, snRNAs, and snoRNAs from the matched sequences through BLASTn search [56] using NCBI Genbank database http://www.ncbi.nlm.nih.gov/blast/Blast.cgi/ and Rfam database http://www.sanger.ac.uk/Software/Rfam/. Mismatches are not allowed in the above two approaches (*see* **Note 17**).

2. Align the unique sequences with known miRNAs from miRBase 19 (http://www.mirbase.org/) [57].
3. Identify potential candidate miRNAs by folding the flanking genome sequences of unique sRNAs using MIREAP (https://sourceforge.net/projects/mireap/) (*see* **Note 18**). Parameters are set based on the criteria for annotation [58].

3.2.3 Target Prediction

Targets of miRNA are predicted using tools of the psRNATarget (http://plantgrn.noble.org/psRNATarget/) and the srna-tools (http://srna-tools.cmp.uea.ac.uk/plant/) [59], by aligning mature miRNA sequences with genome of *M. truncatula* and *A. thaliana*. Target can be validated by 5′ RACE or degradome analysis that also can be processed by high-throughput sequencing [60].

3.3 Analysis of Differential Expression

1. Normalize the frequency of miRNAs of the two libraries to 1 million by total number of miRNAs in each sample (*see* **Note 19**). Normalized expression = actual miRNA count/total count of clean reads × 1,000,000.
2. Calculate the fold change for each miRNA between Sample 1 and Sample 2. If Sample 1 sets as a control, fold-change = $\log_2$ (Sample 2/Sample 1). A fold-change >0 means that the miRNA expression of Sample 2 is higher than that of Sample 1, while a fold-change <0 indicates that the miRNA expression of Sample 2 is lower than that of Sample 1. Perform statistics analysis according to Poisson distribution. If $P \leq 0.05$, the differences are referred to as significant.
3. Validate the differential expression by real-time quantitative PCR or sRNA northern blot.

4 Notes

1. Soak mortars in 0.5 M NaOH for 30 min, and rinse them thoroughly with water. Then bake them at 180 °C for 8 h to inactivate RNases.
2. Wear disposable gloves. Hands often contain bacteria and molds that can contaminate RNA, and introduce RNases.
3. Use disposable RNase-free plasticware and pipettes for RNA work to prevent cross-contamination with RNases.
4. All of the reagents used in isolation and transcriptional of RNA must be RNase-free.
5. Freeze tissue samples immediately with liquid nitrogen upon collection. The samples must be fresh or stored at −70 °C prior to trituration.

6. Trizol reagent is harmful if handled improperly. Work related to Trizol reagent must be done in a fume hood. If Trizol reagent contacts with skin or eyes, wash the exposed area immediately with copious amounts of water for 15 min and seek medical attention if necessary.
7. Avoid drawing the organic layer into the pipette during removing the aqueous phase.
8. Do not allow the RNA to dry completely, because the solubility of the pellet may be lost.
9. Total RNAs should be isolated again if 28S and 18S bands tested by Agilent 2100 are not distinct and sharp.
10. Avoid overloading the pocket of the gel to ensure good fractionation of the RNAs.
11. Leave at least one well empty between the different samples to avoid cross-contamination.
12. Do not run the gel excessively to ensure that the sample contains sRNAs as small as possible.
13. The pH value of cDNA templates should be between 7.0 and 9.5 so that the PCR is not inhibited. If necessary, readjust the pH by adding base or acid.
14. Perform a negative control to check DNA contamination in the reaction mixture.
15. The number of PCR cycles can be adjusted according to gel analysis of different cycles. The gel might appear as a double band with the higher band at 70–80 bp and a lower band at about 50 bp. Never amplify for more than 30 cycles because it will distort the sRNA ratios.
16. Define the optimal cycle number within the exponential amplification phase of the PCR.
17. In the process of miRNA identification, siRNAs, snRNAs, and snoRNAs have to be removed. However, if these sRNAs are studied, they can be used for further analyses following the protocols similar to analysis of miRNAs described in the above section.
18. The new miRNAs can be named after submitting to miRBase (http://www.mirbase.org/).
19. Direct comparison of the sequencing number for a certain miRNA among libraries is not allowed, as the sequencing amounts of each library can be different. Normalization is required prior to comparison of sequencing number among different libraries.

References

1. Lee RC, Feinbaum RL, Ambros V (1993) The *C. elegans* heterochronic gene lin-4 encodes small RNAs with antisense complementarity to lin-14. Cell 75:843–854
2. Jones-Rhoades MW, Bartel DP, Bartel B (2006) MicroRNAs and their regulatory roles in plants. Annu Rev Plant Biol 57:19–53
3. Khraiwesh B, Arif MA, Seumel GI et al (2010) Transcriptional control of gene expression by microRNAs. Cell 140:111–122
4. Bartel DP (2004) MicroRNAs: genomics, biogenesis, mechanism, and function. Cell 116: 281–297
5. Llave C, Xie Z, Kasschau KD et al (2002) Cleavage of Scarecrow-like mRNA targets directed by a class of Arabidopsis miRNA. Science 297:2053–2056
6. Chen X (2004) A microRNA as a translational repressor of APETALA2 in Arabidopsis flower development. Science 303:2022–2025
7. Brodersen P, Sakvarelidze-Achard L, Bruun-Rasmussen M et al (2008) Widespread translational inhibition by plant miRNAs and siRNAs. Science 320:1185–1190
8. Kurihara Y, Watanabe Y (2004) Arabidopsis micro-RNA biogenesis through Dicer-like 1 protein functions. Proc Natl Acad Sci USA 101:12753–12758
9. Reyes JL, Chua NH (2007) ABA induction of miR159 controls transcript levels of two MYB factors during Arabidopsis seed germination. Plant J 49:592–606
10. Palatnik JF, Allen E, Wu X et al (2003) Control of leaf morphogenesis by microRNAs. Nature 425:257–263
11. Liu Z, Jia L, Wang H et al (2011) HYL1 regulates the balance between adaxial and abaxial identity for leaf flattening via miRNA-mediated pathways. J Exp Bot 62: 4367–4381
12. Aukerman MJ, Sakai H (2003) Regulation of flowering time and floral organ identity by a MicroRNA and its APETALA2-like target genes. Plant Cell 15:2730–2741
13. Guo HS, Xie Q, Fei JF et al (2005) MicroRNA directs mRNA cleavage of the transcription factor NAC1 to downregulate auxin signals for arabidopsis lateral root development. Plant Cell 17:1376–1386
14. Wang JW, Wang LJ, Mao YB et al (2005) Control of root cap formation by MicroRNA-targeted auxin response factors in Arabidopsis. Plant Cell 17:2204–2216
15. Boualem A, Laporte P, Jovanovic M et al (2008) MicroRNA166 controls root and nodule development in *Medicago truncatula*. Plant J 54:876–887
16. Li WX, Oono Y, Zhu JH et al (2008) The Arabidopsis NFYA5 transcription factor is regulated transcriptionally and posttranscriptionally to promote drought resistance. Plant Cell 20:2238–2251
17. Trindade I, Capitao C, Dalmay T et al (2010) miR398 and miR408 are up-regulated in response to water deficit in *Medicago truncatula*. Planta 231:705–716
18. Wei LY, Zhang DF, Xiang F et al (2009) Differentially expressed miRNAs potentially involved in the regulation of defense mechanism to drought stress in maize seedlings. Int J Plant Sci 170:979–989
19. Zhao B, Liang R, Ge L et al (2007) Identification of drought-induced microRNAs in rice. Biochem Biophys Res Commun 354:585–590
20. Zhou L, Liu Y, Liu Z et al (2010) Genome-wide identification and analysis of drought–responsive microRNAs in *Oryza sativa*. J Exp Bot 61:4157–4168
21. Lv DK, Bai X, Li Y et al (2010) Profiling of cold-stress-responsive miRNAs in rice by microarrays. Gene 459:39–47
22. Zhang J, Xu Y, Huan Q et al (2009) Deep sequencing of Brachypodium small RNAs at the global genome level identifies microRNAs involved in cold stress response. BMC Genomics 10:449
23. Zhou X, Wang G, Sutoh K et al (2008) Identification of cold-inducible microRNAs in plants by transcriptome analysis. Biochim Biophys Acta 1779:780–788
24. Zhao B, Ge L, Liang R et al (2009) Members of miR-169 family are induced by high salinity and transiently inhibit the NF-YA transcription factor. BMC Mol Biol 10:29
25. Bari R, Pant BD, Stitt M et al (2006) PHO2, microRNA399, and PHR1 define a phosphate-signaling pathway in plants. Plant Physiol 141:988–999
26. Chiou TJ, Aung K, Lin SI et al (2006) Regulation of phosphate homeostasis by MicroRNA in Arabidopsis. Plant Cell 18:412–421
27. Jones-Rhoades MW, Bartel DP (2004) Computational identification of plant microRNAs and their targets, including a stress-induced miRNA. Mol Cell 14:787–799
28. Zeng HQ, Zhu YY, Huang SQ et al (2010) Analysis of phosphorus-deficient responsive miRNAs and cis-elements from soybean (*Glycine max* L.). J Plant Physiol 167:1289–1297

29. Zhou ZS, Zeng HQ, Liu ZP et al (2012) Genome-wide identification of *Medicago truncatula* microRNAs and their targets reveals their differential regulation by heavy metal. Plant Cell Environ 35:86–99
30. Chen L, Wang TZ, Zhao MG et al (2012) Identification of aluminum-responsive microRNAs in *Medicago truncatula* by genome-wide high-throughput sequencing. Planta 235: 375–386
31. Sunkar R, Kapoor A, Zhu JK (2006) Posttranscriptional induction of two Cu/Zn superoxide dismutase genes in Arabidopsis is mediated by downregulation of miR398 and important for oxidative stress tolerance. Plant Cell 18:2051–2065
32. Zhang Z, Wei L, Zou X et al (2008) Submergence-responsive MicroRNAs are potentially involved in the regulation of morphological and metabolic adaptations in maize root cells. Ann Bot 102:509–519
33. Jia X, Ren L, Chen QJ et al (2009) UV-B-responsive microRNAs in *Populus tremula*. J Plant Physiol 166:2046–2057
34. Zhou X, Wang G, Zhang W (2007) UV-B responsive microRNA genes in *Arabidopsis thaliana*. Mol Syst Biol 3:103
35. Bazzini AA, Hopp HE, Beachy RN et al (2007) Infection and coaccumulation of tobacco mosaic virus proteins alter microRNA levels, correlating with symptom and plant development. Proc Natl Acad Sci USA 104:12157–12162
36. He XF, Fang YY, Feng L et al (2008) Characterization of conserved and novel microRNAs and their targets, including a TuMV-induced TIR-NBS-LRR class R gene-derived novel miRNA in Brassica. FEBS Lett 582:2445–2452
37. Axtell MJ, Bartel DP (2005) Antiquity of microRNAs and their targets in land plants. Plant Cell 17:1658–1673
38. Allen E, Xie Z, Gustafson AM et al (2004) Evolution of microRNA genes by inverted duplication of target gene sequences in Arabidopsis thaliana. Nat Genet 36:1282–1290
39. Rajagopalan R, Vaucheret H, Trejo J et al (2006) A diverse and evolutionarily fluid set of microRNAs in *Arabidopsis thaliana*. Genes Dev 20:3407–3425
40. Fahlgren N, Howell MD, Kasschau KD et al (2007) High-throughput sequencing of Arabidopsis microRNAs: evidence for frequent birth and death of MIRNA genes. PLoS One 2:e219
41. Sunkar R, Zhou X, Zheng Y et al (2008) Identification of novel and candidate miRNAs in rice by high throughput sequencing. BMC Plant Biol 8:25
42. Xin M, Wang Y, Yao Y et al (2010) Diverse set of microRNAs are responsive to powdery mildew infection and heat stress in wheat (*Triticum aestivum* L.). BMC Plant Biol 10:123
43. Subramanian S, Fu Y, Sunkar R et al (2008) Novel and nodulation-regulated microRNAs in soybean roots. BMC Genomics 9:160
44. Kwak PB, Wang QQ, Chen XS et al (2009) Enrichment of a set of microRNAs during the cotton fiber development. BMC Genomics 10:457
45. Pantaleo V, Szittya G, Moxon S et al (2010) Identification of grapevine microRNAs and their targets using high-throughput sequencing and degradome analysis. Plant J 62:960–976
46. Barakat A, Wall PK, Diloreto S et al (2007) Conservation and divergence of microRNAs in Populus. BMC Genomics 8:481
47. Kozomara A, Griffiths-Jones S (2011) miRBase: integrating microRNA annotation and deep-sequencing data. Nucleic Acids Res 39:D152–D157
48. Szittya G, Moxon S, Santos DM et al (2008) High-throughput sequencing of *Medicago truncatula* short RNAs identifies eight new miRNA families. BMC Genomics 9:593
49. Jagadeeswaran G, Zheng Y, Li YF et al (2009) Cloning and characterization of small RNAs from *Medicago truncatula* reveals four novel legume-specific microRNA families. New Phytol 184:85–98
50. Lelandais-Briere C, Naya L, Sallet E et al (2009) Genome-wide *Medicago truncatula* small RNA analysis revealed novel microRNAs and isoforms differentially regulated in roots and nodules. Plant Cell 21: 2780–2796
51. Wang TZ, Chen L, Zhao MG et al (2011) Identification of drought-responsive microRNAs in *Medicago truncatula* by genome-wide high-throughput sequencing. BMC Genomics 12:367
52. Chen L, Wang TZ, Zhao MG et al (2012) Ethylene-responsive miRNAs in roots of *Medicago truncatula* identified by high-throughput sequencing at whole genome level. Plant Sci 184:14–19
53. Hafner M, Landgraf P, Ludwig J et al (2008) Identification of microRNAs and other small regulatory RNAs using cDNA library sequencing. Methods 44:3–12
54. Allen E, Xie Z, Gustafson AM et al (2005) microRNA-directed phasing during trans-acting siRNA biogenesis in plants. Cell 121:207–221
55. Li R, Li Y, Kristiansen K et al (2008) SOAP: short oligonucleotide alignment program. Bioinformatics 24:713–714

56. Altschul SF, Gish W, Miller W et al (1990) Basic local alignment search tool. J Mol Biol 215:403–410
57. Griffiths-Jones S, Saini HK, van Dongen S et al (2008) miRBase: tools for microRNA genomics. Nucleic Acids Res 36:D154–D158
58. Meyers BC, Axtell MJ, Bartel B et al (2008) Criteria for annotation of plant MicroRNAs. Plant Cell 20:3186–3190
59. Moxon S, Schwach F, Dalmay T et al (2008) A toolkit for analysing large-scale plant small RNA datasets. Bioinformatics 24:2252–2253
60. German MA, Luo S, Schroth G et al (2009) Construction of parallel analysis of RNA ends (PARE) libraries for the study of cleaved miRNA targets and the RNA degradome. Nat Protoc 4:356–362

Chapter 7

Determining Abundance of MicroRNAs and Other Small RNAs in Legumes

Cecilia Contreras-Cubas, Alejandra A. Covarrubias, and José Luis Reyes

Abstract

High-throughput sequencing techniques have been applied to the discovery of several different regulatory small RNAs, including siRNAs and microRNAs in different plant species. Their subsequent characterization demands the use of sensitive and quantitative methods for their detection. Here we describe the use of northern blot and quantitative PCR for these purposes. We highlight the advantages and shortcomings of each method and offer a detailed description of those techniques that best work in our hands, in particular having in mind their use for the study of several small RNAs in a given sample. We enumerate relevant alternatives as well as cautionary comments in cases where we have detected potential difficulties.

Key words Plant sRNAs, Plant miRNAs, qPCR, Northern blot

1 Introduction

The advent of high-throughput sequencing for small RNAs (sRNAs) has allowed for large-scale identification of microRNAs (miRNAs) and other small regulatory RNAs involved in multiple processes in different plant species [1, 2]. As a consequence of the enormous amount of data generated by these methods, an ensuing challenge is to define the contribution of individual RNA molecules to the process under study and the corresponding need to use an independent method to determine miRNA or other small RNA abundance. Different techniques have been successfully employed to that end, and in particular two have been used extensively and proven to be amenable to accurately determine the abundance of miRNAs and other small RNAs. First, northern blot analysis can be used to conveniently determine miRNA abundance in multiple samples in the same experiment, with the flexibility to reutilize the same blotted membrane to study other RNAs, including constitutively expressed RNAs serving as loading controls. Alternatively, a more sensitive approach consists of the reverse transcription and quantitative PCR amplification (qPCR) of sRNAs to monitor their abundance.

Ray J. Rose (ed.), *Legume Genomics: Methods and Protocols*, Methods in Molecular Biology, vol. 1069,
DOI 10.1007/978-1-62703-613-9_7, © Springer Science+Business Media, LLC 2013

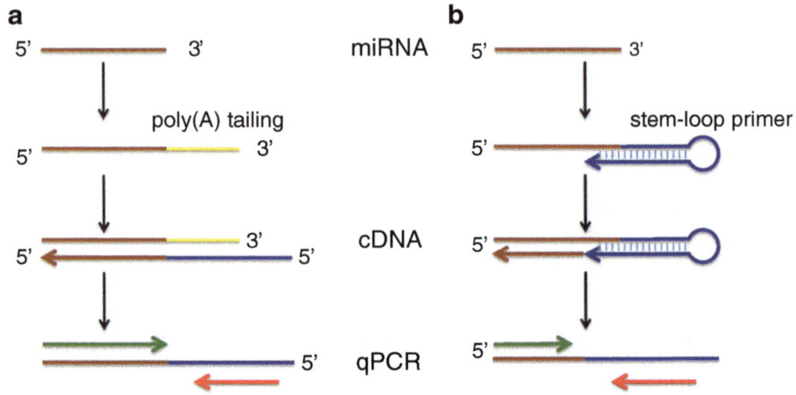

Fig. 1 miRNA qPCR protocols. Both protocols rely on the addition of a known nucleotide sequence to the miRNA so that cDNA synthesis can be performed. (**a**) Poly(A) tailing of total RNA allows for synthesis of the cDNA of the polyadenylated miRNA and the subsequent qPCR reaction. One of the main advantages of this protocol is that a large set of miRNAs and other sRNAs can be analyzed using the same cDNA sample. (**b**) A stem-loop DNA oligonucleotide complementary to the last 4–6 nucleotides at the 3′-end of the miRNA is used as primer for the cDNA synthesis. Separate cDNA synthesis reactions need to be performed per miRNA analyzed. Once the cDNA is synthesized in both protocols, qPCR is accomplished using either a TaqMan probe or SYBR *Green*. The diagrams are taken and modified from the Ncode miRNA First-Strand cDNA Synthesis brochure (Invitrogen) and from Chen et al. [3]

The use of qPCR for miRNA detection is a recurring method used for the detection of both animal and plant miRNAs. The most common protocols are based on the use of a stem-loop primer that base pairs to the 3′-end of the miRNA to prime the cDNA synthesis reaction or alternatively, based on the polyadenylation of sRNAs prior to cDNA production (Fig. 1) [3, 4]. Both protocols consist of a Reverse Transcriptase reaction followed by qPCR amplification for specific miRNA detection. There are currently several commercial kits available, such as the Ncode miRNA First-Strand cDNA Synthesis from Invitrogen or the Mir-X miRNA First-Strand Synthesis from Clontech. Both commercial kits rely on the polyadenylation of sRNAs and the subsequent cDNA synthesis using a complementary oligo-dT primer. qPCR can be performed using TaqMan probe or SYBR Green. Here we describe a protocol using the Ncode miRNA First-Strand cDNA Synthesis kit followed by qPCR using SYBR green.

The choice of northern blot or qPCR methods depends on multiple factors. First, the amount of RNA material needed is very different: in the range of 10–20 μg of total RNA for northern blot analysis as compared to only 100 ng to 2 μg for RT-PCR. Second, under the appropriate conditions qPCR can distinguish individual miRNA family members that may differ by one or two nucleotides.

In contrast, due to cross-hybridization the use of northern blot analysis will, in general, be useful to monitor the combined abundance of various members of a microRNA family present in a given sample. Third, for the analysis of very low abundance microRNAs or other small RNAs, qPCR is better suited due to its higher sensitivity when compared to northern blot.

Because both approaches can be advantageous depending on the scope of the experiment, we will describe both techniques in detail and it is up to the researcher to choose one, based on particular experimental conditions and specific goals to achieve; we also provide guidelines based on our experience as to what kind of alternatives may arise and ways to solve potential obstacles.

2 Materials

2.1 Northern Blot Analysis

2.1.1 Total RNA Preparation

1. Liquid nitrogen.
2. Mortar and pestle (previously autoclaved and stored at −20 °C).
3. Trizol reagent (Life Technologies).
4. Sterile distilled water.
5. 50 % Formamide solution in water, deionized.
6. Low retention 1.5 mL microcentrifuge tubes (Eppendorf).
7. Chloroform.
8. Isopropanol.
9. 100 % ethanol.
10. Tabletop microcentrifuge.

2.1.2 RNA Separation in Polyacrylamide Gels

1. Acrylamide solution: 15 % acrylamide–bis-acrylamide (19:1), 8 M urea, 1× TBE (89 mM Tris-borate pH 8.3, 2 mM EDTA pH 8.0). Store solution in an amber bottle at 4 °C for up to a month or at −20 °C for longer storage.
2. RNA loading buffer: 95 % formamide, 25 mM EDTA pH 8.0, 0.1 % xylene cyanol, 0.1 % Bromophenol Blue.
3. 0.5× TBE running buffer: 45 mM Tris-borate pH 8.3, 1 mM EDTA pH 8.0.
4. Vertical electrophoresis apparatus for small gels.
5. Semidry electrotransfer apparatus.
6. Hybond-N+ membrane (Amersham).
7. 3 MM chromatography paper (Whatman).
8. UV cross-linker.

2.1.3 Probe Hybridization

1. Hybridization bottles and temperature-controlled incubator.
2. ULTRAhyb-Oligo solution (Ambion).

3. Wash solution: 2× SSC, 0.1 % SDS.
4. Stripping solution: 0.1× SSC, 0.1 % SDS. Only if membrane will be reused.
5. PhosphorImager exposure screen and PhosphorImager scanner (Typhoon, Amersham).

2.1.4 5′-End Labeling and Purification of Oligonucleotide Probe

1. DNA oligonucleotide fully complementary to microRNA to analyze.
2. DNA oligonucleotide complementary to a region of a constitutive RNA (U6 snRNA or 5S rRNA are typical loading controls).
3. γP^{32}-ATP (3,000 Ci/mmol, Perkin Elmer).
4. T4 Polynucleotide Kinase (Fermentas).
5. 10× Polynucleotide kinase buffer (provided with enzyme).
6. Sephadex G25 column (mini Quick-spin Oligo column, Roche).
7. STE buffer: 10 mM Tris–HCl pH 8.0, 1 mM EDTA, 100 mM NaCl.
8. Tabletop microcentrifuge.
9. Heat block or incubator for 1.5 mL microcentrifuge tubes set at 37 °C.

2.2 Quantitative PCR for miRNAs

2.2.1 sRNA Polyadenylation

1. 1.5-mL RNase-free microcentrifuge tubes.
2. RNase-free pipette tips.
3. 10 ng to 2.5 μg total RNA.
4. 5× miRNA Reaction buffer (provided with the Ncode miRNA cDNA Synthesis kit).
5. 10 mM ATP (provided with the Ncode kit, Invitrogen).
6. 25 mM $MnCl_2$ (provided with the Ncode kit).
7. Poly-A Polymerase (provided with the Ncode kit).
8. RNase-free H_2O.
9. Heat block or incubator for 1.5 mL microcentrifuge tubes set at 37 °C.

2.2.2 Reverse Transcription

1. Polyadenylated RNA.
2. Annealing buffer (provided with the Ncode kit).
3. Universal RT primer (25 μM) (provided with the Ncode kit).
4. 2× First-Strand Reaction Mix (provided with the Ncode kit).
5. Super Script™ III RT/RNase Out™ Enzyme Mix (provided with the Ncode kit).
6. Heat block or incubator for 1.5 mL microcentrifuge tubes for use at 65, 50, and 85 °C.

2.2.3 qPCR

1. Calibrated micropipettes (a set of pipettes dedicated for qPCR).
2. qPCR tubes or plates.
3. Optical tube-caps or optical adhesive film.
4. qPCR Thermal cycler instrument. Here we present conditions used in the iCycler iQ™5 PCR Thermal Cycler (Bio-Rad) and the StepOne™ Thermal Cycler (Life Technologies).
5. SYBR Green Master Mix (Maxima SYBR Green/ROX qPCR Master Mix 2×, Fermentas).
6. miRNA specific forward primer (10 μM).
7. Universal qPCR primer (10 μM) (provided with the Maxima kit).
8. RNase-Free H_2O.
9. miRNA specific primer. For recommendations on primer design *see* **Note 1**.
10. Reference gene primer. For guidelines on selecting a reference gene *see* **Note 2**.

3 Methods

3.1 Northern Blot Analysis

3.1.1 RNA Purification

1. Grind tissue under liquid nitrogen to a fine powder using a mortar and pestle. Store powdered tissue at −70 °C if not proceeding to RNA preparation.
2. Resuspend as much as 100 mg of tissue in 1 mL of Trizol solution, using low-retention 1.5 mL microcentrifuge tubes. Allow 5 min for complete dissociation of protein–RNA complexes. Centrifuge at 12,000 rpm in a tabletop microcentrifuge for 3 min to pellet fibrous material.
3. Add 100 μL of chloroform and shake by hand for 30 s. Separate phases by centrifugation in a tabletop microcentrifuge, for 10 min at 12,000 rpm.
4. Recover aqueous phase and add 600 μL of isopropanol. Allow for RNA precipitation 10 min at room temperature.
5. Centrifuge 15 min at 12,000 rpm in a tabletop microcentrifuge, remove supernatant with an automatic pipette and wash pellet (often invisible) with 200 μL of 100 % ethanol. There is no need for further sRNA enrichment. *See* **Note 3.**
6. After centrifugation for 3 min at 12,000 rpm in a tabletop microcentrifuge, remove supernatant and repeat centrifugation. The remaining few drops of ethanol can be removed with a 2 or 10 μL micropipette. The RNA pellet is resuspended in 30–50 μL of 50 % formamide solution (*see* **Note 4**). If necessary, the solution can be warmed up to 65 °C for 5 min to help

resuspension and immediately returned to ice. Determine RNA concentration in a NanoDrop and check integrity by RNA electrophoresis in agarose gels or with a Bioanalyzer (Applied Biosystems). Store RNA at −20 °C until use.

3.1.2 RNA Gel Electrophoresis

1. Mix 10–20 μg of total RNA with RNA loading buffer to a convenient volume. Because the sample is already in 50 % formamide the volume can be easily adjusted while maintaining denaturing conditions for the sample.
2. Separate RNA samples using a 15 % polyacrylamide–8 M urea–1× TBE minigel. Electrophorese at 20 mA (~1 h for 8 × 10 cm minigels, *see* **Note 5**).
3. Place a Hybond-N+ membrane (soaked in 0.5× TBE buffer) in direct contact with the polyacrylamide gel, between two layers of Whatman 3 MM paper previously soaked in 0.5× TBE buffer. Assemble semidry apparatus and transfer the resolved RNA to the membrane for 35 min at 100 mA. *See* **Note 6.**
4. Cross-link RNA to the membrane with a UV cross-linker using the standard settings, twice.

3.1.3 Hybridization

1. Prehybridize at least 30 min in Ambion ULTRAhyb-Oligo solution (10 mL for a 10 × 10 cm membrane) in a hybridization bottle at 42 °C with rotation.
2. Add 5′-end labeled oligonucleotide probe (*see* Subheading 3.1.4 for its preparation and *see* **Note 7**) and continue hybridization at 42 °C overnight (at least 10 h).
3. Carefully remove hybridization solution (*see* **Note 8**) and perform two washes in wash solution (2× SSC, 0.1 % SDS) for 30 min at 42 °C.
4. Remove excess liquid and wrap membrane with saran wrap avoiding wrinkles and bubbles as much as possible. Expose to a PhosphorImager screen before scanning in a Typhoon PhosphorImager. *See* **Notes 9** and **10**.
5. Stripping. If rehybridization is to be performed, wash membrane in boiling stripping solution (0.1× SSC, 0.1 % SDS), 15 min while shaking, and repeat wash with fresh boiling stripping solution. Discard used wash solutions as radioactive material. Check membrane by exposing to PhosphorImager before a second hybridization.

3.1.4 5′-End Labeling of Oligonucleotides

1. Set the following reaction on ice:

   3–4 μL γP^{32}-ATP (10–13 pmol of ATP).

   1 μL Oligonucleotide (10 pmol).

   1 μL 10× buffer.

1 μL T4 polynucleotide kinase.

H_2O to 10 μL.

2. Incubate at 37 °C for 1 h.
3. Purify labeled oligonucleotide from unincorporated nucleotide and enzyme using a Sephadex G25 column (Quick-spin mini column, Roche) according to manufacturer directions.
4. Quantify ^{32}P incorporation efficiency using a scintillation counter or similar. Typically 30–40 % incorporation efficiency is adequate.

3.2 Quantitative PCR Analysis for Small RNAs

3.2.1 RNA Purification

1. Purify total RNA using Trizol regent as previously described (Subheading 3.1.1). There is no need for additional sRNA enrichment.

3.2.2 RNA Polyadenylation

2. Set up the sRNA polyadenylation reaction on ice. Mix 10 ng to 2.5 μg of total RNA with the proper concentration of ATP (*see* **Note 11**), 5× miRNA Reaction Buffer, 25 mM $MnCl_2$, Poly A Polymerase to a final volume of 25 μL, according to the manufacturer's directions (Ncode kit). Incubate the reaction in a heating block or incubator at 37 °C for 15 min.
3. After the polyadenylation reaction proceed directly to the first strand synthesis (*see* **Note 12**).

3.2.3 Reverse Transcription

1. For the first strand synthesis mix 4 μL of the polyadenylated RNA with 1 μL of annealing buffer and 3 μL of universal RT primer (25 μM) and incubate the reaction at 65 °C for 5 min.
2. Place the tube on ice for 1 min. Then add 10 μL of the 2× First-Strand Reaction Mix and 2 μL of the SuperScript™ III/RNaseOUT™ Enzyme Mix. Centrifuge briefly and incubate at 50 °C for 50 min (*see* **Note 13**).
3. For inactivation of the SuperScript™ III enzyme incubate at 85 °C for 5 min.
4. Proceed directly to qPCR or store at −20 °C for up to 2 weeks.

3.2.4 Quantitative PCR of Small RNAs

1. For the qPCR amplification, prepare a common mix for as many reactions as possible to minimize individual sample variations (*see* **Note 14**). Add the 2× SYBR green/reference fluorophore mix, the miRNA specific forward primer (10 μM), the Universal qPCR primer (10 μM), cDNA sample and RNase-free water up to 25 μL per individual reaction (*see* **Notes 15** and **16**).
2. Distribute each reaction volume to the qPCR tubes or plates and carefully cap each tube or place the adhesive film to the plate.

3. For particular qPCR thermal cycler an optimal qPCR run method is specified in the Ncode™ miRNA First-Strand cDNA Synthesis brochure. The optimal annealing temperature is 60 °C (*see* **Note 17** and **18**).

4. The following are miRNA qPCR run methods for the iCycler iQ™5 PCR Thermal Cycler (Bio-Rad) and for the StepOne™ Thermal Cycler (Life Technologies), which we have successfully used.

iCycler iQ5 PCR Thermal cycler	StepOne™ Thermal cycler
95 °C for 10 min	95 °C for 2 min
35 Cycles 95 °C for 15 s 55–60 °C (*see* **Note 7**) for 15 s	40 Cycles 95 °C for 15 s 55–60 °C (*see* **Note 7**) for 30 min
Melting curve 95 °C for 1 min 55 °C for 1 min 55–95 °C (ramp 0.5 °C) each 10 s	Melting curve 95 °C for 15 s 60–95 °C (ramp 0.5 °C)

4 Notes

1. For the initial design of the miRNA specific forward primer we suggest to use a DNA oligonucleotide with the sequence of the full-length mature miRNA. Nevertheless, in some cases truncating the primer sequence by one or two nucleotides at the 3′-end can result in more specific and accurate results. For discrimination between two or more miRNAs whose mature sequence only varies by a few nucleotides, raising the annealing temperature up to 65 °C is suggested, though there might be a slight loss in sensitivity. In our experience, changing the melting curve temperature transition rate with the lowest ramp (up to 0.1 °C/s) allows to define conditions that will differentiate two or more miRNAs whose sequence varies in one or two nucleotides.

2. Choosing the qPCR reference gene(s). For a reference gene for miRNA qPCR the conventional sRNA constitutive genes can be used such as U6 snRNA. Even so, there are some experimental works that demonstrate that some miRNAs are more constant normalizers than other constitutive sRNAs used as loading controls such as U6 snRNA for northern blots for sRNAs described above [5–7]. Thus, using those miRNAs that can serve as the best reference genes in the studied conditions is suggested. For some plant species several miRNAs have already been reported as reference genes [7, 8].

3. RNA pellet is washed with 100 % ethanol rather than 70 % ethanol to minimize resuspension of small RNAs.
4. Use of 50 % formamide solution to resuspend RNA increases yield and does not interfere with subsequent enzymatic reactions.
5. RNA molecules of ~20 nucleotides migrate between the xylene cyanol and bromophenol blue dyes in 15 % polyacrylamide gels.
6. Extreme care should be taken not to extend times of semidry transfer of RNA to the membrane, as small molecules tend to pass through and end in the buffer tank.
7. We have tested the use of alternative labels such as digoxigenin but results do not compare in sensitivity to the use of ^{32}P. Standard procedures for handling radioactive materials should be familiar to the experimenter and are not covered here.
8. The hybridization solution containing radiolabeled oligonucleotide can be recovered and reused to hybridize another membrane. In this case, perform pre-hybridization step with fresh ULTRAhyb solution as indicated and replace with probe-containing hybridization solution. Thereafter, proceed as indicated.
9. The time of exposure to a PhosphorImager screen depends on various factors such as the amount of RNA blotted, abundance of the miRNA, or the specific activity of the probe, among others. Expose for 6–8 h first to estimate band intensity, and determine if longer exposures are required. Extended exposure times (~3 days) will be expected for low abundance RNAs.
10. Expected results. In our hands, this procedure is highly sensitive and the hybridization signal is very clear. Blots contain very low levels of nonspecific signal and low background noise. We normalize our blots by performing a second hybridization against U6 snRNA or 5S rRNA using the same procedure and 20-nt long oligonucleotides complementary to conserved regions in these small RNAs. Because of the high abundance of these RNA molecules we perform oligonucleotide labeling with 30 pmol of oligonucleotide and only 1 μL of ^{32}P-γ-ATP (3–4 pmol). Exposure times are often a few hours in this case. In addition, due to the larger size of these RNA molecules, hybridization can be performed in the same blots after microRNA detection without removing microRNA probes. All other steps are carried out as described for microRNA detection.
11. The Ncode First Strand Synthesis kit includes a 10 mM ATP solution. The ATP dilution factor for the used RNA concentration is determined by the quotient of 5,000 and the RNA concentration used. The ATP is diluted in 1 mM Tris, pH 8.0.

12. Because only 4 μL of the polyadenylated RNA are used for the first strand cDNA synthesis, the rest of the polyadenylated RNA can be stored at −20 °C or −70 °C for up to 2 weeks if additional cDNA synthesis reactions are needed. The final volume of the polyadenylation reaction is 25 μL, thus it can be used for up to six first strand cDNA reactions.
13. For the RT negative control, use 1 μL of RNase-free water and 1 μL of RNaseOUT™ Recombinant Ribonuclease Inhibitor instead of the Enzyme Mix, as recommended in the Ncode kit.
14. To minimize sample variation, prepare a common qPCR mix containing as many components as possible, i.e., if different cDNA samples will be tested for the same miRNA, include primers in the mix, separate into individual reactions and add cDNA sample at the end. Alternatively, if the presence of different miRNAs in a given condition will be tested, prepare a mix containing the cDNA sample, aliquot into individual reactions and add miRNA primer last.
15. The recommended primer final concentration is 200 nM. Nevertheless up to 600 nM can provide acceptable results.
16. Always consider an additional reaction volume as pipetting can change the final volume for each reaction. A dilution factor of 10 is suggested for the cDNA sample. Even though a standard dilution curve most be performed to determine the best cDNA amount to be used for each microRNA, lower dilutions for low-abundant microRNAs (1:5 or 1:2) are recommended in order to avoid high Ct values.
17. However, performing a gradient of the annealing temperature might be necessary to choose the optimal Tm for each tested primer.
18. Expected results. In our experience, we have obtained very satisfactory results with the qPCR for miRNAs protocol described here. We find the presented miRNA qPCR protocol advantageous compared to the stem-loop qPCR protocol because we achieved quantitation of several miRNAs from the same cDNA sample. In contrast, when using the stem-loop miRNA qPCR protocol, the cDNA sample is exclusively employed for detection of a single miRNA.

    Here we present results that we obtained using the Bio-Rad iCycler iQ5 PCR Thermal Cycler for the characterization of the pvu-miR-159a primer using samples from adult leaves of common bean (*Phaseolus vulgaris*) from the Pinto Villa cultivar, in order to point out three of the main features that have to be considered for the detection of miRNAs using this qPCR method (Fig. 2). The dynamic range of the primer was

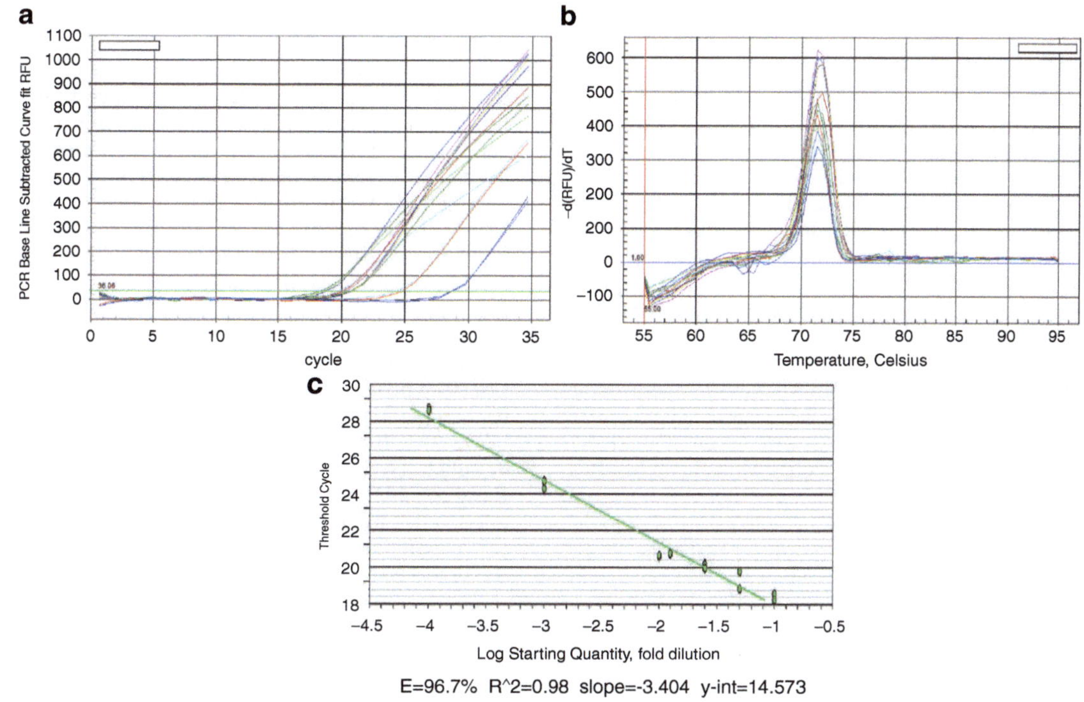

Fig. 2 Dynamic range of the pvu-miR-159a primer. (**a**) Amplification chart of the standard curve of the pvu-miR159a primer. Base line subtracted the relative fluorescence units (RFU) is plotted against the quantification value (Cq). (**b**) Melt peak chart for the pvu-miR-159a primer. A single peak is obtained for this primer at 72 °C confirming that a single qPCR product is detected. The quotient of the negative derivative of RFU and time is plotted against the temperature value in °C. (**c**) Efficiency chart of the pvu-miR-159a primer using six serial tenfold dilutions of the cDNA sample. The efficiency score obtained for this primer is $E = 96.7$ % and $R^2 = 0.98$. The threshold cycle is plotted against the logarithm of the starting quantity fold dilution. The qPCR was performed using the iCycler iQ5 PCR Thermal Cycler (Bio-Rad)

determined testing serial tenfold dilutions of the cDNA template sample. This primer had an efficiency value very close to 100 % and only one peak was detected in the melting curve analysis, thus we have used this primer obtaining reliable results.

Acknowledgments

C. C-C. received a postdoctoral fellowship from CONACyT (CB2010-151571). This work was partially supported by grants from UNAM (PAPIIT IN-205112) and from CONACyT (CB2010-151571) to J.L.R.

References

1. Rajagopalan R, Vaucheret H, Trejo J, Bartel DP (2006) A diverse and evolutionarily fluid set of microRNAs in *Arabidopsis thaliana*. Genes Dev 20:3407–3425
2. Fahlgren N, Howell MD, Kasschau KD, Chapman EJ, Sullivan CM, Cumbie JS, Givan SA, Law TF, Grant SR, Dangl JL, Carrington JC (2007) High-throughput sequencing of Arabidopsis microRNAs: evidence for frequent birth and death of MIRNA genes. PLoS One 2:e219
3. Chen C, Ridzon DA, Broomer AJ, Zhou Z, Lee DH, Nguyen JT, Barbisin M, Xu NL, Mahuvakar VR, Andersen MR, Lao KQ, Livak KJ, Guegler KJ (2005) Real-time quantification of microRNAs by stem-loop RT-PCR. Nucleic Acids Res 33:e179
4. Shi R, Chiang VL (2005) Facile means for quantifying microRNA expression by real-time PCR. Biotechniques 39:519–525
5. Peltier HJ, Latham GJ (2008) Normalization of microRNA expression levels in quantitative RT-PCR assays: identification of suitable reference RNA targets in normal and cancerous human solid tissues. RNA 14:844–852
6. Chang KH, Mestdagh P, Vandesompele J, Kerin MJ, Miller N (2010) MicroRNA expression profiling to identify and validate reference genes for relative quantification in colorectal cancer. BMC Cancer 10:173
7. Kulcheski FR, Marcelino-Guimaraes FC, Nepomuceno AL, Abdelnoor RV, Margis R (2010) The use of microRNAs as reference genes for quantitative polymerase chain reaction in soybean. Anal Biochem 406:185–192
8. Feng H, Huang X, Zhang Q, Wei G, Wang X, Kang Z (2012) Selection of suitable inner reference genes for relative quantification expression of microRNA in wheat. Plant Physiol Biochem 51:116–122

Chapter 8

Forward Genetics Screening of *Medicago truncatula Tnt1* Insertion Lines

Juan Carlos Serrani Yarce, Hee-Kyung Lee, Million Tadege, Pascal Ratet, and Kirankumar S. Mysore

Abstract

A large population of *Medicago truncatula* insertion lines has been generated using the *Tnt1* retrotransposon. More than 21,000 insertion lines have been generated, representing more than 500,000 insertion events. This mutant population is being used by the legume research community to screen for various different mutants using a forward genetics approach. Some of the phenotypes that have been screened using this population include developmentally abnormal phenotypes in leaves, stem, flowers, and roots. In addition to these, mutants with defects in symbiosis with *Rhizobium* and mycorrhiza, mutants with altered nonhost resistance against Asian Soybean Rust and switch grass rust pathogens, mutants with altered lignin content, mutants with altered cell wall structure, etc. have been identified. Here, we describe the high throughput methodology that is being used to identify these *M. truncatula* mutants.

Key words *Medicago truncatula*, *Rhizobium*, *Tnt1*, Forward genetics, Mutant screening

1 Introduction

Legumes are the second most important source of human food and animal feed after grasses contributing 33 % of the dietary protein needs of humans [1]. *Medicago truncatula* has become a model plant because of its small and diploid genome, self-fertility, short life cycle and ease of transformation. Furthermore, it forms symbioses with nitrogen-fixing rhizobia (*Sinorhizobium meliloti* and *Sinorhizobium medicae*) and arbuscular mycorrhizal fungi. The sequenced genome and the bioinformatic resources generated provide powerful tools for genetic, metabolomic, proteomic, and genomic studies of *M. truncatula*.

The generation of mutant collections is necessary to determine plant gene functions. Both forward and reverse genetics strategies can be used to identify desired mutants. Generating a large number of T-DNA tagged lines is a challenge in *M. truncatula* due to unavailability of a reproducible *in planta* transformation method.

Ray J. Rose (ed.), *Legume Genomics: Methods and Protocols*, Methods in Molecular Biology, vol. 1069,
DOI 10.1007/978-1-62703-613-9_8, © Springer Science+Business Media, LLC 2013

Transposable elements represent an attractive alternative for generating large numbers of mutants because the number of transgenic lines to be produced is not a limiting factor when using these mobile elements. Compared to other methods of mutagenesis, the use of retrotransposons presents some advantages since more mutations can be generated with a considerably smaller population. In addition, unlike DNA transposons, the mutations are stable over generations.

Large-scale mutagenesis in *M. truncatula* has been done using the *Tnt1* retrotransposon [2]. *Tnt1* is a long terminal repeat (LTR) retrotransposon of the Ty1-copia type isolated from tobacco [3]. Retrotransposons transpose through a copy and paste mechanism [4]. Two identical direct repeats of LTRs flank two major genes called *gag* and *pol* that produce the necessary proteins for transcription and transposition [4]. It has been demonstrated that the *Tnt1* transposition occurs during tissue culture and/or somatic embryogenesis [5]. It has also been shown that some signals such as different abiotic (hormones) or biotic stresses (pathogen attack) can also induce transcription of *Tnt1* [6]. The efficiency of transposition in *M. truncatula* resulted in a wide range of insertion events. Thus, from 4 to 50 insertions can be found per genome in different regenerated lines [2]. Retrotransposon-mediated mutagenesis has also been used in other plant species such as rice [7], lettuce [8] and soybean [9].

At the Samuel Roberts Noble Foundation, we have generated a population of over 21,000 *Tnt1* insertion mutant lines, each with an average of 25 insertions per line [2, 10]. The population contains approximately 500,000 *Tnt1* insertions randomly located in the *M. truncatula* genome. This represents near saturation mutagenesis; we have a success rate of 85 % in finding insertions in the 900 genes that have been targeted for reverse genetics (requests from 34 different research groups), using a PCR-based approach [11]. Members of 20 different laboratories from around the world have participated in annual *Tnt1* mutant screening workshops over the past 8 years. Using forward genetics screening hundreds of interesting mutants have been identified. For example, screening 9,300 *Tnt1* insertion lines yielded 317 lines with putative defects in nodule development and/or nitrogen fixation. After rescreening, 156 lines were confirmed to be defective in symbiotic nitrogen fixation [12]. A Web site has been established (http://bioinfo4.noble.org/mutant/), which contains information for ~11,000 *Tnt1* insertion lines including images of different phenotypes, as well as more than 30,000 flanking sequence tags from ~2,700 lines. Here, we describe the protocol to screen *M. truncatula Tnt1* insertion lines to identify mutants that show abnormal development phenotypes and altered symbiosis with *Rhizobium*.

2 Materials

2.1 Plant Material

The *M. truncatula Tnt1* lines were obtained through *Agrobacterium*-mediated transformation of *M. truncatula* R108 accession as previously described [2, 5]. We have chosen a *Tnt1* line (tnk88-7-7) that has five copies of *Tnt1* as mother plant for all the mutagenesis work. Around 21,000 plants were regenerated from the same mother plant and we designate these lines as R0. R1 seeds from R0 plants are used for forward genetics screening as described below.

2.2 Scarification and Germination

1. Sulfuric acid (H_2SO_4).
2. 15 ml falcon tubes.
3. Rounded rack for 15 ml falcon tubes.
4. Bleach.
5. Petri dishes (60 × 15 mm).
6. Filter paper.
7. Plastic pipettes.
8. Rounded glass beaker
9. 2.25 × 3.5 Kraft coin envelopes.

2.3 Rhizobia Inoculation

1. *Sinorhizobium meliloti* strain 1021 [13].
2. TY media.

   Tryptone peptone 5 g.

   Yeast extract 3 g.

   Adjust pH to 7.2 by adding 10 ml of sterile 1 M $CaCl_2$ after autoclaving.
3. Low nitrogen nutrient solution.

   This solution is made of three components (A, B, C):

   Component A

A1. KNO_3	20.2 g/l
A2. KH_2PO_4	27.2 g/l
A3. $CaCl_2{\cdot}2H_2O$	73.0 g/l
A4. $MgSO_4{\cdot}7H_2O$	24.6 g/l
A5. K_2SO_4	43.5 g/l

Dissolve each component (A1–A5) separately in 1 l of deionized water.

Component B

EDTA 2Na2Fe (Sigma E6760)	8.2 g/1 l of deionized water

Component C

H_3BO_3	11 g
$MnSO_4{\cdot}H_2O$	6.2 g
KCl	10 g
$ZnSO_4{\cdot}7H_2O$	1 g
$(NH_4)_6Mo_7O_{24}{\cdot}4H_2O$	1 g
$CuSO_4{\cdot}5H_2O$	0.5 g
H_2SO_4	0.5 ml

Dissolve the components in the given order in deionized water in 1 l of final volume. Heating facilitates the solubility of components.

Stock solution

Mix in a >10 l container with deionized water up to 10 l of final volume:

250 ml of solution A1.

250 ml of solution A2.

250 ml of solution A3.

250 ml of solution A4.

250 ml of solution A5.

250 ml of solution B.

13.5 ml of solution C.

For the watering solution add 200 ml of stock to 7.8 l of deionized water.

2.4 Plant Growth, Screening, and Harvest

Plants were grown in a greenhouse under cooling/heating setpoints of 18.3/19.4. The light threshold was 250 W/m^2 along with a 16 h light photoperiod and the average relative humidity was 58.6%.

1. Plastic seed starting trays (8 × 12 = 96 cells).
2. Peter's professional water soluble fertilizer (20–10–20) at 100 ppm.
3. 20 oz dart styrofoam cups.
4. Metro-Mix 830 peat mixture.
5. 1 gal plastic pots.
6. Dissection microscope.
7. Plastic pollination bags.

3 Methods

3.1 Seed Scarification

1. 24 Seeds of each line are separated in KRAFT coin envelopes.
2. Seeds per line are individually poured in 15 ml Falcon tubes perforated in the bottom with holes smaller than seeds. Eighteen of such falcon tubes are attached to a plastic rack with holder that supports the tubes (*see* **Note 1**).
3. Tubes with seeds are immersed in 100% sulfuric acid for 6–8 min in a rounded glass beaker (*see* **Note 2**).
4. After sulfuric acid treatment, seeds are washed with running tap water for 1 min.
5. The washed seeds are immersed in 30% bleach for 2 min.
6. Seeds are washed with running tap water two times, 1 min each.
7. Seeds are then placed in 60×15 mm Petri dishes using a distilled water hose to wash them down onto the Petri dish. The seeds will remain overnight immersed in water. The Petri dishes are labeled according to the line numbers and contain autoclaved filter paper.
8. Excess water is removed from Petri dishes with plastic pipettes 1 day after scarification.

3.2 Seed Germination

The scarified seeds in Petri dishes will remain at 4 °C for 1 week. Plates are sealed with surgical tape. Some seeds may start to germinate from the second day. This vernalization will help for uniform germination. The germination rate is almost 90% for each line.

3.3 Transplanting

1. After 1 week, germinated seeds are planted in the plastic seed starting trays. Each tray holds eight lines and all germinated seeds of each line are planted individually in 12 different cells (two seedlings per cell). Later, only 12 plants per line are kept for screening.
2. The growth media in the trays contain a mixture of perlite and sand (3:1). This mixture will help in minimizing the damage to the roots when the seedlings are pulled out of the tray for screening.
3. Once transplanted, in addition to regular watering, the seedlings are fed with the low nitrogen nutrient solution every other day.

3.4 Rhizobium Inoculation

1. 1 Week after transplanting, *S. meliloti* is grown overnight at 28 °C in TY medium until OD 600 = 1.0.
2. Bacterial suspension is centrifuged at 8,000×*g* and supernatant is discarded.

3. The pellet is resuspended in diluted low nitrogen nutrient solution and adjusted to OD 600 = 0.015–0.03.
4. Plants are sufficiently watered with the inoculum solution.
5. Some more inoculum solution is poured in the tray at the bottom to insure that the roots get in contact with the bacteria.
6. 5–7 Days after the first inoculation, a second inoculation is recommended.

3.5 Mutant Screening

1. After 5–6 weeks, plants are ready for screening. Alterations in shoot development are recorded. Shoot samples are collected for screening other defects such as lignin content, cell wall composition, etc.
2. For symbiotic mutant screening, roots are washed with tap water for all plants of a single line. Perlite and sand residues are removed. A labeled plastic stick for every line is placed in the styrofoam cup.
3. Half of the cup is filled with water and plants are placed inside the cup leaving the leaves and stems out of the water while roots are immersed.
4. The plants are removed from the cup to take individual pictures for recording both shoot and root phenotypes. Plants are now ready for nodulation mutant screening (*see* **Note 3**).
5. Roots are observed under a dissection microscope to record the defects in nodulation.
6. Plants are then transferred to the peat mixture in 1 gal pots. Each pot contains six plants (two pots per line). A labeled plastic stick is placed in every pot. These plants are regularly watered and treated with the fertilizer solution. Further visual phenotypic screening can be performed if needed (*see* **Note 4**).

3.6 Seed Harvest

1. As the plant grows, a 4 ft bamboo stake is inserted vertical into the 1 gal pot and the plant is tied up to the stake with vinyl tape. Routinely the plant is tied up to keep upright and not interfere with neighboring plants.
2. Once there are several pods on the plant that are starting to mature by turning dark (takes approx. 4–5 months) the plant is bagged and the water is withheld. The bags are perforated plastic pollination bags to allow the plant to breath but yet contain the pods. The labeled bagged plant is placed in a dry area to fully mature the pods for about a month.
3. When the plant is completely dried, the seed pods are extracted by cutting the plant stems from the pot. The bag is gently beaten with another bamboo pole to dislodge all pods. Stems are pulled out of the bag and the remaining chafe and pods stay in the bag along with the label.

4. Pods are stored in brown paper bags. If there is more than one bag with the same line number, they are pooled together at this time.
5. The paper bags are then stored in water proof containers in numerical order until ready to be threshed.
6. When ready for threshing, the bags are placed in a drying oven at 36 °C for approximately 24 h for ease in crushing the pods to remove the seed.
7. Seed pods are placed in a seed thresher to crush the pods and allow the seed to fall out.
8. Clean seeds are placed in paper envelopes with all information printed on the envelope and stored in a 4 °C cold room.

4 Notes

1. The rounded rack used to place the falcons for seed scarification is used in convenience for the high number of lines chosen for the screening. It is also possible to use squared racks or other kinds but resistant to acid corrosion.
2. It is necessary to wear gloves, lab coat, lab mask and protective glasses when working with sulfuric acid. The acid can be used several times but is recommended to change after 15 passes of the rack. Same works for the bleach solution.
3. At the screening stage we normally observe genetic segregation of the phenotype. If it is a single gene recessive mutant, one out of four plants shows mutant phenotype.
4. Once the mutant phenotypes of interest are identified, genomic DNA will be extracted and flanking sequence tags (FSTs; plant DNA flanking *Tnt1*) can be identified to determine the mutated gene. Several PCR based approaches including transposon display (TD)-PCR, thermal asymmetric interlaced (TAIL)-PCR and inverse (I)-PCR can be used to identify FSTs [11, 14]. Co-segregation analyses will be performed to correlate FSTs with the phenotype. Putative gene candidate will be further characterized by complementation analyses to confirm that a particular mutated gene is the cause of the desired phenotype.

Acknowledgments

This work was supported by the Samuel Roberts Noble Foundation and NSF Plant Genome Program Grants (DBI-0703285 and IOS-1127155).

References

1. Graham PH, Vance CP (2003) Legumes: importance and constraints to greater use. Plant Physiol 131:872–877
2. Tadege M et al (2008) Large-scale insertional mutagenesis using the Tnt1 retrotransposon in the model legume *Medicago truncatula*. Plant J 54:335–347
3. Kumar A, Bennetzen JL (1999) Plant retrotransposons. Ann Rev Genet 33:479–532
4. Perez-Hormaeche J et al (2008) Invasion of the Arabidopsis genome by the tobacco retrotransposon Tnt1 is controlled by reversible transcriptional gene silencing. Plant Physiol 147:1264–1278
5. d'Erfurth I et al (2003) Efficient transposition of the Tnt1 tobacco retrotransposon in the model legume *Medicago truncatula*. Plant J 34:95–106
6. Grandbastien MA (1997) The expression of the tobacco Tnt1 retrotransposon is linked to plant defense responses. Genetica 100:241–252
7. Greco R et al (2001) Transposon insertional mutagenesis in rice. Plant Physiol 125:1175–1177
8. Mazier M et al (2007) Successful gene tagging in lettuce using the Tnt1 retrotransposon from tobacco. Plant Physiol 144:18–31
9. Cui Y et al (2013) Tnt1 retrotransposon mutagenesis: a tool for soybean functional genomics. Plant Physiol 161:36–47
10. Tadege M et al (2009) Mutagenesis and beyond! Tools for understanding legume biology. Plant Physiol 151:978–984
11. Cheng X et al (2011) Reverse genetics in *Medicago truncatula* using Tnt1 insertion mutants. Meth Mol Biol 678:179–190
12. Pislariu CI et al (2012) A *Medicago truncatula* tobacco retrotransposon insertion mutant collection with defects in nodule development and symbiotic nitrogen fixation. Plant Physiol 159:1686–1699
13. Mulligan JT, Long SR (1985) Induction of *Rhizobium meliloti* nodC expression by plant exudate requires nodD. Proc Nat Acad Sci USA 82:6609–6613
14. Ratet P et al (2010) Tnt1 induced mutations in Medicago: characterization and applications. In: Meksem K, Kahl G (eds) The handbook of plant mutation screening: mining of natural and induced alleles. Wiley-Blackwell, Chichester, pp 83–99

Chapter 9

Reverse Genetics in *Medicago truncatula* Using a TILLING Mutant Collection

Maria Carelli, Ornella Calderini, Francesco Panara, Andrea Porceddu, Ilaria Losini, Pietro Piffanelli, Sergio Arcioni, and Carla Scotti

Abstract

Medicago truncatula is one of the model species for legume molecular genetics. In the last decade different types of mutant populations have been created in this species that can be screened by forward and reverse-genetic approaches to identify and functionally characterize genes of interest. TILLING is a reverse-genetic method combining random chemical mutagenesis and a PCR-based screen to identify point mutations in regions of interest. The different steps of the TILLING analysis are described in a mutant collection of ~2,300 M2 individuals for which genomic DNA and M3 seed were obtained. A two-dimensional DNA pooling strategy was adopted to reduce the number of PCR reactions necessary to screen the collection and to unambigously identify the individual M2 plant carrying the mutation. The genotypic and phenotypic analyses of the mutant M3 progeny provide the possibility to study the gene function. In spite of its reduced size, this mutant collection has proved valid in the study of the biosynthetic pathway of a class of secondary metabolites present in the genus *Medicago*, the triterpenic saponins.

Key words *Medicago truncatula*, TILLING, Reverse genetics, PCR-based screening, Mutants

1 Introduction

Genetic variation is the basis of breeding selection. Variation can be produced by natural or artificial mutations as well as through sexual crossing. Natural mutations have been instrumental for production of varieties with suitable traits for cultivation and in later years they have enabled discovery of the genes influencing the domestication traits. Similarly artificial mutation was employed for creation of new genetic variability to be used for crop improvement but also it has been essential in understanding gene function [1, 2].

Legumes are a worldwide major source of proteins for human and animal nutrition. Use of legume proteins is characterized by a high sustainability; in fact food from animal origin is far more land and energy intensive compared to food of vegetable origin [3].

Ray J. Rose (ed.), *Legume Genomics: Methods and Protocols*, Methods in Molecular Biology, vol. 1069,
DOI 10.1007/978-1-62703-613-9_9, © Springer Science+Business Media, LLC 2013

In addition livestock production takes up to 80 % of global agriculture land and is estimated to be responsible for more than half of the greenhouse gas (GHG) emissions due to agriculture [4, 5].

The presence of legumes in the crop rotation brings many environmental benefits including reduced energy use, global warming potential, ozone formation, acidification and ecotoxicity [6, 7]. The main reason for the positive environmental performance of legumes is their ability to fix nitrogen and thereby to reduce the dependence on mineral fertilizers [7]. Legumes are characterized by a high content of protein, fiber, and micronutrients and are poor in fats, therefore they are recommended for improving the health quality of the human diet, especially in those countries where dietary patterns are too high in energy dense food and too low in fiber rich food, resulting in an excessive intake of total fat, saturated fat, free sugars and salt, leading to risks of metabolic syndrome and cardiovascular diseases [8].

The renewed interest in legumes brought about by interest in functional foods/feeds and sustainable cropping systems boosted research for legume crops and also led to the adoption of model legume species (*Medicago truncatula* and *Lotus japonicus*). Models are being used to perform genetic analyses that enable the understanding of key traits with agronomical importance whose results can be translated to crops [9].

Medicago truncatula has reached the status of the main model species for legumes in recent years. The main reasons to use *M. truncatula* as a model species were: diploidy, relatively small genome (450–500 Mb), self-fertility, short generation time and relative ease of genetic transformation [10]. The genome sequence version 3.5 has recently been published providing 94 % of expressed genes and a highly informative platform for analysis of euchromatin [11]. Subsequent sequencing of 26 accessions of *M. truncatula* is giving a basis for a plethora of information for descriptions of how nucleotide diversity varies along chromosomes, on the importance of both background and positive selection in shaping that diversity, on the extent of linkage disequilibrium or evolutionary independence of genes in different chromosomal regions and on the relative importance of recombination and mutation in introducing variation. Besides, whole-genome sequence data for 26 accessions will enable genome-wide association studies and candidate genes to be discovered underlying natural occurring variation [12, 13]. The genomic effort in *M. truncatula* was nicely mirrored by functional genomics initiatives that have released a variety of mutant collections [10, 14]. Basically three main types of collections are available to the scientific community:

(a) Transposon tagged collections which made use of the ability of the tobacco *Tnt1* transposon to be activated in *M. truncatula* tissue culture [15]. A *Tnt1* collection amenable for both forward and reverse screenings is based at the Noble Foundation and counts 15,000 lines; this collection has been recently surveyed for nodulation mutants [16] and proved efficient for analysis of other traits as reported in several publications [17]. A multinational European effort also produced 6,000 *Tnt1* lines [10] and 1,000 lines were obtained by an Italian functional genomic initiative [18].

(b) Fast neutron based collections. Genomic deletions were obtained for two available populations. In one case a population of 156,000 plants was arranged for reverse screening with a sensitive polymerase chain reaction-based detection, allowing recovery of mutants with a 29 % rate using five targets per gene [14]; such a population is available for screening through RevGenUK at the John Innes Center. A second population was established by R. Chen at the Noble Foundation [19]; in this case the size of deletions does not allow reverse screening and the population is being characterized with a forward approach. Gene cloning by mapping in such a population can be substantiated by parallel *Tnt1* allele discovery [20–22]. Development of tiling arrays that might help gene cloning is in progress (R. Chen pers.comm.).

(c) TILLING collections. A mutagenized population was first created by EMS treatment and it was screened for the presence of developmental mutants [23]; the population was successively arranged for TILLING analysis via a LICOR based platform [10]. Another population was established by Porceddu et al. [18] and it will be reported in the present paper. A third population is available through RevGenUK and it is also arranged for reverse genetic screening via capillary sequencer [24].

Chemical mutagenesis is relatively easy to perform in many plant species; it is based on the treatment of seeds with a mutagen solution, ethylmethanesulfonate (EMS) being the most reported in the literature. A main advantage of chemical mutagenesis is that it does not involve any GM technique, with no laborious tissue culture procedures and no need for containment and control measures provided for by law in many countries for GM plants. Mutant plants can be directly developed in pure line varieties or used in breeding programs for cultivar development. TILLING approach investigates the chemically induced mutation via reverse PCR-based screenings in region of interest (candidate genes). However, recent developments of TILLING-by-sequencing strategies based on next-generation sequencing [25] could allow the sequencing of the entire mutant population bringing a great aid to gene cloning in forwards screenings [10]. Chemical mutagens induce single base-pair changes producing different types of alleles, not only loss-of-function ones due to stop codon introduction, but also missense mutations resulting in conservative or nonconservative amino acid substitutions. A typical mutagenesis protocol is based on: (1) seed treatment and subsequent development of M1 and M2 generations; (2) DNA extraction from individual M2 plants and pooling for TILLING analysis; (3) set up of an M3 seed bank. TILLING analysis involves the detection of mutations in the gene-of-interest via PCR amplification of a ~1,000-bp fragment and detection of heteroduplex by cleavage and electrophoresis [26]. The following protocol reports a case example for *M. truncatula* [18] concerning chemical mutagenesis and TILLING analysis for genes putatively involved in triterpenic saponin biosynthesis.

2 Materials

2.1 *M. truncatula* TILLING Population and Mutagenesis

1. Seeds used for chemical mutagenesis belong to the *M. truncatula* genotype 2HA10-9-3. This genotype was the genetic background used in the large scale *Tnt1* insertional mutagenesis effort carried out during the completion of the European Union Grain Legumes Integrated Project (FOOD-CT-2004-506223). It is characterized by a higher regeneration ability compared to the genotype Jemalong and a close genetic similarity to the sequenced genotype A17 [27]. Mutants in 2HA10-9-3 can contribute allelic series for the EU *Tnt1* functional genomics platform but can also be of use for other genetic backgrounds, such as A17, the sequenced genotype, and R-108.
2. Solution 0.15 % EMS.
3. 70 % H_2SO_4.
4. 20 % NaOCl.
5. 10 N NaOH.

2.2 DNA Extraction and Pooling

1. GenElute DNA isolation kit (Sigma-Aldrich Corporation, St. Louis, MI, USA) or similar kit.
2. 96-Well plates.

2.3 PCR Amplification and Digestion of the PCR Products

1. Primers: unmodified (Sigma-Aldrich Corporation, St. Louis, MI, USA) and 5′-labelled (Applied Biosystems [ABI], Forster City, CA, USA) primers in lyophilized form. The fluorophores used for 5′-labelling are 6-FAM (forward primer) and VIC (reverse primer).
2. HotStartTaq Master Mix (Qiagen).
3. PCR reaction mixture: 3.75 μl HotStartTaq Master Mix; 0.2625 μl primers from primer work solution (ratio 3 vol:2 vol labelled–unlabelled).
4. Surveyor® Mutation Detection Kit (Transgenomics, Omaha, NE, USA).

5. Digestion buffer (10×): 0.1 M $MgSO_4$, 0.1 M Hepes, 0.1 M KCl, 0.02 % Triton X-100, and 2 μg/ml BSA.
6. EDTA 150 mM.
7. 100 % ethanol.
8. 70 % ethanol.
9. Ultrapure DNAse/RNAse-free water.

2.4 Electrophoresis and Analysis of Data

1. Applied Biosystems 3730 DNA analyzer (Applied Biosystems [ABI], Forster City, CA, USA).
2. ABI GS 1200 LIZ size standard (20–1,200 bp) (Applied Biosystems).
3. GeneMapper® 4.0 program (Applied Biosystems).

2.5 Genotypic Analyses of the Mutant Plant and Progeny Genotyping

1. Big Dye Terminator sequencing kit, version 3.1 (Applied Biosystems).
2. Mutation Surveyor software (SoftGenetics LLC., PA, USA).
3. PARSESNP (Project Aligned Related Sequences and Evacuate SNPs) http://www.prowb.org/parsesnp/.
4. QIAQuick PCR purification kit (Qiagen) or similar.
5. Real time rotary analyzer Rotor-Gene 6000 (Corbett Research, NSW, Australia).
6. EVAGreen supermix (Bio-Rad) or similar.

3 Methods

3.1 Chemical Mutagenesis

The EMS concentration used (0.15 %) provides the highest mutation rate compatible with germination and plant development for *M. truncatula* as confirmed in different mutagenesis experiments [23, 24].

1. Seeds were scarified by treatment for 7 min in 70 % H_2SO_4 and sterilized in 20 % NaOCl for 2 min then washed five times in sterile double-distilled water.
2. Pre-scarified seeds were treated with 0.15 % EMS and kept overnight at 24 °C with gentle agitation.
3. The mutagen was removed and disactivated in 10 N NaOH. Mutagenized seeds were washed three times with sterile double-distilled water and vernalized 36 h at 4 °C.
4. Seeds subjected to the same protocol but treated with 0 % EMS were used as control (not mutagenized).

3.2 Creation of a Mutagenized Population

Approximately 2,500 M1 plants and 65 control plants (0 % EMS) were grown and seed was harvested from 2,281 plants. M2 generation was obtained by growing 1–5 plants per fertile M1 individual

Table 1
Phenotypic screening of the M2 generation (5,000 seeds examined)

Phenotypic class	Percentage of individuals		
	Porceddu et al. [18] (unpublished data)	Penmetsa et al. [23]	LeSignor et al. [24]
Not germinated seed	7.5 ± 9	30	
Embryo lethality	15.5 ± 1.2		15–19
Albino	0.6 ± 0.2	0.4 ± 0.7	1.2–1.4
Chlorotic	0.9 ± 0.2	3.5 ± 1.8	
Pale green	0.7 ± 0.2	2.0 ± 1.7	

(1,658 families and 2,560 total plants) together with 34 control families. DNA was extracted from 2,300 plants and M3 seed was harvested.

Seed germination rate, embryo lethality and mutant phenotypes at the seedling stage were recorded in the M2 generation to evaluate the efficiency of the EMS treatment. As reported in Table 1 the frequency of mutant phenotypes is comparable for the reported mutagenesis experiments.

M2 generation was not obtained under a strict single-seed descent procedure (*see* **Note 1**); this was done in order to limit the loss of mutations still segregating in M1 generation and the loss of lines due to mortality or infertility of individual M2 plants. Pedigree data carrying the information on the kinship among lines are recorded. The estimated rate of mutation based on the TILLING screening for six genes of interest was 1 mutation/Kbp/500 plants.

3.3 Screening DNA Pools

Gene-specific primers are used to selectively amplify a region of ~1,000 bp in the gene of interest from DNA pools. Genomic DNA is extracted from individual M2 plants and pooled together following a two-dimensional (2D) strategy based on eightfold (8×) and 12-fold (12×) pools obtained by pooling a column (8×) and a line (12×) of the original DNA 96-well plates. The 2D DNA pooling strategy enables the identification of a single mutant DNA at the intersection of a specific 8× and 12× pool. A single mutant will then be detected both in a 8× and in a 12× pool and the coordinates of the two pools will allow its unambiguous identification. This strategy allows to considerably reduce the size of the screening (~2,300 M2 plants are screened in our case by means of 480 PCR reactions) and to focus the sequencing reactions necessary to confirm the mutation to few and precisely identified DNA samples. The outline of the TILLING analysis procedure is summarized in Fig. 1.

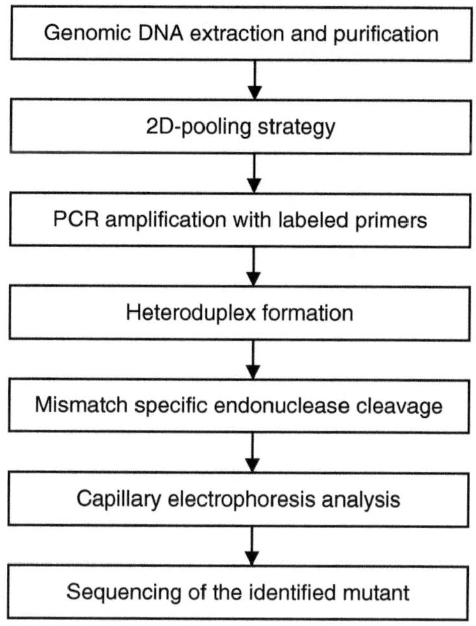

Fig. 1 TILLING analysis procedure flow chart

3.3.1 Genomic DNA Isolation

1. 0.2 g of fresh leaf tissue are collected from each M2 plant in 2 ml Eppendorf tubes, frozen in liquid nitrogen and powdered with glass beads.
2. DNA is extracted using the GenElute DNA isolation kit (*see* Subheading 2.2, **item 1**) according to the manufacturer's protocol (any other protocol for genomic DNA extraction can be used as well).
3. DNA quality and concentration are checked on 1 % agarose gel; DNA is stored at −20 °C.
4. DNA working solutions are prepared at the concentration of 1.5 ng/μl in water and transferred in 96-well plates with barcode identification; the 96-well plates are stored at −20 °C.

3.3.2 Genomic DNA Pooling

Figure 2 shows the schematic diagram of 2D-pooling strategy. DNA pools are created as follows:

1. 8× pools: 10 μl of genomic DNA from each of 8 individual samples of column 1 are placed in the 1A well of the 8×-pool plate, etc.
2. 12× pools: 10 μl of genomic DNA from each of 12 individual samples of line A are placed into the A1 well of the 12×-pool plate, etc.
3. 8× and 12× plates are stored at 4 °C for working dilutions and at −20 °C for the undiluted DNA 96-well plates.

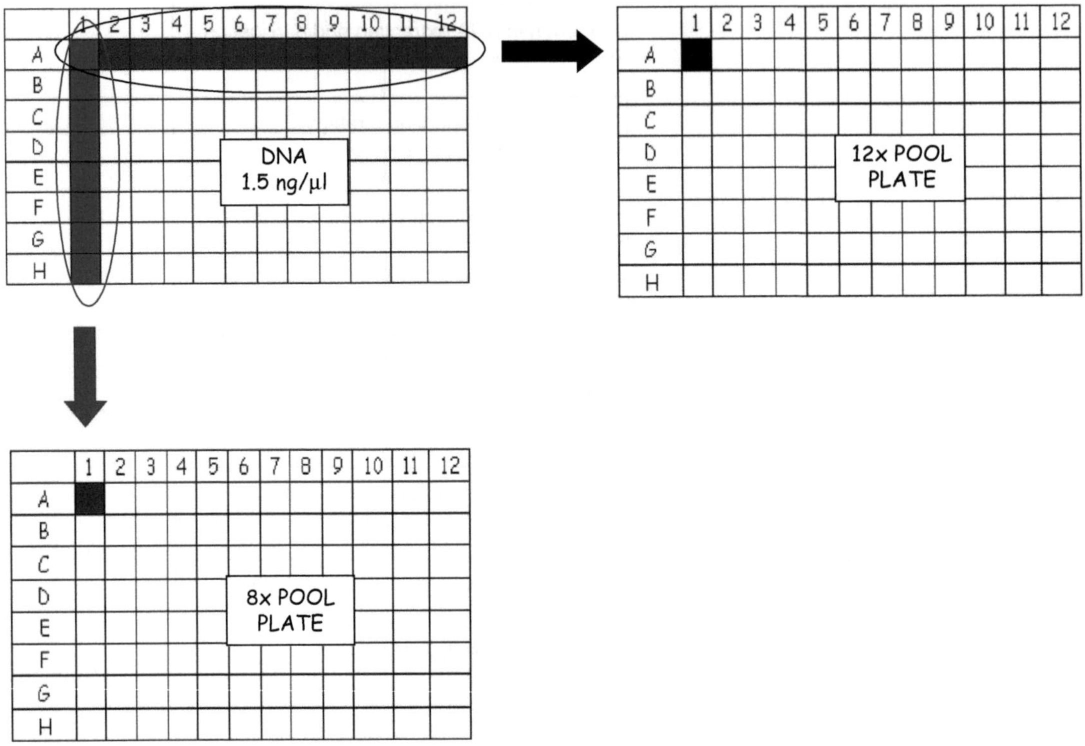

Fig. 2 Genomic DNA 2D-pooling diagram of the M2 plants

3.3.3 Primer Design

In the case of genes for which genomic sequences are available, primers are designed using the program CODDLE (Codon Optimized to Discover Deleterious Lesions) by selecting the TILLING w/EMS (plant) option and provided an amplicon of max 1,200 bp is generated. Otherwise, primers are designed using PRIMER3 program (*see* **Note 2**). Primers are designed to match melting temperatures (Tm) from 60 to 70 °C.

1. Each pair of primers is ordered in two forms, unmodified (not labelled) and 5′-labelled with two different fluorophores, 6-FAM for the forward primer and VIC for the reverse one (*see* Subheading 2.3, **item 1**). Labelling of the primers with different dyes will produce—after denaturation, renaturation, and cleavage—DNA fragments labelled with one of the two fluorophores respectively that will be run simultaneously in the electrophoresis capillary gel.
2. Primer stock solution: Both unmodified and 5′-labelled lyophilized primers, are resuspended in ultrapure water DNA/RNAase-free to a final concentration of 100 µM.
3. Primer work solution: Primers are used in the PCR reaction at the final concentration of 0.35 µM in the ratio 3:2 labelled–unlabelled.

3.3.4 Amplification of the Target Region

1. The reaction is performed in 96-well plates in a final volume of 7.5 μl. The reaction mixture (*see* Subheading 2.3, **items 2** and **3**) is as follows: 2 μl pooled DNA; 3.75 μl HotStartTaq Master Mix; 0.2625 μl primers (*see* **Note 3**). PCR program is as indicated in Colbert et al. [28]: 95 °C for 15 min (Taq activation); 8 cycles using a touchdown program: 95 °C for 1 min (denaturation), from Tm +3 °C to Tm −4 °C with a decrease of 0.7 °C per cycle for 30 s (annealing), and 72 °C for 1 min (extension); 95 °C 1 min, Tm −5 °C 30 s, 72 °C 1 min, 35 cycles; 72 °C for 10 min; 99 °C for 10 min (inactivation); 35 cycles of 30 s starting from 70 to 49 °C with a decrease of 0.6 °C per cycle (re-annealing allowing mutated-wt DNA hybridization).
2. PCR products are analyzed to verify the correct amplification of the target region by loading 2 μl of product on 2 % agarose gel.

3.3.5 Digestion of the PCR Products

PCR products are processed using Surveyor® Mutation Detection Kit (*see* Subheading 2.3, **items 4** and **5**) to cleave at base mismatches, according to Colbert et al. [28].

1. 5 μl of PCR products from the 96-well plate are digested with 0.0625 μl Surveyor nuclease, a member of the CEL mismatch-specific endonuclease, in 1× digestion buffer (*see* Subheading 2.3, **item 5**). The digestion total volume is 15 μl.
2. The PCR products are incubated at 45 °C for 15 min; the reaction is stopped by adding 2.5 μl EDTA 150 mM.
3. The digested PCR products are purified by adding 30 μl cold ethanol 100 % and incubated for 15 min at room temperature; after centrifugation at 2,000 × *g* for 45 min the plate is put in an inverted position on a paper towel and centrifuged at 180 × *g* for 10 s to allow ethanol to drain away.
4. The samples are washed by adding 30 μl cold ethanol 70 % and centrifuged at 2,000 × *g* for 15 min; the fluid is removed as previously indicated and the plate is allowed to dry in the air for 15–30 min.

3.3.6 Electrophoresis and Analysis of Data

Capillary electrophoresis is performed on a Applied Biosystems 3730 DNA analyzer (*see* Subheading 2.4, **item 1**).

1. The digested samples are resuspend in 12 μl Hi-Di formamide and 0.05 μl ABI GS 1200 LIZ size standard (20–1,200 bp) are added (*see* Subheading 2.4, **item 2**).
2. Run conditions are set as follows: pre-run voltage 15 V, pre-run time 180 s, injection voltage 2 kV, injection time 20 s, run voltage 8 kV.

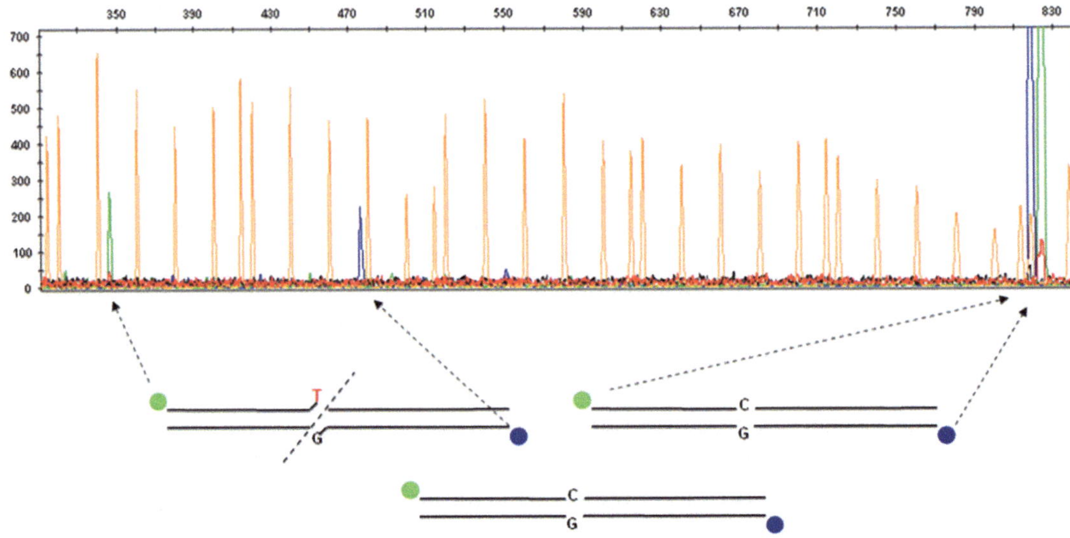

Fig. 3 Electropherogram of a DNA pool containing a mutant. The products of the cleavage by the endonuclease are identified as single peaks of *blue* or *green* color (*left side* of the image); the whole amplicon is identified by the presence in the same position of two peaks of the two colors (*right side* of the image). In *orange* the peaks corresponding to the DNA size standard (ABI GS 1200 LIZ 20-1200)

Examination of the electropherograms:

1. Identification of a mutant in a DNA pool: Primers used for the amplification of the target region are labelled with two different fluorophores: forward primer is labelled with the fluorophore 6-FAM of blue color and the reverse primer with the green fluorophore VIC. The product of the reaction of amplification is visualized in the electropherogram by the concomitant presence in the same position of two peaks, one green and the other one blue (Fig. 3, right side). The presence of additional single peaks, green and blue respectively, in different positions in the electropherogram corresponds to two fragments derived from the cleavage by the endonuclease as the fluorophores remain associated each to one of the two products of the cleavage (Fig. 3, left side).
2. Control of the size of the PCR products obtained by cleavage: The size of the PCR products depends on the position of the mutation in the amplicon determining the point where the cleavage occurs. Consequently the sum of the sizes of the fragments obtained by the endonuclease cut need to correspond to that of the entire (uncut) PCR amplicon. To establish the size of the obtained PCR products and to check if their sum corresponds to that of the uncut amplicon the DNA size marker ABI GS 1200 LIZ (20–1,200) is analyzed by capillary electrophoresis together with each DNA pool sample. The mentioned size

standard, especially conceived for TILLING analyses, generates a series of orange peaks of known size (Fig. 3).

3. Because of the 2D pooling strategy adopted, a single mutant is present in one and only one 8× and 12× pool. The output obtained from the two pools must then be comparable in term of position and size of the detected fragments.
4. *See* **Note 4** for testing the efficiency of the endonuclease digestion.

Data analysis: The data analysis is performed using the GeneMapper® 4.0 program (*see* Subheading 2.4, **item 3**).

1. The analysis is carried out starting from the 8×-pool plates and separately for each fluorophore: by overlaying all the 96 samples, the mutant peak is easily identified as it will appear above the background and it will be the only single peak as no overlay with any other peak should occur.
2. By clicking the position of the peak, all the corresponding data are visualized in a summary table (position in the plate, molecular size and height). The same analytical procedure is repeated for the second fluorophore. If for both fluorophores a mutant peak is identified in the same DNA sample (same position in the plate) the size of the two DNA fragments is calculated and should correspond to that of the entire amplicon. When this situation occurs a putative mutant is identified.
3. To confirm the mutant and to identify the individual plant carrying the mutation, the same analysis is performed on the 12×-pool plates. The same peaks, in term of molecular size and height, must be found in only one 12× pool; the sum of the two peaks must again correspond to that of the entire amplicon. By crossing the coordinates corresponding to the two putative mutant pools it is possible to identify the DNA sample (M2 plant) that is likely to carry the mutation.

Mutations occurring in the ~80 bp at the two ends of the amplicon are probably missed [29] because of possible artifacts in these areas due to PCR priming and technical limitations of the sequencing by capillary electrophoresis itself.

3.3.7 Sequence Analysis of the Putative Mutant Plant

The putative mutant M2 plant identified by the TILLING 2D-analysis undergoes a PCR amplification for the gene of interest; the product of the reaction is then sequenced in both forward and reverse directions.

1. Sequencing is performed using the ABI BigDye Terminator V3.1 Cycle Sequencing Kit (*see* Subheading 2.5, **item 1**) according to the manufacturer's protocol; the products of the sequencing reaction are purified by precipitation with ethanol–EDTA and run on a Applied Biosystems 3730 DNA analyzer.

Table 2
Example of the final report of the TILLING analyses for the genes of interest (amplicons of ~1,000 bp)

Gene	Sequence mutation/position	Amino acid mutation
CYP72A68	Not found	–
CYP93E2	G>A/1099	Asp397/Asn
CYP716A12	G>GA/1457 C>T/1174 G>GA/1727 C>CT/1354	Trp449/STOP Pro355/Leu None 415Thr/Ile

2. The sequencing data analysis is carried out by the Mutation Surveyor software (*see* Subheading 2.5, **item 2**). PARSESNP (Project Aligned Related Sequences and Evacuate SNPs, http://www.prowb.org/parsesnp/) (*see* Subheading 2.5, **item 3**) can be further used to analyze the sequenced mutations and to predict the likely consequences potentially affecting the gene function using a protein homology model.

3.4 Progeny Genotyping

The identified mutant plants, confirmed by sequencing, can show the mutation in homozygous or heterozygous state; in the latter case M3 progeny genotyping is necessary to identify the segregating classes before phenotyping the mutation.

Table 2 reports an example of the results of TILLING analyses for three cytochromes P450 on the *M. truncatula* collection of Porceddu et al. [18]; the P450s are putatively involved in the biosynthesis of triterpenic saponins, a class of secondary metabolites with important biological properties [30].

Heterozygous mutations (i.e., G>GA in Table 2) ask for the genotyping of the segregating M3 progeny.

3.4.1 Seed Germination and Growth Conditions

1. M3 seeds are chemically scarified by soaking in concentrated H_2SO_4 for 10 min and washed three times with water.
2. Seeds are germinated in Petri dishes with water-imbibed filter papers.
3. The viable seedlings are transferred at the stage of fully open cotyledons in individual tube-plots. Plants are grown in greenhouse or alternatively in growth chamber depending on phenotype screening.

3.4.2 Sequence-Based Genotyping

1. DNA is extracted from each individual M3 plant as described in Subheading 3.3.1.
2. Amplification of the gene of interest: Primers and PCR mix composition are the same previously used for TILLING

screening; 20 ng of genomic DNA from each M3 plant are used as template. The PCR program is: 95 °C for 15 min; 8 cycles using a touchdown program: 95 °C for 1 min (denaturation), from Tm +3 °C to Tm –4 °C with a decrease of 0.7 °C per cycle for 30 s (annealing), and 72 °C for 1 min (extension); 95 °C 1 min, Tm –5 °C 30 s, 72 °C 1 min, 35 cycles; 72 °C for 10 min. The PCR products are analyzed to verify the correct amplification of the target region by loading 2 μl of the products on 1 % agarose gel.

3. The PCR products are purified by the QIAQuick PCR purification kit (*see* Subheading 2.5, **item 4**) according to manufacturer's protocol and final elution is made in water.
4. Sequence reaction is performed on a ABI 310 analyzer. Sequence PCR products using the most suitable primer and the Big Dye Terminator, version 3.1, sequencing kit (*see* Subheading 2.5, **item 1**).
5. The genotyped plants are used for phenotypic analyses and seed production.

3.4.3 HRM-Based Genotyping

High Resolution Melt (HRM) analysis is a post-PCR method for the detection of polymorphisms in double-stranded DNA samples. The technique requires a specific instrument (e.g., Corbett Rotor-Gene 6000, Subheading 2.5, **item 5**) and standard samples (wt and/or mutant in homozygous state).

1. Primers for PCR reaction are designed using PRIMER3. Usually, a short segment around the SNP of interest is amplified; the amplicon is typically 50–100 bp in length (*see* **Note 5**).
2. DNA is extracted from each progeny plant as previously described.
3. The PCR reaction is performed using a commercial reaction mix containing a specialized double-stranded DNA binding dye (e.g., SYBR Green, EVA Green, *see* Subheading 2.5, **item 6**) on standard and unknown samples; PCR conditions depend on the primers, the PCR reaction mix and the instrument used.
4. The HRM analysis is performed on the amplified fragments. HRM settings are instrument-dependent, however typical conditions are indicated: melt curve from 70 to 90 °C with 0.01–0.1 °C increments at each step. The melt curves of the unknown and standard samples are compared to identify the different classes (Fig. 4).

3.5 Progeny Phenotyping

EMS-mutagenesis creates mainly G-A and C-T transitions ([2], Table 2) often resulting in amino acid substitutions or in the introduction of novel stop codons. This can generate a loss-of-function protein as well as a protein with modified enzymatic activity or a regular function protein. The phenotypic analysis is then aimed to

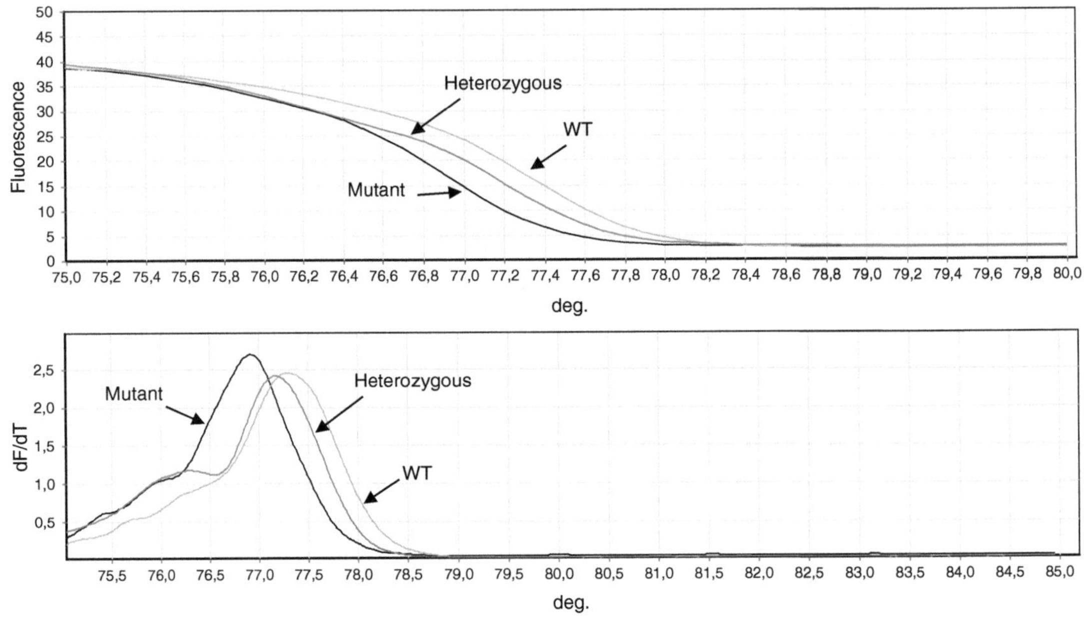

Fig. 4 Example of results obtained by HRM analysis on the progeny of a M2 plant carrying a mutation in heterozygous state in a cytochrome P450 gene

verify if the TILLING mutation produces a detectable phenotype. In the case of genes listed in Table 2, for instance, the content and composition of sapogenins, the aglyconic moieties of saponins, were chemically analyzed in the mutant progenies.

1. In the case of a mutation in heterozygous state, the M3 progeny is genotyped as previously indicated and phenotypic analyses are performed on the different classes: homozygous mutant/wild type (wt) and heterozygous individuals. This situation is of particular interest as it gives the opportunity to evaluate the effect of a mutation within the same full sib line, *M. truncatula* being an autogamous species. The mutant and wt phenotypes can then be compared in the same genetic background, i.e., in the presence of the other unknown point mutations induced by the mutagen EMS treatment (*see* **Note 6**).
2. In the case of a mutation in homozygous state, the phenotypic control (wt) is represented by the M3 progeny of plants treated with 0 % EMS, grown together with the mutant lines. The comparison between mutant and wt is less stringent as it is not possible to estimate the phenotypic effects of the unknown point mutations outside the gene of interest.
3. TILLING can produce an allelic series of mutants in a gene of interest as shown in Table 2 for CYP716A12; the presence of a same mutant phenotype in independent allelic mutant lines is a strengthening evidence of the functional relationship between gene and phenotype.

3.6 Maintenance of the TILLING Lines

The actual seed (M3 generation) availability of the TILLING lines can be hampered by the low seed setting of the M2 individuals. In our collection, for instance, about 35 % of the M2 individuals showed a seed production ≤50 seeds. Seed increase is obtained by germinating M3 seeds and growing up M3 plants to produce pods and seeds (M4 generation). Due to the reduced size of our collections (about 2,000 M2 plants), single seed descent does not seem the most suitable procedure as it involves the loss of part of the mutations present in the M1 [2]. On the contrary our objective is to maintain, as far as possible, the entire range of mutations carried by the M1 plant even if in a segregant M4 generation. For this purpose the following procedures are adopted: (1) use of a maximum of 20 M3 plants per line to obtain the M4 generation, (2) growing techniques limiting among-plants competition and avoiding stressing conditions, and (3) pod collection in bulk (*see* **Note 1**).

We briefly describe the protocol we adopt in semi-controlled conditions for a sub-continental climate (45° 19′ N, 9° 30′ E, 81 m asl; average annual rainfall of 800 mm; average annual mean daily temperature of 12.5 °C, January being the coldest month,1.1 °C, and July the hottest one, 22.9 °C) where *Medicago truncatula* is not an endemic species.

1. M3 seeds, typically 20 per line, are germinated in Petri dishes with mechanical scarification to ensure contemporary germination. All the viable seedlings are transplanted in plug trays with cells 5 cm diameter (two seedlings/cell) at the end of January and grown for about 2 months in a greenhouse where minimum temperature is set at 8 °C. The slow growth of plantlets allows for the investigation of morpho-physiological mutations at young stages.
2. In the second half of March, plugs are transferred in tube-plots 5 cm diameter × 80 cm height with a layer of expanded clay in the bottom and filled with a mixture vol:vol 46 % soil–46 % peat–8 % expanded clay sand grade, P- and K-supplied; no mineral N is used. The tube-plots are put on the outside under a rain shelter structure and provided with individual automatic irrigators; lines are kept apart by tissue separators. Pods are collected in bulk at maturity (June–July); typically each line, represented by 15–20 M3 adult plants, can produce 100–400 pods.

4 Notes

1. The general strategy to create a EMS mutagenized population in a diploid species [2] indicates that a large number (around 20,000, depending on the genome size) of M2 lines are necessary for the saturated coverage of the genome. Successive generations (M3–M6) obtained by self-fertilization under single-seed descent result in the "fixation" of about half of all

the mutations present in M1, the remaining being lost through segregation in the process. The reduced size of our EMS mutagenized population (2,000 M2 lines) requires a different strategy in order to limit the loss of mutations. For this reason, generations after the M0 (seed exposed to EMS treatment) are reduced to M4; 1–5 M1 plants/line are used to obtain M2 generation; a maximum of 20 M2 and M3 plants/line are used to obtain M3 and M4 generation respectively and the seed of each line is collected in bulk. As a consequence M4 lines are still largely segregant but likely to maintain a higher proportion of the mutation present in M1 generation than under single-seed descent procedure. Pedigree notations allow identification of the mutations 'not unique' that are carried by related lines.

2. For TILLING analysis if the sequence of gene of interest is larger than 1,200 bp, a target region of ~1 kb can be chosen within the gene or the gene can be split into fragments. In the case where a large intron is inserted in the expressed target region, the gene can be split into two fragments and forward/reverse primers for each fragment should be designed based on the intron sequence close to the intron–exon boundaries.
3. In PCR amplification mix, $MgCl_2$ concentration , that is 1.5 mM in the HotStartTaq Master Mix, can be adjusted, if necessary, to improve gene amplification.
4. Test of the efficiency of the endonuclease digestion: The visual examination of the background in the electropherograms of the DNA pools containing or not containing mutants can be used to estimate the efficiency of the endonuclease digestion. The occurrence of the same "background noise" between different DNA pools is indicative of a correct efficiency of the nuclease cutting. Moreover, the low intensity of the "background noise" when compared to that of the two peaks occurring when a mutation is present indicates a good efficiency of the purification of the PCR products following endonuclease digestion.
5. In HRM analysis amplicons longer than 50–100 bp make more difficult to distinguish the different genotypes and increase the chance of including unwanted SNPs that will affect discrimination. Ideally, the SNP is located close to the middle of the amplicon.
6. As an example of the interest of mutations in heterozygous state, we observed a retarded-in-growth phenotype associated to a mutation in the CYP716A12 gene (Table 2, Trp449/STOP) resulting in the absence of hemolytic saponins [30]. The phenotyping of the three classes (homozygous mutant/wild type and heterozygous individuals) for the presence/absence of both hemolytic saponins and retard in growth allowed to formulate the hypothesis of an association between the two phenotypes [30].

Acknowledgments

This work was supported by the Italian Ministero dell'Istruzione dell'Università e della Ricerca (MIUR FIRB—Post genomica di leguminose foraggere).

References

1. Doebley JF, Gaut BS, Smith BD (2006) The molecular genetics of crop domestication. Cell 127:1309–1321
2. Parry MAJ, Madgwick PJ, Bayon C, Tearall K, Hernandez-Lopez A, Baudo M, Rakszegi M, Hamada W, Al-Yassin A, Ouabbou H, Labhilili M, Phillips AL (2009) Mutation discovery for crop improvement. J Exp Bot 60:2817–2825
3. Garnett T (2009) Livestock-related greenhouse gas emissions: impacts and options for policy makers. Environ Sci Pol 12:491–503
4. FAO (2009) The state of food and agriculture: livestock in the balance. Food and Agriculture Organization, Rome
5. Steinfeld H, Gerber P, Wassenaar T, Castel V, Rosales M, Haan CD (2006) Livestock's long shadow: environmental issues and options. Renew Resour J 24:15–17
6. Kopke U, Nemecek T (2010) Ecological services of faba bean. Field Crop Res 115: 217–233
7. Nemecek T, von Richthofen JS, Dubois G, Casta P, Charles R, Pahl H (2008) Environmental impacts of introducing grain legumes into European crop rotations. Eur J Agron 28:380–393
8. Hallström E (2011) Examination paper SASH 37 Gastronomy-Sustainable Eating 2011
9. Cannon SB, May GD, Jackson SA (2009) Three sequenced legume genomes and many crop species: rich opportunities for translational genomics. Plant Physiol 151:970–977
10. Tadege M, Wang TL, Wen J, Ratet P, Mysore KS (2009) Mutagenesis and beyond! Tools for understanding legume biology. Plant Physiol 151:978–984
11. Young N, Debelle F, Oldroyd GE, Geurts R, Cannon SB, Udvardi MK et al (2011) The *Medicago* genome provides insight into the evolution of rhizobial symbioses. Nature 480:520–524
12. Branca A, Paape TD, Zhou P, Briskine R, Farmer AD, Mudge J, Bharti AK, Woodward JE, May GD, Gentzbittel L, Ben C, Denny R, Sadowsky MJ, Ronfort J, Bataillon T, Young ND, Tiffin P (2011) Whole-genome nucleotide diversity, recombination, and linkage disequilibrium in the model legume *Medicago truncatula*. Proc Natl Acad Sci U S A 108:864–870
13. Paape T, Zhou P, Branca A, Briskine R, Young N, Tiffin P (2012) Fine-scale population recombination rates, hotspots, and correlates of recombination in the *Medicago truncatula* genome. Genome Biol Evol 4:726–737
14. Rogers C, Wen J, Chen R, Oldroyd G (2009) Deletion-based reverse genetics in *Medicago truncatula*. Plant Physiol 151:1077–1086
15. d'Erfurth I, Cosson V, Eschstruth A, Lucas H, Kondorosi A, Ratet P (2003) Efficient transposition of the *Tnt1* tobacco retrotransposon in the model legume *Medicago truncatula*. Plant J 34:95–106
16. Pislariu CI, Murray J, Wen J, Cosson V, Duvvuru Muni RR, Wang M, Benedito V, Andriankaja A, Chen X, Torres JI, Mony S, Zhang S, Taylor M, Tadege M, Ratet P, Mysore KS, Chen R, Udvardi MK (2012) A *Medicago truncatula* tobacco-retrotransposon (Tnt1)-insertion mutant collection with defects in nodule development and symbiotic nitrogen fixation. Plant Physiol 159:1686–1699
17. Panara F, Calderini O, Porceddu A (2012) *Medicago truncatula* functional genomics. An invaluable resource for studies on agriculture sustainability. In: Meroni G (ed) Functional genomics, In Tech ISBN 978-953-51-0727-9, doi:10.5772/51016. Available from: http://www.intechopen.com/books/functional-genomics/medicago-truncatula-functional-genomics-an-invaluable-resource-for-studies-on-agriculture-sustainabi
18. Porceddu A, Panara F, Calderini O, Molinari L, Taviani P, Lanfaloni L, Scotti C, Carelli M, Scaramelli L, Bruschi G, Cosson V, Ratet P, de Larembergue H, Duc G, Piano E, Arcioni S (2008) An Italian functional genomic resource for *Medicago truncatula*. BMC Res Notes 1:129
19. Wang H, Li G, Chen R (2006) Fast neutron bombardment (FNB) mutagenesis for forward and reverse genetic studies in plants. In: Teixeira da Silva JA (ed) Floriculture, ornamental and plant biotechnology: advances and topical issues. Global Science Books, Isleworth, UK, pp 629–639

20. Chen J, Yu J, Ge L, Wang H, Berbel A, Liu Y, Chen Y, Li G, Tadege M, Wen J, Cosson V, Mysore KS, Ratet P, Madueño F, Bai G, Chen R (2010) Control of dissected leaf morphology by a Cys(2)His(2) zinc finger transcription factor in the model legume *Medicago truncatula*. Proc Natl Acad Sci U S A 107:10754–10759
21. Peng J, Yu J, Wang H, Guo Y, Li G, Bai G, Chen R (2011) Regulation of compound leaf development in *Medicago truncatula* by fused compound leafl, a class M *KNOX* gene. Plant Cell 23:3929–3943
22. Chen J, Moreau C, Liu Y, Kawaguchi M, Hofer J, Ellis N, Chen R (2012) Conserved genetic determinant of motor organ identity in *Medicago truncatula* and related legumes. Proc Natl Acad Sci U S A 109:11723–11728
23. Penmetsa RV, Cook DR (2000) Production and characterization of diverse developmental mutants of *Medicago truncatula*. Plant Physiol 123:1387–1398
24. Le Signor C, Savois V, Aubert G, Verdier J, Nicolas M, Pagny G, Moussy F, Sanchez M, Baker D, Clarke J, Thompson R (2009) Optimizing TILLING populations for reverse genetics in *Medicago truncatula*. Plant Biotechnol J 7:430–441
25. Tsai H, Howell T, Nitcher R, Missirian V, Watson B, Ngo KJ, Lieberman M, Fass J, Uauy C, Tran RK, Khan AA, Filkov V, Tai TH, Dubcovsky J, Comai L (2011) Discovery of rare mutations in populations: TILLING by sequencing. Plant Physiol 156:1257–1268
26. Henikoff S, Comai L (2003) Single-nucleotide mutations for plant functional genomics. Annu Rev Plant Biol 54:375–401
27. Trinh H, Barker D, Ratet P (2001) EMBO Practical course on the New Plant Model System *Medicago truncatula* Nov 19–Dec 1st 2001. http://www.isv.cnrs-gif.fr/embo01/manuels/pdf/module2.pdf
28. Colbert T, Till BJ, Tompa R, Reynolds S, Steine MN, Yeung AT, McCallum CM, Comai L, Henikoff S (2001) High-throughput screening for induced point mutation. Plant Physiol 126:480–484
29. Talamè V, Bovina R, Sanguineti MC, Tuberosa R, Lundquist U, Salvi S (2008) TILLMore, a resource for the discovery of chemically induced mutants in barley. Plant Biotechnol J 6:477–485
30. Carelli M, Biazzi E, Panara F, Tava A, Scaramelli L, Porceddu A, Graham N, Odoardi M, Piano E, Arcioni S, May S, Scotti C, Calderini O (2011) *Medicago truncatula* CYP716A12 is a multifunctional oxidase involved in the biosynthesis of hemolytic saponins. Plant Cell 23: 3070–3081

Chapter 10

High-Throughput and Targeted Genotyping of *Lotus japonicus* LORE1 Insertion Mutants

Dorian Fabian Urbański, Anna Małolepszy, Jens Stougaard, and Stig Uggerhøj Andersen

Abstract

The Lotus Retrotransposon 1 (LORE1) is used for genome-wide mutagenesis of the model legume *Lotus japonicus*. Characterization of the LORE1 insertion sites in individual mutant lines is critical for development and use of the resource. Here we present guidelines for use of the LORE1 reverse genetics resource and provide detailed protocols for insertion site identification and validation. For high-throughput identification of insertions in up to 9,216 pooled lines, the FSTpoolit protocol takes advantage of Splinkerette adapters, molecular barcoding, 2D pooling, Illumina sequencing, and automated data analysis using the freely available FSTpoolit software. Complementing the high-throughput approach, we describe a simplified sequence-specific amplification polymorphism (SSAP) protocol well suited for quick identification of insertion sites in a limited number of lines. Both the FSTpoolit and simplified SSAP protocols are generally applicable to insertion site identification in any insertional mutagenesis setup.

Key words LORE1, FSTpoolit, Insertion mutagenesis, Next-generation sequencing, Reverse genetics, FST, Retrotransposon

1 Introduction

The recent accumulation of transcriptome and proteome data for *Lotus japonicus* (Lotus) and other plants opens the possibility for rapid generation of hypotheses for candidate gene function. These high throughput approaches therefore accentuate the need for reliable and easily applicable reverse genetic techniques that are adequate for experimental hypothesis testing. To be efficient, a platform for reverse genetics in a model plant like Lotus should provide well-annotated knock-out mutations for the majority of the known genes. Among the existing approaches, introduction of large deletions or insertions is the most reliable method for generating knock-out mutations. To date however, only insertions can be systematically traced in a large mutant population. A new reverse genetics resource for Lotus was founded in 2011. This resource

Ray J. Rose (ed.), *Legume Genomics: Methods and Protocols*, Methods in Molecular Biology, vol. 1069, DOI 10.1007/978-1-62703-613-9_10, © Springer Science+Business Media, LLC 2013

takes advantage of the random integration of Lotus endogenous retrotransposon 1 (LORE1) in the Lotus genome [1, 2]. LORE1 mutant lines have a relatively low number of insertions (~3), which can be annotated using an efficient next-generation sequencing-based high throughput protocol. The following chapter is an introduction to the LORE1 resource and comprises detailed protocols for LORE1 insertion site identification and validation.

1.1 LORE1: An Endogenous Retrotransposon with Germ Line-Specific Activity

A novel Ty3-*gypsy* type retrotransposon, 5,041 bp in length, was discovered in the ecotype Gifu, whilst screening a tissue-culture derived Lotus population for symbiotically impaired individuals [3–5]. This retrotransposon was designated LORE1 and the high mutation frequency caused by the activated LORE1 element motivated the investigation of its activity and origin. A Southern blot analysis, followed by cloning from bacterial artificial chromosomes (BACs), led to identification of eight members of the LORE1 family [6]. Interestingly, all new mutant alleles were derived from a single family member, namely LORE1a. Promoter fusion constructs with a β-glucuronidase reporter gene revealed that, in both Lotus and Arabidopsis, expression driven from the LORE1 promoter was high in pollen and pollen tubes [7]. The hypothesis that LORE1a could be active in germ lines was tested by crossing a Lotus Gifu line containing an active LORE1a element to the MG20 ecotype that is devoid of LORE1a copies. New insertions occurred predominantly in crosses where Gifu was used as a pollen donor, and to a much lesser extent in crosses with Gifu as the female partner [7]. New insertions of LORE1a were found in plants regenerated from tissue culture, but not in the sporophyte or callus tissue itself [6, 7]. LORE1 de-repression occurred stochastically among the tissue culture regenerants [2, 7].

1.2 LORE1 as a Gene-Targeting Insertion Mutagen Suitable for Large Scale Mutagenesis

Two large scale experiments, which provided data on almost 13,000 LORE1 insertions, indicated a strong LORE1 preference for genic insertions, while maintaining an even distribution across Lotus chromosomes [1, 2]. Similar to tobacco *Tnt1* and unlike rice-derived *Tos17*, LORE1 did not show a strong preference for specific gene families [8, 9]. The even genomic distribution was confirmed by the observation that among 1,798 unique genes with exonic insertions, 1,586 were targeted by a single LORE1 insertion. In agreement with the observed random gene targeting, no conserved insertion-site motif was found when comparing LORE1 insertion sites [1].

The unique activation mechanism and the germinal transposition differentiate LORE1 from other active retrotransposons used for mutagenesis, allowing easy and ample generation of mutant lines simply by harvesting and propagating seeds from a founder line carrying a de-repressed LORE1a element. At the same time, LORE1 shares features of other retro-element-based mutagenesis tools, which are stable insertions via a copy-and-paste mechanism, and ease of detection by flanking sequence tag (FST) amplification. Given the small LORE1 family, LORE1a is well suited for efficient high-throughput insertion identification by next-generation sequencing techniques. As a result, systematic LORE1 insertion identification can be carried out on a large scale in a short period of time.

1.3 The LORE1-Based Mutant Collection

The Centre for Carbohydrate Recognition and Signalling (CARB—Denmark) and the National Institute of Agrobiological Sciences (Japan) have launched a joint initiative for genome-wide LORE1 mutagenesis of Lotus. The aim of the project was to establish large populations of LORE1-tagged lines and develop protocols for robust and high-throughput insertion annotation. As a result, two protocols for amplification of insertion border fragments were developed, both taking advantage of NGS technologies. The entire population established in Denmark, and a large part of the Japanese population, were developed from Lotus Gifu line G329-3. This line was regenerated from callus meaning that all LORE1 mutants provided by the resource are non-transgenic [2]. The R1 generation of the line G323-3 was chosen as a founder for the population because of its frequent germinal transpositions (2.9 new insertions in R2), an important factor influencing the size of the mutant population required for saturation mutagenesis. The G329-3 line had accumulated three insertions in the R1 generation: two in exons of putative genes and one in an intron of cinnamoyl-CoA reductase, an enzyme involved in the early stages of lignin biosynthesis (unpublished data) [1, 10, 11]. We have confirmed that the presence of background mutations in the G329-3 line has no influence on plant morphology and does not alter symbiotic processes after rhizobial inoculation, when compared to the Gifu control.

Together, work by Fukai et al. and Urbanski et al. [1, 2] delivered approximately 6,000 LORE1-tagged lines in 2011, laying the foundation for a new Lotus reverse genetics resource. The resource is being expanded, and more than 60,000 lines are currently scheduled for sequencing.

Requests for lines from the National Institute of Agrobiological Sciences, Japan, can be made through Legume Base (http://www.legumebase.brc.miyazaki-u.ac.jp). Line name coding format: Pxxxx. To order lines developed by CARB (IDs: 3xxxxxxx), visit http://carb.au.dk/resources. If available, ~40 seeds from each requested line will be sent.

2 Materials

Kazusa DNA Research Institute, The *Lotus japonicus* genome browser, http://www.kazusa.or.jp/lotus/.

The Centre for Carbohydrate Recognition and Signalling (2007), http://carb.au.dk/resources.

The National BioResource Project, Legume Base (2004) http://www.legumebase.brc.miyazaki-u.ac.jp.

2.1 FSTpoolit Laboratory Protocol

2.1.1 Reagents

Amounts shown are for 200 samples.

1. CTAB buffer (1,000 ml).
2. TE buffer pH 8.0 (100 ml).
3. Tris–HCl 10 mM, pH 8.0 (>100 ml).
4. Isopropanol (600 ml).
5. Chloroform–isoamyl alcohol (24:1, 1,000 ml).
6. 70 % EtOH (1,200 ml).
7. 10× Annealing buffer (>15 μl).
8. 10× Extra buffer (1,000 μl).
9. 20 mM ATP (300 μl).
10. 25 mM $MgCl_2$ (1,500 μl).
11. 20 mM dNTPs mix (30 μl).
12. 20 mM dATP (40 μl).
13. 5 U/μl T4 DNA polymerase (60 μl).
14. 10 U/μl T4 Polynucleotide Kinase (PNK, 600 μl).
15. 5 U/μl Taq polymerase (100 μl).
16. 5 U/μl T4 DNA Ligase (200 μl).
17. 10× T4 DNA Ligase buffer (1,000 μl).
18. KAPA2G Fast Multiplex PCR Kit (5,000 μl); KAPABIOSYSTEMS Cat. no. KK5801.
19. 10 mg/ml RNAse H (1 ml).
20. λ bacteriophage DNA (1 μg).
21. Quant-iT PicoGreen (50 μl) Invitrogen Cat. no. P7581.
22. Oligomers: 2.5 μM P2, P3, Splink 1, Splink 2 (500 μl).

2.1.2 Equipment

1. 9.3 mm ceramic beads, reusable (four per sample).
2. Vortexer.
3. 65 °C incubator.
4. Centrifuge for 13 ml (round bottom) tubes. Up to 10,000 rcf. With cooling (Sorval).
5. Agarose gel electrophoresis apparatus.
6. Covaris S2 or E210 shredder (Covaris).
7. NanoDrop spectrophotometer (Thermo Scientific).
8. Table centrifuge.
9. Thermocycler.
10. Centrifuge for 96× plates. Up to 4,000 rcf.
11. QIAvac 96 vacuum system (Qiagen Cat. no. 19504).
12. Vacuum pump with a vacuum controller.

2.1.3 Consumables

1. 30 ml tubes, round bottom + plug caps, chloroform resistant, (Sarstedt Cat. no: 55.517; art. no: 65.791).
2. 13 ml round bottom tubes + screw-caps, chloroform resistant, 400/200 samples (Sarstedt Cat. No. 60.541.545).
3. 1.5 ml tubes snap-cap tubes, 200/200 samples.
4. 96× well plates 300–500 μl.
5. 96× well PCR plates.
6. PCR plate aluminum cover (VWR, Cat. no. 391-1282).
7. COVARIS S2 microTUBES crim-cup 130 μl or COVARIS 96× microTUBE plates (KBIOscience (96 microTUBE Plate 520051 or 520054)).
8. QIAquick 96 PCR Purification Kit (Qiagen Cat. no. 28181 or 28183).
9. Homemade spin column, 200/200 samples (Urbanski et al. 2012 [1] supplemental materials).
10. Agarose.
11. Ice.
12. Liquid nitrogen.
13. Autoclaved ddH_2O.

2.1.4 Primers and Oligonucleotides

SPL short	CCACTAGTGTCGACACCAGTCTCTAATTTTTTTTTT CAAAAAAA
SPL long	CGAAGAGTAACCGTTGCTAGGAGAGACCGTGGCTG AATGAGACTGGTGT CGACACTAGTGGT
Splink 1	CGAAGAGTAACCGTTGCTAGGAGAGACC
Splink 2	NNGTGGCTGAATGAGACTGGTGTCGAC
P2	CCATGGCGGTTCCGTGAATCTTAGG
P3	CCAACTAACAACCCAGCTAACAGT

2.1.5 Buffers

10× Extra buffer.

100 mM Tris–HCl pH 8.3, 500 mM KCl, 15 mM $MgCl_2$, 1 % Triton X-100.

10× Annealing buffer.

100 mM Tris–HCl, 500 mM NaCl, 10 mM EDTA.

CTAB buffer 200 ml.

4 g CTAB (cetyl-trimethyl-ammonium-bromide, do not autoclave CTAB solutions), 16.36 g NaCl,10 ml 0.4 M EDTA pH 8.0 (autoclaved separately), 20 ml 1 M Tris–HCl pH 8.0 (autoclaved separately), ddH_2O to 200 ml.

PE and EB buffers.

Supplied with the QIAquick 96 PCR purification kit (Qiagen).

TE buffer.

10 mM Tris–HCl pH 8, 1 mM EDTA.

10× Taq buffer.

Use the buffer supplied with your Taq polymerase of choice.

2.2 Simplified SSAP Protocol for LORE1 and LORE2 Detection

2.2.1 Reagents

1. CTAB buffer.
2. TE buffer.
3. Tris–HCl 10 mM, pH 8.0.
4. Chloroform–isoamyl alcohol (24:1).
5. 70 % EtOH.
6. 96 % EtOH.
7. Autoclaved ddH_2O.
8. dNTPs mix (20 mM of each).
9. 5 U/μl Taq polymerase.
10. 10× Annealing buffer.
11. T4 DNA Ligase.
12. 10× T4 DNA Ligase buffer.
13. Ethidium bromide.

2.2.2 Equipment

1. Table centrifuge.
2. Vortexer.
3. Apparatus for vertical polyacrylamide gel electrophoresis.
4. Apparatus for agarose gel electrophoresis.

2.2.3 Consumables

1. 200 μl PCR tubes.
2. 1.5 ml snap-cap tubes.
3. PCR purification kit.
4. Polyacrylamide gel components: 30 % Acrylamide–Bis-acrylamide (37:1).
   Tetramethylethylenediamine (TEMED), Ammonium persulfate (APS).
5. Agarose.

2.2.4 Primers

Splink 1 5′ CGAAGAGTAACCGTTGCTAGGAGAGACC

Splink 2 5′ NNGTGGCTGAATGAGACTGGTGTCGAC

P2 5′ CCATGGCGGTTCCGTGAATCTTAGG

P3 5′ CCAACTAACAACCCAGCTAACAGT

LORE2_SSAP1 5′ TCCCTTTTTGGATCGGGAGCTCTG
LORE2_SSAP2 5′ CTGCCACCTCACCAGGTGGTGGTTAT
P4 5′ GAGGACAAGGAAATTGTTAAGGGGAGGGATA
P5 5′ CTTCATCTCTTCTTCTTCTTCAACCTTCTGTCC

3 Methods

3.1 LORE1 Resources

3.1.1 Quick Search for Insertions in a Candidate Gene Using BLAST

To facilitate the search for insertions in candidate genes, BLAST tools were established at the CARB Center (http://carb.au.dk/resources/) and KAZUSA DNA Research Institute (http://www.kazusa.or.jp/lotus/blast.html). The CARB LORE1 BLAST database contains 2,000 bp long flanking sequences; 1,000 bp upstream and 1,000 bp downstream of each LORE1 insertion site. To extract flanking sequences, *Lotus japonicus* MG20 genome release 2.5 was used as reference (available at http://www.kazusa.or.jp/lotus). The coding sequences of genes of interest can be used as queries in the BLAST search, and multiple sequences can be submitted using multifasta format. If an insertion in an exon of the gene of interest is found, the BLAST search result will span position 1,000 of the database subject (Fig. 1).

3.1.2 Common Methods for LORE1 Insertion Site Validation

A LORE1 insertion site is presented as the first base upstream from the LORE1 5′ long terminal repeat (LTR). It is therefore always described by a contig/chromosome number, position and orientation. Information on all LORE1 insertions identified by *FSTpoolit* analysis (described in detail below) is available at http://carb.au.dk/resources/. See also Table 1.

Seeds delivered by the LORE1 resource are the R3 generation of line G329-3 and are therefore a segregating population for the insertions that occurred in R2. Individuals homozygous for mutant alleles have to be identified from the pool by genotyping PCR. To facilitate the genotyping step, a set of PCR primers will automatically be generated for the majority of the insertion sites [1, 12]. When designing genotyping primers for insertions in the reverse orientation, the reverse complemented sequence is used. Therefore the primer, marked as "Forward" should always be used together with primers "P2" or "Reverse." Use primers "Forward" and "Reverse" to detect wild type alleles and "Forward" and "P2" for alleles with an insertion (Fig. 2). Primer sets were designed to yield PCR products ranging from 300 to 700 bp. Optimized PCR conditions are presented below (Subheading 3.1.3).

```
Query: 2272 aagatttattggatgaactgggtggagctaaaatttttc taagctggacctaagagctg 2331
            ||||||||||||||||||||||||||||||||||||||| ||||||||||||||||||||
Sbjct: 1040 aagatttattggatgaactgggtggagctaaaatttttc taagctggacctaagagctg 981
                                                    ↑
                                Insertion site at subject position "1000"
```

Fig. 1 BLAST search results for LORE1 insertion in a coding sequence. Significant alignments will be also found by coding sequence searches if an insertion is present in introns or in 5′ or 3′ UTRs in close proximity to the coding sequence. In this case, the match will not span position 1000 of the subject, and the search will have to be repeated with the genomic sequence of gene of interest to identify the precise location of the insertion site. Unlike the CARB BLAST tool, BLAST searches on the KAZUSA Web page will indicate the insertion site at position "1" of the search subject. For both search tools it is suggested to restrict the E-value of the BLAST search to 1e^{-5} or lower (advanced option-e) to limit false positives. If successful, a BLAST search outputs a list of insertion sites that can be linked to a correct Lotus mutant line by viewing the Lotus genome browser http://www.kazusa.or.jp/lotus/ or by searching the LORE1 insertion database at the CARB Web page http://carb.au.dk/resources/

Table 1
LORE1 insertion details

Batch	Experimental batch
Chromosome	Lotus chromosome or assembly contig
Position	Position on the chromosome. 1st base upstream from LORE1 5′ LTR
Orientation	Orientation of retrotransposon insertion. Please refer to Fig. 2a
Column_barcode	Multiplexing barcode number from the pooling array
Row_barcode	Multiplexing barcode number from the pooling array
Plant_ID	Plant line identifier
Seed_stock	Seed availability
Pool_coordinates	Barcodes describing single insertions
Coordinate_abundances	Abundance of sequencing reads for individual barcode describing single insertions
Abundances_sum	Sum of sequencing reads for insertion
Forward_primer	Forward genotyping primer (Fig. 2b)
Reverse_primer	Reverse genotyping primer (Fig. 2b)
PCR_product_insertion_positive	PCR product size for locus with insertion if primers "For" and "P2" were used
PCR_product_wt	PCR product size for wild type locus if primers "For" and "Rev" were used
±1,000_bp_insertion_flanking	insertion flanking region sequence

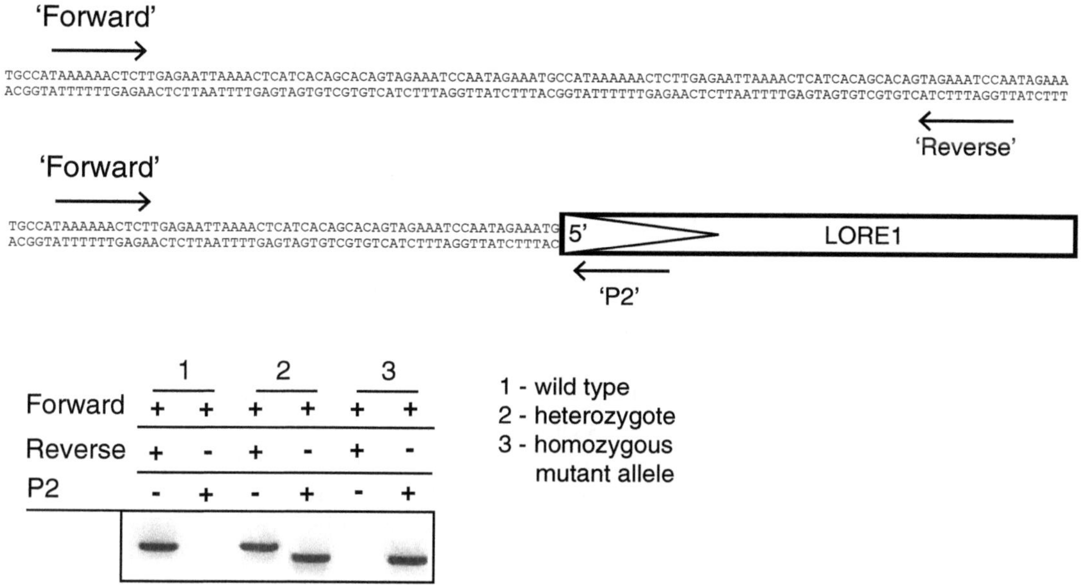

Fig. 2 Genotyping PCR for LORE1 insertions. Two combinations of primers have to be used to characterize a locus for wild type or LORE1 insertion alleles. The "Forward" and "Reverse" primers were custom designed for every LORE1 insertion identified and are suitable for the generic PCR conditions described

3.1.3 Protocol for the LORE1 Genotyping PCR

PCR reaction mix.

Reagent	Volume (μl)	Final concentration
Primer "Forward" (2.5 μM)	2	250 nM
Primer P2 or "Reverse" (2.5 μM)	2	250 nM
10× Taq buffer	2	1×
dNTPs (20 mM)	0.1	0.1 mM
Taq polymerase (5 U/μl)	0.1	0.5 U
gDNA template (100 ng/μl)	1	5 ng/μl
ddH$_2$O	To 20	

PCR conditions.

95 °C	3′	
95 °C 72 °C	30″ 1′15″	5×
95 °C 72–68 °C 72 °C	30″ 30″ 45″	10× touchdown

(continued)

(continued)

95 °C	30″	20×
68 °C	30″	
72 °C	45″	
72 °C	10′	

For troubleshooting, *see* **Note 1**.

3.1.4 Validation of Gene Function by Mutant Complementation

To obtain strong evidence for linkage between mutation and phenotypic changes, null alleles should be complemented with a wild type allele. For genes involved in root morphology, the transgene can be introduced via hairy roots transformation [13, 14] and stable transformants can be obtained using previously published protocols [15, 16]. In cases where another mutant allele is available for a gene of interest, complementation crosses between two mutant lines can be used to confirm allelism.

3.1.5 FSTpoolit Protocol for High-Throughput LORE1 5′ Border Fragment Amplification

The *FSTpoolit* (flanking sequence tag pooling and identification) protocol was designed for high-throughput identification of LORE1 insertions by Illumina sequencing. The *FSTpoolit* protocol consists of a laboratory protocol for sample pooling, multiplexing and Illumina library preparation, as well as custom made software for Illumina sequencing data analysis. *FSTpoolit* enables simultaneous analysis of material from 9,216 individuals pooled in a 2D 96×96 array. Plant material from each row and column of the array is collected independently, which creates 192 samples for individual treatment. From each of these sub-pools gDNA is extracted and mechanically sheared on a COVARIS Adaptive Focused Acoustics shredder (Fig. 3). Fragmented DNA is subjected to end-repair and terminal adenylation treatments, purified, and subsequently ligated with Splinkerette-based intermediate adapters (IA) [17]. Next, nested PCR reactions are conducted with IA and LORE1 LTR-specific primers and the products are then separated by electrophoresis and size selection steps. The second PCR step introduces a unique 7 bp-long molecular barcode for each sub-pool. Barcoded samples are mixed in equimolar amounts and pooled as a template for Illumina standard paired-end library preparation. The library is sequenced in one Illumina HiSeq2000 flow cell lane. Due to the uneven base distribution caused by the presence of IA and LORE1 sequences, the library has to be sequenced together with at least 1 % of PhiX control. We recommend making your sequencing provider aware of the unusual base distribution to ensure optimal sequencing results.

The sequencing output can be directly converted into a list of LORE1 insertions in individual plants by the *FSTpoolit* software package. The separate modules *FSTpoolit_map*, *FSTpoolit_filter* and *FSTpoolit_primer* are responsible for identification of correct

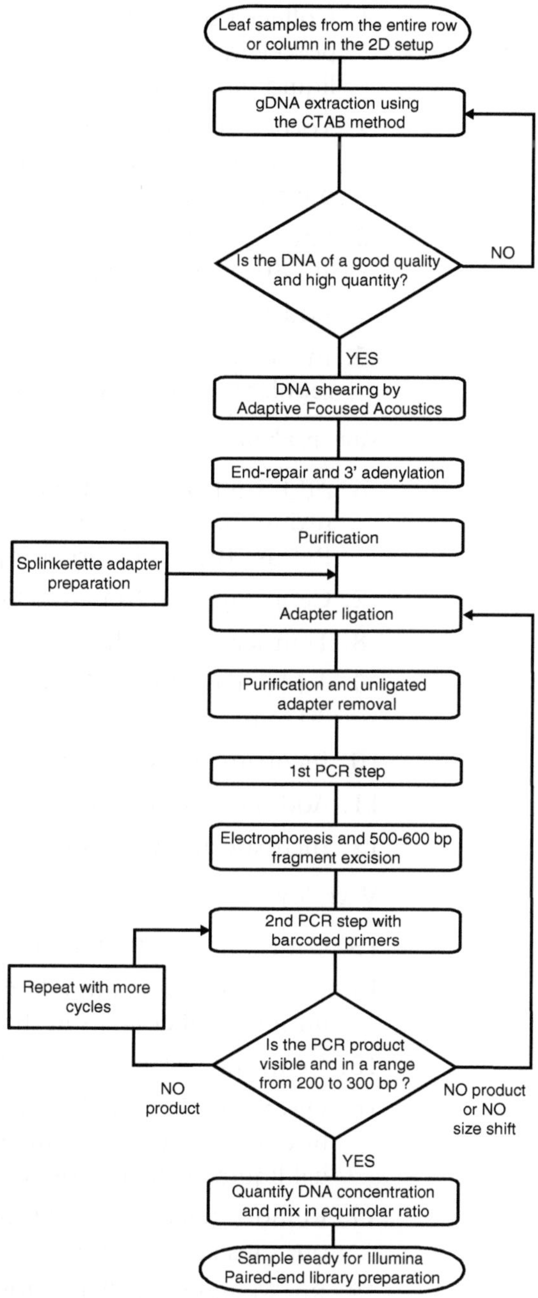

Fig. 3 *FSTpoolit* protocol flowchart

read pairs containing LORE1 flanking sequence information, mapping them to genome reference, filtering data noise, and designing genotyping primers. A detailed description of the *FSTpoolit* software can be found in the manuscript by Urbanski et al. [1] and the software package including documentation can be downloaded from http://carb.au.dk/resources/.

3.1.6 FSTpoolit Laboratory Protocol

DNA Extraction

1. Collect the plant tissue in a 30 ml centrifuge tube. Place 4 porcelain beads (9.3 mm) in the tube with tissue and freeze in liquid nitrogen.
2. Vortex three times for 30 s to crush tissue completely. Vortex at maximum speed. Pour the beads out (be careful not to pour out the tissue).
3. Pour 1/3 of the powder into a new 13 ml tube. Leave the rest at −80 °C as a backup.
4. Add 5 ml of CTAB buffer preheated to 60 °C.
5. Incubate at 60 °C for 30 min with caps closed. Mix content regularly during incubation.

Now work in a fume hood

6. Add 5 ml of chloroform–isoamyl alcohol (24:1), mix by vortexing for at least 20 s or until a uniform emulsion is formed. Release the gas by briefly opening the cap.
7. Centrifuge for 20 min, 6,000 rcf at 4 °C.
8. Transfer 4 ml of the upper aqueous phase to a new 13 ml tube.
9. Add 5 μl of RNAse H (10 mg/ml stock), incubate at 37 °C for 30 min.
10. Incubate on ice for 5 min.
11. Add 3 ml of ice-cold isopropanol and mix by vortexing.
12. Incubate at −20 °C overnight.

Next day

13. Centrifuge at 10,000 rcf for 20 min at 4 °C.
14. Discard supernatant, add 6 ml of 70 % EtOH, and mix gently until the pellet floats in the ethanol.
15. Centrifuge again at 10,000 rcf for 10 min at 4 °C.
16. Discard EtOH carefully so as not to lose the pellet. Centrifuge briefly (until the centrifuge reaches 4,000 rcf) to collect residual liquid at the bottom of the tube.
17. Remove traces of EtOH using a pipette, incubate tubes without caps at 37 °C for 15 min to dry the pellet.
18. Resuspend the pellet in 300 μl of TE and incubate at 65 °C for 20 min then at 4 °C over night.
19. Transfer the supernatant to one well in a 96-well plate. Avoid touching the insoluble pellet.
20. Check DNA quality on a 0.8 % agarose gel and measure DNA concentration spectrophotometrically (e.g. with a NanoDrop).

DNA Shearing

1. Prepare 130 μl of a 150 ng/μl DNA dilution in 10mM Tris–HCl.
2. *For Covaris S2 series.* Turn on the Covaris S2, fill the water container with fresh ddH_2O up to level 12 and set the temperature

of the thermostat to 3 °C. This will result in a water bath temperature of around 6 °C. *It takes about 1 h to cool down to the appropriate temperature.* Adjust the settings to:

Duty cycle	5 %
Intensity	3
Cycles per burst	200
Time	40
Temperature	6–8 °C
Power mode	Frequency Sweeping
Degassing mode	Continuous

Transfer 130 μl of DNA solution into Covaris Snap-cap microTUBE and insert the tube into a holder. Submerge the holder into a water bath and close the instrument cover. Run the instrument.

Transfer sheared DNA from the microTUBE into a fresh 1.5 ml tube.

3. *For Covaris E210 series.* Turn on the Covaris E210, fill water container with fresh ddH_2O up to level 6 and set the temperature of the thermostat to 3 °C. That would result in a water bath temperature of around 6 °C. *It takes about 2 h to cool down to the appropriate temperature.* Adjust the settings to:

Duty cycle	5 %
Intensity	3
Cycles per burst	200
Time	40
Temperature	6–8 °C
Power mode	Frequency sweeping
Degassing mode	Continuous

Transfer 130 μl of DNA solution into a 96 microTUBE plate and cover the plate with aluminum PCR plate cover. Insert the plate into a holder. Follow the manufacturer's instructions regarding instrument use.

Spin the plate at 180 rcf for 1 min. Slowly remove the sticky aluminum foil cover (*see* **Note 2**).

Transfer sheared DNA from the 96 microTUBE plate into a fresh 96× well plate.

End Repair and Adapter Ligation (*See* **Note 3**)

1. Intermediate Adapter preparation.

Prepare the Intermediate Adapter the same day as it will be ligated. Prepare enough to have ten times more moles of adapter than gDNA.

Component	For 100 samples of 2 μg DNA each[a] (μl)	Final conc.
SPL long (100 μM)	30	42.8 μM
SPL short (100 μM)	30	42.8 μM
10× Annealing buffer	7	1×
H_2O	3	

[a]2 μg of DNA equals 3 pmol provided that the DNA is sheared to, on average, 1,000 bp-long dsDNA fragments. In this case 30 pmol of adapter is needed for each reaction.

Insert tube with mixed oligos and buffer into a heating block (95 °C) and incubate for 3 min.

Remove the heating block with the tubes in it, cover it with another hot block and leave until temperature drops below 30 °C (around 1.5–2 h). Put on ice.

2. End repair.

Prepare the end repair mix.

	1× (μl)	100× (μl)	Final concentration in 30 μl
10× Extra buffer	3	300	1×
ATP (20 mM)	1.5	150	1 mM
$MgCl_2$ (25 mM)	7.5	750	6.25 + 1.5 mM from the buffer
dNTPs mix (20 mM of each)	0.15	15	0.1 mM
T4 DNA polymerase	0.3	30	1.5 U
T4 DNA Polynucleotide kinase (PNK)	3	300	30 U

Distribute the reaction mix into a 96-well PCR plate and add 14.55 μl DNA to reach a final volume of 30 μl. Vortex and centrifuge briefly.

Incubate for 30 min in 37 °C—end repair reaction.

Incubate for 15 min in 75 °C—inactivation.

3. Adenylation.

Prepare the Adenylation mix.

	1× (μl)	100× (μl)	Final conc. in 50 μl
Extra buffer	2	200	1×
dATP (20 mM)	0.2	20	>0.08 mM
Taq polymerase	0.5	50	25 U
H_2O	17.3	1,730	

- Add 20 μl of the adenylation mix to 30 μl of heat inactivated "end-repair" reaction.
- Incubate at 72 °C for 30 min.
- Purify DNA. Purification step example based on QIAquick 96 PCR Purification Kit.
- Prepare the manifold by placing the waste tray inside the QIAvac base and the QIAvac top on the base. Attach the QIAvac to a vacuum source and insert the QIAquick 96 PCR plate.
- Add 150 μl of Buffer PM to 50 μl of adenylation reaction and mix by pipetting or vortexing, followed by a brief centrifugation. Add the mixture to the QIAquick 96 PCR Purification Kit directly on the column material.
- Apply vacuum (~200–600 mbar) until the membrane is dry.
- Wash the plate with 900 μl PE buffer and switch on the vacuum. *Repeat this step once.*
- Apply ~600 mbar vacuum for 10 min to remove residual PE buffer. Switch off vacuum and ventilate the plate slowly. To remove residual Buffer PE, lift the top plate together with QIAquick 96 plate and vigorously tap the top plate on a stack of absorbent paper until no drops come out. Blot the nozzles of the QIAquick plate with clean absorbent paper.
- Replace the waste container with a 96× PCR plate placed in a holder so that the nozzles of QIAquick plate are inserted into the tubes of the PCR plate.
- Place the top plate of the manifold back on the base. Add 65 μl of EB buffer (or 10 mM TRIS pH 8) to every well and incubate for 1 min. Apply ~600 mbar vacuum for 5 min. Ventilate slowly.

4. Ligation of Intermediate Adapter.

   Prepare ligation reaction mix.

	1× (μl)	100× (μl)	Final conc. in 50 μl
10× T4 DNA Ligase buffer	5	500	1×
T4 DNA Ligase	1	100	5 U
Intermediate Adapter[a] (42.8 μM)	0.7	70	30 μM

[a]Adjust the volume of IA to 10× higher concentration than dsDNA.

Add ligation mix to a fresh PCR plate and add 43 μl of adenylated DNA. Incubate at 22 °C for 1 h. Add 7 μl of the ligation mix.

Inactivate the T4 DNA ligase by incubation at 70 °C for 5 min or at 65 °C for 10 min.

Purify DNA using the QIAquick 96 PCR Purification Kit (like **step 3** in Subheading 3.1.6.3). Use the same plate as in **step 3** but wash it two times with 200 μl of TE before use.

Elute with 80 μl of 10 mM Tris–HCl or EB buffer (*see* **Note 4**).

PCR Amplification of 5′ LORE FSTs

1. First PCR reaction. Prepare PCR master mix.

	For 1× (μl)	100× (μl)	Final conc. in 25 μl
Primer "Splink 1" (2.5 μM)	2.5	250	250 nM
Primer "P2" (2.5 μM)	2.5	250	250 nM
KAPA multiplex polymerase mix	12.5	1,250	
DNA	7.5		

PCR conditions.

95 °C	3′	
95 °C	30″	Touchdown ten cycles
72–68 °C	30″	
72 °C	1′30″	
95 °C	30″	Ten cycles
68 °C	30″	
72 °C	1′30″	
72 °C	7′	

2. Resolve the reaction products on a freshly prepared 1.5 % agarose gel.

3. Cut out approximately 500–600 bp-long fragments. Spin the gel fragments on home-made columns [1] for 10 min, 14,000 rcf at room temperature.
4. Prepare the nested PCR reaction mix. *It is very important to do this step the same day as the gel extraction step. DNA samples are very unstable in the flow-through.*

	For 1× (μl)	100× (μl)	Final conc. in 25 μl
Primer "Splink 2" (2.5 μM)	2.5	250	250 nM
KAPA multiplex polymerase mix	12.5	1,250	
ddH_2O	5.5	550	

5. To PCR master mix add 2 μl of eluate from the home made column in **step 3** and 2.5 μl of the appropriate 2.5 μM primer P3 with molecular barcode tail.
6. Run the second (nested) PCR reaction.

95 °C	3′	
95 °C	30″	Touchdown ten cycles
72–68 °C	30″	
72 °C	30″	
95 °C	30″	Ten cycles
68 °C	30″	
72 °C	30″	
72 °C	7′	

DNA Concentration Quantification and Pooling

1. Make the following standards of DNA, diluting λ DNA or any other short length DNA with TE-buffer, using the following table.

Conc. (pg/μl)	Dilution
1,000	2 μl (0.5 μg/μl λ-DNA) + 998 μl TE
600	300 μl (1,000 pg/μl) + 200 μl TE
400	200 μl (1,000 pg/μl) + 300 μl TE
200	200 μl (400 pg/μl) + 200 μl TE
100	200 μl (200 pg/μl) + 200 μl TE
50	200 μl (100 pg/μl) + 200 μl TE
25	200 μl (50 pg/μl) + 200 μl TE
0	Only TE

2. Transfer 5 μl of each sample to 145 μl of TE (30× dilution).
3. Prepare 200× dilution of PicoGreen stock solution. For 100 samples: add 25 μl of PicoGreen to 4,975 μl TE.
4. Pipette 50 μl of diluted sample and 50 μl of diluted PicoGreen to a black microtiter plate. Be sure that standards are included on every plate.
5. Measure emission on a microtiter-plate reader. Excitation: 480 nm. Emission: 520 nm.
6. Calculate sample concentration from the standard curve equation. Remember to amplify by the 60× dilution factor.
7. Combine all samples in one tube by adding 70 ng of each sample to one 1.5 ml tube.

Prepare a Paired-End Library

Prepare a paired end library according to the standard Illumina protocol or ship the sample to your sequencing provider.

3.2 LORE1 as a Tool in Forward Genetic Screens

3.2.1 Setup for Forward Genetics Screens

The high tagging efficiency makes LORE1 mutant populations suited for forward genetic screens [1, 2]. LORE1 mutants have the additional advantage that the retrotransposon insertions can be rapidly localized in the genome via polymerase chain reactions with target-specific primers, eliminating the need for map based cloning. To conduct forward genetic screens using LORE1 lines, we use the procedure described below.

- Develop appropriate screening conditions that allow a clear distinction of a mutant phenotype from the wild type.
- Germinate 40–60 R3 generation seeds separately for each progenitor (Family screening—Fig. 4). Make sure that the ID of the progenitor is retained.
- Screen for an adequate phenotype that should occur in approximately 25 % of the siblings.
- Propagate all the individuals that show a phenotype of interest and at least three siblings with the wild type phenotype. If the causative mutation decreases fertility, the mutant allele can be retained in heterozygous individuals.
- When possible, collect leaf samples for DNA extraction.
- If the progenitor comes from the population that was annotated for LORE1 insertions by high-throughput methods, the search for candidate causative insertions can be done in silico. Perform the genotyping PCR described earlier in this chapter for all the insertions identified in a progenitor to confirm co-segregation of the LORE1 insertion with the mutant phenotype.
- If insertion data is not available, there are too many insertions in the line, or the genotyping PCR did not show positive results, the causative mutation has to be identified by other means. For this purpose, use the simplified Sequence Specific Amplification

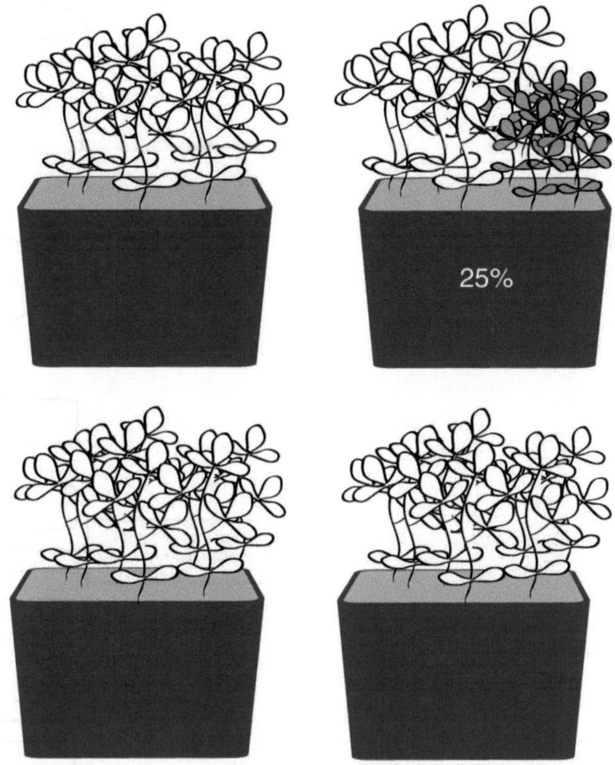

Fig. 4 LORE1 family screening. Progeny of the self pollinated R2 generation, which was genotyped for LORE1 insertions by high-throughput methods, is grown under screening conditions. The single gene mutations that occurred in the R2 generation should display Mendelian segregation resulting in a 1:3 ratio of null homozygotes to heterozygous and wild type alleles

Polymorphism (SSAP) protocol described below. The technique was chosen for its simplicity, robustness and possibility of marker sequence identification. In the forward genetic screens, the principle of identifying causative LORE1 insertions is to show that an SSAP band not present in the founder line co-segregates with an observed phenotype. Because SSAP markers are dominant, 10–16 plants with mutant phenotypes should be screened to reliably score co-segregation.

- Rescreen the offspring from candidate mutants to confirm phenotype and eliminate false positives.

If you wish to estimate the total number of new LORE1 insertions in the mutant line, perform Southern blot analysis following the protocol proposed by Madsen et al. [6].

3.2.2 Simplified SSAP Protocol for LORE1 and LORE2 Detection

The simplified sequence specific amplification polymorphism (SSAP) protocol takes advantage of the same Splinkerette adapters used for high-throughput LORE1 insertion detection [17]. The adapters were modified to contain cohesive ends complementary to the ones produced by five different restriction enzymes (MboI, MspI, Csp6I,

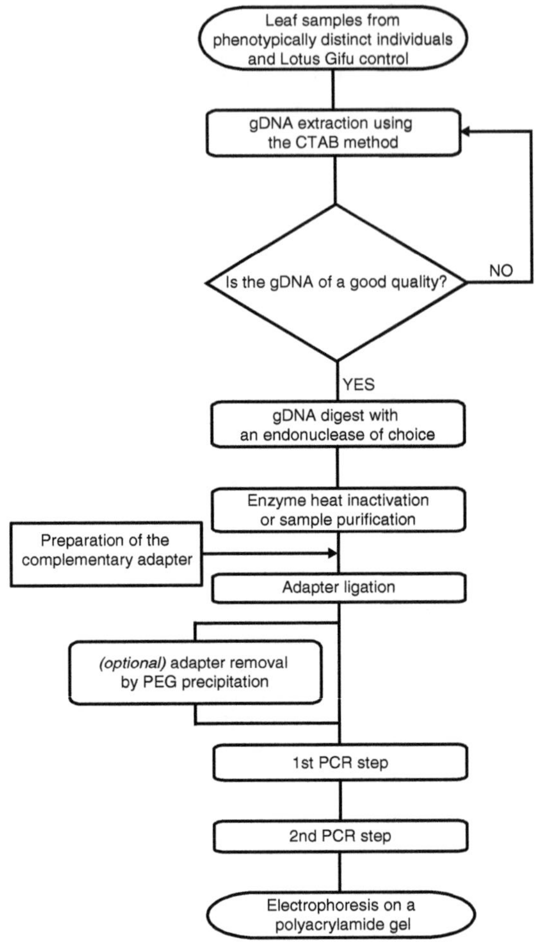

Fig. 5 Flowchart of simplified SSAP procedure for LORE1 insertion site detection

TaqI, BstNI) that may be used in this protocol. In short, high quality DNA (1 μg) is digested by a single endonuclease (Fig. 5). The enzyme has to be inactivated prior to the adapter ligation step. Removal of the unbound adapters is optional and can be done by PEG precipitation [18]. The border fragments of insertions are subsequently amplified in nested PCR reactions using primers specific for the adapter and the 5′ or 3′ ends of the LORE1 insert. PCR products are resolved on a polyacrylamide gel, which provides increased sensitivity and higher resolution as compared to an agarose gel (Fig. 6). Polymorphic bands can be extracted and used for re-amplification and sequencing of candidate fragments. Because line G329-3 exhibited low LORE2 activity, the following protocol includes a set of primers that enables detection of LORE2 insertions [2, 19].

The protocol for the simplified SSAP procedure is designed for robust amplification of LORE1 elements. Five different frequently cutting restriction enzymes, MboI, MspI, Csp6I, TaqI, and BstNI, can be used followed by amplification of fragments 5′ or 3′ of

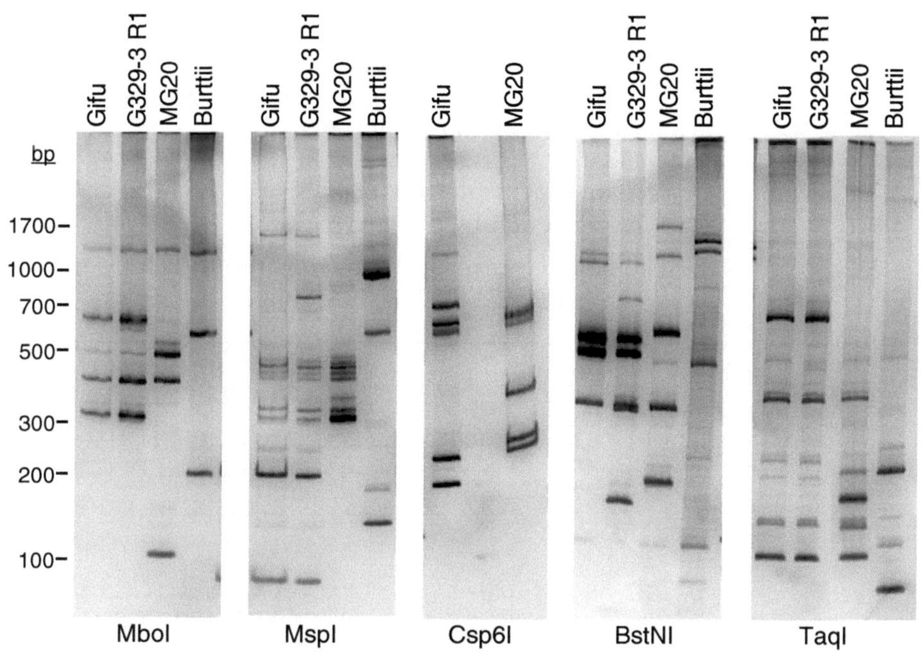

Fig. 6 SSAP bands of LORE1 family members resolved on polyacrylamide gels. Polymorphisms in LORE1 element numbers and genomic locations among different Lotus ecotypes and species are apparent. Repeating the SSAP analysis with multiple restriction enzymes increases the chance for detecting polymorphisms

LORE1 insertions. Only MspI can be used for amplification of LORE2 border regions if the presented primer set is used.

Use DNA of high purity, preferably extracted using the CTAB protocol [20]. 1 μg of gDNA per sample per enzyme is sufficient. Use more DNA if restriction enzymes that cannot be heat inactivated are used.

Remember to include the control DNA from Gifu and/or the G329-3 line, preferably in duplicates.

Thermo-inactivated enzymes: MboI, MspI, Csp6I.

Not thermo-inactivated enzymes: TaqI, BstNI.

1. Digest sample in 20 μl of the appropriate buffer for the enzyme of choice. Enzyme: 1 μl; time: 3–4 h.

   For FERMENTAS buffers.

Enzyme 1 μl	Buffer (2 μl) 10×	Digest temp. (°C)
MspI	Tango	37
MboI	R	37
Csp6I	B	37
TaqI	Taq buffer	65
BstNI (MvaI)	R	37

2. If appropriate, heat inactivate the restriction enzyme.

   If heat inactivation is not possible, purify DNA using a PCR clean-up kit following the manufacturer's instructions. Elute in 45 μl of Tris–HCl pH 8.0.

3. Prepare the adapter with cohesive ends complementary to the ones generated by the enzyme used.

   The adapter is made by mixing the SPL long oligo with an appropriate LORE1_SSAP oligo.

LORE1_SSAP_MboI	GATCACCACTAGTGTCGACACCAG TCTCTAATTTTTTTTTTCAAAAAAA
LORE1_SSAP_TaqI + MspI	CGACCACTAGTGTCGACACCAGTC TCTAATTTTTTTTTTCAAAAAAA
LORE1_SSAP_BstNI	WACCACTAGTGTCGACACCAGTCT CTAATTTTTTTTTTCAAAAAAA
LORE1_SSAP_Csp6I	TAACCACTAGTGTCGACACCAGTC TCTAATTTTTTTTTTCAAAAAAA
SPL long	CGAAGAGTAACCGTTGCTA GGAGAGACCGTGGCTGAATGA GACTGGTGTCGACACTAGTGGT

4. Adapter preparation.

   Mix oligos with each other in equimolar concentrations and add annealing buffer to 1× final concentration. For 1 μg of digested sample 15 pmol of ready adapter is needed.

   Insert tube with mixed oligos and buffer into a heating block (95 °C) and incubate for 3 min.

   Remove the heating block with the tubes in it, cover it with another hot block and leave until temperature drops below 30 °C (around 1.5–2 h).

   Transfer tubes to ice and keep on ice from now on.

5. Adapter ligation.

   If you have heat-inactivated the enzyme, mix:

Digested DNA	1 μg
Adapter	10× higher molar concentration than DNA
ATP (20 mM)	2 μl
T4 DNA ligase	1 μl
ddH_2O	To 50 μl

   If DNA was purified after digestion, mix:

Digested DNA (approx. 1 μg)	43 μl
Adapter (10× higher molar concentration than DNA)	1 μl
T4 DNA ligase Buffer (10×)	5 μl
T4 DNA ligase	1 μl

6. Vortex, briefly centrifuge and incubate for 1 h at RT.
7. Inactivate ligase by incubating samples at 65 °C for 20 min.
8. PEG precipitate samples to remove unligated adapters.

   ! This step is not critical. The PEG precipitation step removes unspecific, low molecular mass bands from the resolving gel.

   Transfer ligation reaction to 150 μl PEG selection solution.

   - PEG 8000 Final conc. 9 % (51 μl of 35 % stock).
   - 10 mM $MgCl_2$ (2 μl of 1 M stock).
   - 97 μl H_2O.

     Vortex well and leave in RT for 10 min.

     Centrifuge with maximum speed for 30 min.

     Carefully remove supernatant without dislodging the pellet.

     Add 195 μl 70 % EtOH and vortex on low speed until the pellet dislodges from the wall of the tube.

     Centrifuge for 15 min at maximum speed, remove EtOH, and leave until dry.

     Redissolve the pellet in 50 μl H_2O; vortex on low speed to solubilize most of the DNA.

9. PCR Amplification.

   Use primer Splink 1 together with: P2 for LORE1 5′ end, P4 for LORE1 3′ end, and LORE2_SSAP1 for LORE2 5′ end.

   First reaction. Prepare PCR mix.

	For 1 reaction (μl)	Final conc.
Primer "Splink 1" (2.5 μM)	2	250 nM
Primer "P2", "P4" or "LORE2_SSAP1" (2.5 μM)	2	250 nM
dNTP mix (20 mM)	0.1	0.1 mM
10× Taq polymerase buffer	2	1×
Taq polymerase (5 U/μl)	0.1	0.02 U
DNA	3–6	
ddH_2O		To final 20 μl

PCR conditions.

95 °C	3′	
95 °C	30″	Touchdown 10 cycles
72–68 °C	30″	
72 °C	2′	
95 °C	30″	15 cycles
68 °C	30″	
72 °C	2′	
72 °C	7′	

Dilute each sample 100× with ddH_2O.

Prepare the nested PCR reaction mix.
Use primer Splink 2 together with: P3 for LORE1 5′ end, P5 for LORE1 3′ end, and LORE2_SSAP2 for LORE2 5′ end.

	For 1 reaction (μl)	Final conc.
Primer "Splink 2" (2.5 μM)	2	250 nM
Primer "P3," "P5" or "LORE2_SSAP2" (2.5 μM)	2	250 nM
dNTP mix (20 mM)	0.1	0.1 mM
10× Taq polymerase buffer	2	1×
Taq polymerase (5 μ/l)	0.1	0.02 U
DNA	1	
ddH_2O	12.8	

Run the nested PCR.

95 °C	3′	
95 °C	30″	Touchdown 10 cycles
72–68 °C	30″	
72 °C	2′	
95 °C	30″	20 cycles
68 °C	30″	
72 °C	2′	
72 °C	7′	

10. Depending on the well size, run 5–15 μl of the sample on a 5 % polyacrylamide gel made from a 37:1 Acrylamide–Bis-acrylamide stock.
11. Stain the gel for 10 min in 0.5 μg/ml ethidium bromide solution in TAE, or using another appropriate gel dye. Flush the gel with water to remove excess dye.
12. Photograph the gel and analyze the band segregation. Retain the gel.
13. Identify the sequences of amplified fragments.

    Cut the band with a clean scalpel and put it into a clean 1.5 ml tube.

    Crush the polyacrylamide with a pipette tip, add 100 μl TE buffer and vortex thoroughly.

    Incubate for 10 min in a 95 °C heating block.

    Vortex the tube, cool the sample on ice and spin for 2 min at maximum speed to precipitate gel fragments.

    Use 1 μl of supernatant as a template for a PCR reaction identical to the the nested PCR from **step 9**.

    Resolve a fraction of the PCR reaction on an agarose gel. If a single band was amplified, sequence the PCR product directly or clone it into a vector. All true LORE1 insertion border fragments should contain the 5′ LTR sequence of LORE1:

    LORE1 sequence—CCAACTAACAACCCAGCTAACA GTAAACATCTGTAACA—genomic sequence

    Or the 3′ LTR sequence of LORE1:

    LORE1 sequence—CTTCATCTCTTCTTCTTCTTCAA CCTTCTGTCCTATCA—genomic sequence.

    All fragments with LORE2 insertions should contain the common LORE2A and LORE2B 5′LTR sequence:

    LORE2 sequence—TTTGGGATCCTAAGCT TCTAGAATGTTGGCCTCCAGCTGTGGCT TGGGCCTCCTAGCATAGACGTTGCCAAACC TGGGCCTGGAGCTTGGGTTTGCAACA —genomic sequence.

3.2.3 Design of a Suppressor Screen Experiment

In an activated founder line, LORE1a transposes in pollen and the actively transposing element can be transmitted by crossing, using the founder line as pollen donor. This makes LORE1a suitable mutagen for suppressor screens aimed at identifying genetic interactors that modify the phenotype of the original mutant. The procedure starts with crossing a mutant line that has an aberrant phenotype to a LORE1 active line such as G329-3 in order to generate a large-scale LORE1 mutagenized population (Fig. 7).

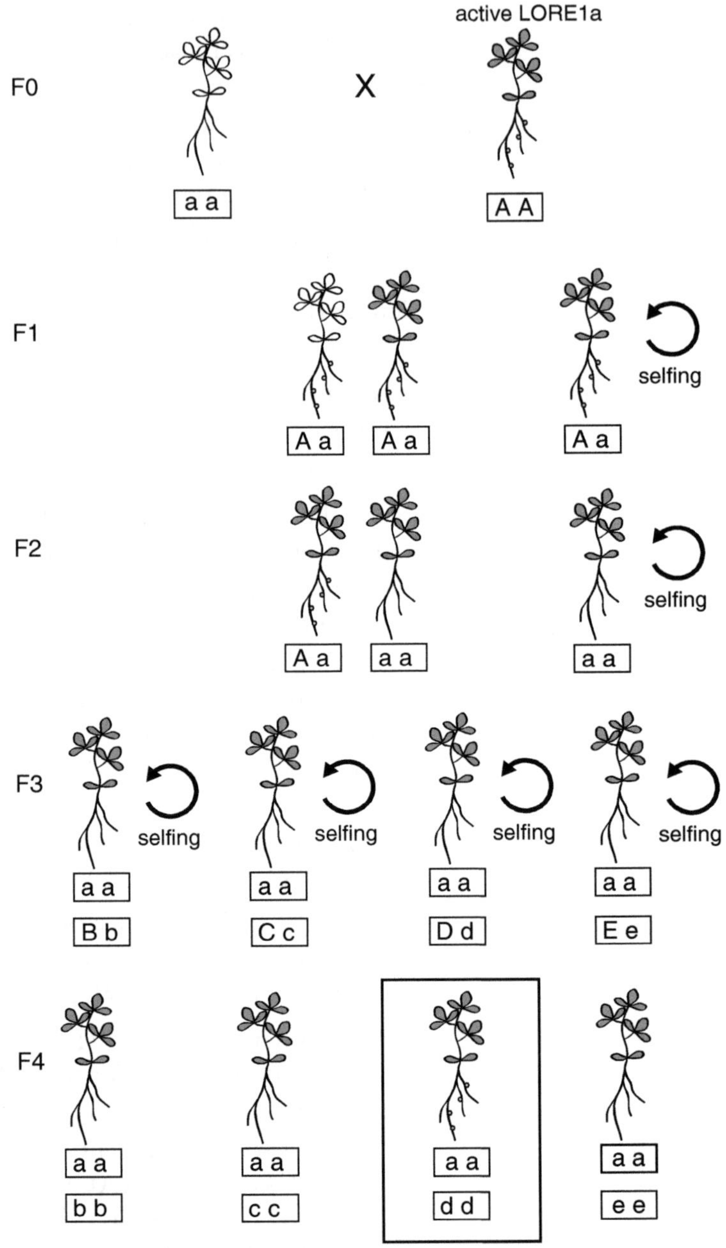

Fig. 7 The principle of LORE1-based suppressor screening. A plant with a mutation altering root symbiosis is shown in *white*. To identify possible genetic interactors, the *line* should be crossed to an individual carrying an activated LORE1 element (*grey* plant, F0). A plant heterozygous for the allele and showing high LORE1 activity has to be chosen from the F1 population and propagated by self pollination. F2 individuals homozygous for the original mutation and carrying active LORE1 elements can be used as founders for a large F3 population. Family screening conducted on the F4 generation may lead to identification of new mutations responsible for reversion or modification of the original mutant phenotype

New LORE1 insertions responsible for phenotype modification can then be identified using either the *FSTpoolit* or the SSAP protocol presented above.

4 Notes

1. Always use good quality gDNA in sufficient amounts. Preferably 100 ng per reaction. In case neither set of primers yields a PCR product, try to decrease the annealing temperature or to increase the Mg^{2+} concentration.
2. Be very careful not to cross-contaminate the samples.
3. To avoid mistakes it is advised to process only one 96-well plate at a time.
4. After this step samples can be stored at –20 °C.

References

1. Urbański DF, Małolepszy A, Stougaard J, Andersen SU (2012) Genome-wide LORE1 retrotransposon mutagenesis and high-throughput insertion detection in *Lotus japonicus*. Plant J 69:731–741
2. Fukai E, Soyano T, Umehara Y, Nakayama S, Hirakawa H, Tabata S et al (2012) Establishment of a *Lotus japonicus* gene tagging population using the exon-targeting endogenous retrotransposon LORE1. Plant J 69:720–730
3. Yokota K, Fukai E, Madsen LH, Jurkiewicz A, Rueda P, Radutoiu S et al (2009) Rearrangement of actin cytoskeleton mediates invasion of *Lotus japonicus* roots by Mesorhizobium loti. Plant Cell 21:267–284
4. Stracke S, Kistner C, Yoshida S, Mulder L, Sato S, Kaneko T et al (2002) A plant receptor-like kinase required for both bacterial and fungal symbiosis. Nature 417:959–962
5. Kanamori N, Madsen LH, Radutoiu S, Frantescu M, Quistgaard EM, Miwa H et al (2006) A nucleoporin is required for induction of Ca2+ spiking in legume nodule development and essential for rhizobial and fungal symbiosis. Proc Natl Acad Sci U S A 103: 359–364
6. Madsen LH, Fukai E, Radutoiu S, Yost CK, Sandal N, Schauser L et al (2005) LORE1, an active low-copy-number TY3-gypsy retrotransposon family in the model legume Lotus japonicus. Plant J 44:372–381
7. Fukai E, Umehara Y, Sato S, Endo M, Kouchi H, Hayashi M et al (2010) Derepression of the plant chromovirus LORE1 induces germline transposition in regenerated plants. PLoS Genet 6:e1000868
8. Miyao A, Tanaka K, Murata K, Sawaki H, Takeda S, Abe K et al (2003) Target Site Specificity of the Tos17 retrotransposon shows a preference for insertion within genes and against insertion in retrotransposon-rich regions of the genome. Plant Cell 15:1771–1780
9. D'Erfurth I, Cosson V, Eschstruth A, Lucas H, Kondorosi A, Ratet P (2003) Efficient transposition of the Tnt1 tobacco retrotransposon in the model legume *Medicago truncatula*. Plant J 34:95–106
10. Sarni F, Grand C, Boudet AM (1984) Purification and properties of cinnamoyl CoA reductase and cinnamyl alcohol dehydrogenase from poplar stems (Populus X euramericana). Eur J Biochem 139:259–265
11. Lacombe E, Hawkins S, Doorsselaere J, Piquemal J, Goffner D, Poeydomenge O et al (1997) Cinnamoyl CoA reductase, the first committed enzyme of the lignin branch biosynthetic pathway: cloning, expression and phylogenetic relationships. Plant J 11: 429–441
12. Rozen S, Skaletsky H (2000) Primer3 on the WWW for general users and for biologist programmers. Methods Mol Biol 132:365–386
13. Hansen J, Jørgensen JE, Stougaard J, Marcker KA (1989) Hairy roots—a short cut to transgenic root nodules. Plant Cell Rep 8:12–15

14. Stougaard J (1995) *Agrobacterium rhizogenes* as a vector for transforming higher plants. Application in *Lotus corniculatus* transformation. Methods Mol Biol 49:49–61
15. Handberg K, Stougaard J (1992) *Lotus japonicus*, an autogamous, diploid legume species for classical and molecular genetics. Plant J 2: 487–496
16. Thykjær T, Schauser L, Danielsen D, Finneman J, Stougaard J (1998) Transgenic plants: agrobacterium-mediated transformation of the diploid legume *Lotus japonicus*. In: Celis JE (ed) Cell biology: a laboratory handbook, vol 3, 2nd edn. Academic, New York, pp 518–525
17. Devon RS, Porteous DJ, Brookes AJ (1995) Splinkerettes–improved vectorettes for greater efficiency in PCR walking. Nucleic Acids Res 23:1644–1645
18. Paithankar KR, Prasad KSN (1991) Precipitation of DNA by polyethylene glycol and ethanol. Nucleic Acids Res 19:1346
19. Fukai E, Dobrowolska AD, Madsen LH, Madsen EB, Umehara Y, Kouchi H et al (2008) Transposition of a 600 thousand-year-old LTR retrotransposon in the model legume *Lotus japonicus*. Plant Mol Biol 68:653–663
20. Murray MG, Thompson WF (1980) Rapid isolation of high molecular weight plant DNA. Nucleic Acids Res 8:4321–4326

Chapter 11

Isolation and Forward Genetic Analysis of Developmental Genes in Pea

James L. Weller, Valérie F.G. Hecht, and Frances C. Sussmilch

Abstract

Understanding of developmental processes relies heavily on isolation and functional characterization of relevant genes. The garden pea (*Pisum sativum* L.) is one of the classic model species in plant genetics and has been used for a wide range of physiological and molecular studies of plant development. Here we describe the resources and approaches available for isolation of genes and genetic characterization of loci affecting development in pea.

Key words Garden pea, Gene isolation, Mapping, Markers, Plant development, Gene function

1 Introduction

Recent development of genomics and transcriptomic resources in a range of legume species are improving ease and efficiency of isolating genes and cloning genetically defined loci, in other less well resourced species. In addition, comparative mapping has documented extensive synteny among different model and crop legumes, allowing genomic information from more advanced systems to be harnessed for mapping efforts in less developed systems.

Despite its importance as a model for physiological and developmental genetics, the large genome size of pea (approx. 4.5 Gb) has so far effectively excluded it as a priority for genomics initiatives. However, the availability of sequence resources from *Medicago truncatula* has dramatically improved the efficiency of pea gene isolation over the past few years. In addition, with the advent of low-cost transcriptome sequencing, pea transcript sequence collections are beginning to appear in publicly accessible databases (e.g., [1, 2]). Overall, these developments have lead to a dramatic acceleration in progress in pea molecular genetics over the past few years, with hundreds of developmental genes brought under study and many loci controlling developmental processes already (or in the process of being) cloned. This chapter will highlight useful

Ray J. Rose (ed.), *Legume Genomics: Methods and Protocols*, Methods in Molecular Biology, vol. 1069,
DOI 10.1007/978-1-62703-613-9_11, © Springer Science+Business Media, LLC 2013

approaches in gene isolation and in cloning of genetically defined loci, as in most cases these two components are both important for functional characterization of genes involved in development.

2 Materials

2.1 Isolation of Pea Homologs of Genes of Interest

1. Access to NCBI BLAST (http://blast.ncbi.nlm.nih.gov/Blast.cgi).
2. Access to Legume Information System (LIS) site and other online legume data bases.

General	http://comparative-legumes.org/ (LIS)
	http://www.phytozome.net/
	http://compbio.dfci.harvard.edu/tgi/plant.html
Medicago	http://www.jcvi.org/cgi-bin/medicago/overview.cgi
Soybean	http://soybase.org/
Lotus	http://www.kazusa.or.jp/lotus/index.html

3. Software for constructing multiple sequence alignments and phylograms.

   Relevant software can be found as component of many sequence analysis software packages and are also available online. One simple and widely used program is Clustal and recent versions can be downloaded from http://www.clustal.org/ or accessed online, for example through EMBL-EBI (http://www.ebi.ac.uk/Tools).
4. Primer design programs, e.g., Primer 3 for simple primers (http://primer3.wi.mit.edu/) or CODEHOP for degenerate primers (http://bioinformatics.weizmann.ac.il/blocks/codehop.html).
5. Long-range Taq polymerase or polymerase mix, and systems for conducting 5′ and 3′ RACE and genome walking. Available from major international molecular biology suppliers, e.g., Clontech (www.clontech.com).

2.2 Mapping of Genes

Access to latest version of Medicago genome (e.g., through www.medicago.org, www.jcvi.org, www.phytozome.net).

2.3 Forward Genetics

A wide range of genetic material including wild accessions, land races, cultivars and induced mutant lines can be obtained from major international stock centers such as at the John Innes Centre (Norwich, UK; www.jic.ac.uk) and USDA (Pullman, WA, USA; http://www.ars.usda.gov/Main/docs.htm?docid=15144).

3 Methods

3.1 Isolation of Pea Homologs of Genes of Interest

When it becomes of interest to isolate a pea homolog of a particular gene that has been isolated and functionally characterized in another system, the initial approach is to identify and design primers on similar sequences from related legumes and attempt to amplify partial sequence from pea cDNA. These sequences can then be extended by standard methods, and the corresponding genomic sequence also amplified (*see* **Note 1**).

3.1.1 Identify Legume Homologs Through BLAST Searches

In order to identify the closest pea homolog of a given gene from another (non-legume) species, it is useful to have as clear a picture as possible of the phylogenetic relationships of that gene to those present in legumes. This can be also be important in determining whether the gene in question actually exists in legumes or, in the absence of a true ortholog, which is the most closely related homolog. It also is important for identifying the degree of sequence conservation in different regions of the gene, which informs the design of primers used to isolate the gene.

To do this, BLAST searches (usually blastp or tblastn) are conducted across a range of legume species. The Legume Information System (LIS; http://comparative-legumes.org/) is a useful access point for sequenced legume genomes, and also provides a compendium of legume genomics resources, many of which are also available on stand-alone sites (Subheading 2.1, **item 2**).

The cross-species sequence search is a particularly useful tool provided on LIS.

In the case of so-called single-copy genes BLAST results are (almost by definition) unambiguous and clearly indicate whether the gene in question is present in the database or not. For a gene that is part of a gene family and/or has a complex modular structure, BLAST results may feature a large number of similar sequences and thus be unclear and potentially misleading. It then becomes more important to refer to a phylogenetic analysis (*see* **Note 2**).

3.1.2 Identify Legume Orthologs or Closest Homologs Through Phylogenetic Analyses

A simple multiple sequence alignment and phylogenetic tree can then be constructed using Clustal, which is included in most sequence analysis software suites and available online. While this may not be sufficient to resolve complex phylogenetic questions, it is usually adequate to give an overall view on gene family structure in legumes relative to other model species, to identify legume orthologs of the target gene is they exist, and if not, to identify the next most similar gene or genes.

3.1.3 Clarify the Structure of the Gene in Question and its Close Legume Relatives

Comparative analysis also gives information about gene structure that can be useful in strategies for gene isolation. In general, intron positions in a given group of legume orthologs or close homologs are well conserved, and can be inferred with a high degree of confidence from the well-characterized species (but *see* **Note 3**). Thus, it is often useful to map intron positions onto cDNA sequences as a guide for subsequent primer design. This can help ensure, for example, that primers are not located over a splice site and can thus be used for both cDNA and genomic isolation. It may also help ensure that the maximal proportion of intron sequence is amplified, where the main objective is isolation of noncoding sequence for polymorphism screening.

3.1.4 Design Primers and Amplify from cDNA

Obviously, the simplest and ideal approach for isolation of pea genes is first to identify as much of the sequence as possible directly by BLAST in pea transcript databases, as this allows design of specific primers and amplification of cDNA. The main limitation with this direct approach is that publicly available pea sequence databases are not yet as comprehensive as those for other legume species, and many genes are not represented. Although this is likely to change dramatically as more resources from next-generation sequencing approaches become available, in the short term it may still be necessary to isolate pea genes without access to any pea sequence for that gene. This is particularly likely to be a problem in the case of genes with very low, localized or otherwise specific expression patterns, as these are less likely to be represented in transcript databases.

If the gene of interest cannot be identified in pea sequence databases, or only a small fragment is represented and a more complete sequence is needed, then the next option is to make use of related legume species included in the phylogenetic analysis (Subheading 3.1.2). The nucleotide similarity of coding sequences in pea and Medicago varies between around 75 and 95 %, and even at the lower end of this range, regions of much higher conservation generally exist. This high degree of sequence conservation means that primers designed against Medicago sequence in these regions can usually be used with success on pea template.

At this point it may be useful to construct a new alignment of the target and related sequences in Medicago using nucleotide sequences. This can then be used to guide the design of primers for the Medicago ortholog of the gene of interest, focusing on sites of high conservation and/or to regions that discriminate it from its closer homologs. In general, primers designed to conserved regions of a Medicago gene work reasonably well on pea template, and where an initial attempt is unsuccessful it may simply be necessary to try other different primer combinations or PCR conditions. Where amplification of closely related family members is a potential problem, it may also be useful to use a standard nesting approach where the product from a first PCR is used as template in a second PCR with primers internal to the first pair.

A small proportion of target sequences may not be represented in either Medicago or pea databases, and it may then be necessary to rely on sequence information from more distant legume relatives. While amplification with heterologous primers may still be successful, their efficiency may decline as sequences diverge with taxonomic distance. To account for this, more refined approaches have been developed that include the use of degenerate primers [3, 4]. In this approach, primers are designed from an alignment of legume (and other) sequences in such a way that multiple variants of each primer accommodate a range of potential sequence differences at variable sites.

Following these approaches, it is relatively straightforward in most cases to obtain at least a partial cDNA sequence for pea orthologs/homologs of a gene of interest. Complications can be encountered in the case of weakly expressed genes or multiple highly similar genes (*see* **Note 3**).

3.1.5 Extending Partial Sequence

Depending on the structure of the gene in question, a partial cDNA and/or genomic sequence may be adequate for mapping or expression analysis (see below), but it may also be necessary to obtain full-length coding sequence if the purpose is to screen for potential causal mutations or conduct functional studies.

If orthologous genes appear in sequenced genomes of Medicago or other legumes, it may be possible to amplify a significant portion of most pea genes by testing a number of heterologous primers designed to the 5′ or 3′ ends of the gene, in combination with specific primers designed on the known pea sequence. Alternative approaches to extend a partial sequence include RACE (Rapid Amplification of cDNA Ends), in which products containing the 3′ or 5′ end of the transcript are generated by PCR with a gene-specific primer and an adaptor primer specific for either the 3′ (poly T) or a homopolymeric extension of the 5′ end introduced during reverse transcription.

3.1.6 Amplifying gDNA

In most cases where a pea cDNA sequence has been isolated and is predicted to contain one or more introns, these introns can then be amplified from genomic DNA, although the large size of some pea introns may require the use of long-range Taq polymerase and persistence (*see* **Note 4**). Where promoter or intergenic sequences are required, homology-based approaches are generally not useful, due to much lower conservation in these noncoding regions. In such cases, a "genome walking" approach may be used, in which libraries of uncloned, adaptor-ligated genomic DNA fragments are generated and screened by nested PCR using specific primers from the known sequence in combination with adaptor primers.

Another approach involves the screening of large-insert libraries. Two bacterial artificial chromosome (BAC) libraries exist in pea, and may be available for use through consultation with their developers [5, 6]. Once again, a detailed description is outside the scope of this chapter, but in essence clones positive for the gene of interest are identified by PCR screening. They are then isolated and fingerprinted, and the relevant fragments containing the gene can be subcloned and sequenced.

3.2 Mapping of Genes

3.2.1 Comparative Mapping Using Medicago and Other Legumes

Recent developments in comparative mapping have established extensive colinearity between the pea and Medicago genomes [7]. This means that the rough map position of a given gene in pea can often be inferred from the position of its Medicago ortholog with a high degree of confidence. For example, we recently reported the identification in Medicago of two candidate genes for the pea *HR* locus and subsequently confirmed conserved map locations for their pea orthologs [8]. However, there are still some genes for which comparisons with Medicago provide no useful mapping information, or may even be misleading (*see* **Note 5**). For this reason, is generally advisable to confirm the location of the pea gene experimentally.

Where an ortholog of a gene of interest cannot be identified in pea or Medicago, but is present in another legume species, it may be possible to infer its location in these two species from broader comparative mapping. The obvious limitation with this is that blocks of collinear regions become smaller with taxonomic distance. The Legume Information System provides a convenient portal for identifying and visualizing comparative maps between Medicago and other sequenced legume genomes.

3.2.2 Mapping Resources in Pea: Markers and Populations

Mapping a gene or locus in pea, as in any other system, requires three fundamental components: a segregating population, a marker system (or existing genome-wide marker genotype information for that progeny), and a polymorphism for the gene of interest. Mapping a gene based on sequence polymorphism can be performed in existing populations for which genome-wide marker genotype information has already been generated (see below). In contrast, mapping of a locus defined by an induced mutation requires the construction of a new progeny or population (*see* Subheading 3.3.2).

A wide range of different markers for mapping in pea have been developed as marker technology has evolved. These have included RAPD, SSR, AFLP and retrotransposon-based insertion polymorphism (RBIP) markers [9]. However more recent mapping studies are converging on the use of gene-based EST-SSR and SNP markers [10] as these are the most useful for comparative mapping across different populations within pea and across other legume species.

Table 1 lists populations and associated marker sets that have been generated in a number of different groups working on pea. Parents of these populations may be genetically very similar ("narrow," i.e., between cultivars or other highly domesticated types), or highly diverged ("wide," often involving wild or primitive accessions as at least one of the parents). In general, the SNP density is much higher in wide crosses, and the chance of finding a

Table 1
Examples of populations used for mapping in pea

Cross type	Parents	Population type	Reference
Narrow	cv. Térèse × cv. Torsdag (K586)	RILs	[11]
	Primo × OSU442-15	F_2	[12]
	Shawnee × Bohatyr	RILs	[13]
	cv. Carneval × MPO1401	RILs	[14]
	cv. Cameor × cv. Ballet	RILs	[15]
Intermediate	JI281 × cv. Cennia (JI399)	RILs	[16]
	cv. Térèse × cv. Champagne	RILs	[17]
	cv. Chemin Long × DP	RILs	[18]
Wide	JI1794 (var. *humile*) × cv. Slow	RILs	[19]
	JI1794 (var. *humile*) × cv. Torsdag (NGB5839)	F_2	[8]

suitable polymorphism within a small gene fragment is therefore greater. In contrast, the low level of polymorphism in some narrow crosses can make it difficult to find useful markers for a specific gene, particularly where the gene is short and/or has no introns. In such a case it may be necessary to extend the sequence into promoter or intergenic regions to identify suitable polymorphism. In addition, common recent ancestries of certain cultivars may be reflected in significant islands of homozygosity that inherently limit mapping resolution in certain regions.

These populations have been generated for different purposes and characterized with different marker sets, and although there has been some degree of integration through the use of common markers, detailed comparisons between maps are so far not generally possible. The most comprehensive effort to date has involved integration of mapping data including large numbers of sequence-characterized markers across six different populations [7], of which three include cv. Cameor as a parent. Although none of these populations are in the public domain, several may be available for community use through contact with the groups concerned.

Most recently, NGS-based approaches such as RAD-seq [20] and DArT-seq [21] are making possible the simultaneous development and genotyping of large numbers of sequence-characterized markers that are to a varying extent enriched for genic sequences. These approaches mean that it may be feasible to construct dense genetic maps for any cross, without needing to rely on existing resources. Whether this is practical in any specific case will depend on the nature of the study. For mapping a small number of sequence-defined genes, it may still be preferable to arrange access to existing populations. However, for mapping a locus defined by an induced mutant, or for QTL analysis in novel material, these newer approaches have major advantages and should rapidly become more cost-effective.

3.3 Identification of Developmental Genes Through Forward Genetics

Discovery of important developmental genes in pea has come largely through isolation and analysis of mutants. A wide variety of spontaneous pea mutants have accumulated over centuries of cultivation and are represented in major *Pisum* stock centers such as those at the John Innes Centre (Norwich, UK) and USDA (Pullman, WA, USA; http://www.ars.usda.gov/Main/docs.htm?docid=15144). In addition, numerous mutagenesis programs have added new loci (Table 2). To date, genetic backgrounds most systematically used for mutant isolation include Torsdag and its derivative NGB5839 [22–25], Terese [26] and most recently Cameor [27]. One advantage of conducting further mutant screens within these backgrounds is the ease of creating isogenic double mutants, and the availability of genetic maps and markers.

3.3.1 Genetic Analysis of Mutants

Following identification of a potentially novel pea mutant of interest, all standard genetic tests should be conducted (*see* ref. 32). These include.

- Crossing to the progenitor line and characterizing the mode of inheritance by observing phenotypes in the F_1, F_2 and F_3 generations.
- Back-crossing to the progenitor line to isolate the mutation of interest away from independent mutations occurring in the original M_1 plant.
- Allelism testing with other similar mutants.

For some pea loci, the mutant phenotype alone may be sufficiently distinctive to move directly to molecular analysis of a

Table 2
Examples of mutagenesis and mutant screening in pea

Line	Mutagen	Purpose	Reference
cv. Frisson	EMS	Symbioses	[28]
cv. Sprint		Symbioses	[29]
SGE	EMS	Symbioses	[30]
cv. Torsdag	EMS/gamma	Symbioses	[22]
	EMS	Photomorphogenesis	[25]
cv. Terese	EMS	Branching	[26]
	FN		[31]
NGB5839	EMS	Flowering	[23]
JI2822	FN	Leaf development	[6]
cv. Cameor	EMS	TILLING	[27]

specific functional candidate (e.g., [24, 33, 34]). For others, it may be necessary to first narrow the field of functional candidates by mapping the locus (e.g., [23, 35]). In some cases, cloning of the gene responsible for a pea mutant phenotype has relied on characterization of a similar locus with conserved location in another model legume (Medicago or Lotus) that has been cloned by a positional or tagging approach [31, 36, 37]. Insertional mutation platforms have not yet been developed for pea, and therefore for pea mutants defining novel loci with no obvious counterparts in other legumes, information about chromosomal position may be the only route to identifying the underlying gene.

3.3.2 Mapping Loci Defined by Mutants

Wide and Narrow Crosses

Unlike mapping a gene, which can be performed in an existing population provided sequence polymorphism for the gene can be identified, mapping a mutant locus involves generation of a segregating progeny from a cross between the mutant and another genetically distinct (usually wild-type) line. This cross can be considered narrow (e.g., between two cultivars, or other closely related lines) or wide (e.g., between a domestic and a wild line), depending on the degree of genetic difference between the parents. Although the effective marker density is higher in a wide cross, for obtaining a rough location of a mutant locus it is preferable in practice to use a relatively narrow cross. This is because it is important to be able to reliably follow phenotypic segregation at the locus in question, and relatively similar parent lines are less likely to carry additional genetic variation that could interfere with this process (*see* **Note 6**).

Mapping Strategies

To be effective, a mapping exercise requires adequate marker coverage, and where single marker genotyping is planned, a judicious choice of markers is therefore important. The density of markers required to detect linkage is influenced by population size and whether heterozygous segregants can be distinguished. In practice, in a standard F_2 population of 100–120 individuals, an interval between markers of around 30–35 cM intervals should be sufficient to locate the majority of genes. Recent maps of pea have a total length of approximately 1,200–1,400 cM, which would require around 40 such markers for full map coverage (5–7 per chromosome). Phenotyping the F_3 to distinguish heterozygotes for the locus in question will also improve the efficiency.

As an alternative to single marker analysis, it is increasingly feasible to convert SNPs for rapid, high-throughput genotyping using systems such as the Illumina GoldenGate assay and several groups are now developing marker panels for pea. Alternatively, as mentioned above, genotyping by sequencing approaches can rapidly provide information of hundreds to thousands of gene-based (or gene-enriched) markers in one analysis.

Another consideration is how the mapping process is handled. The approach of Bulk Segregant Analysis (BSA) aims in effect to reduce the amount of work in the initial stages of mapping by combining mutant segregants and scanning for markers that are homozygous in these pools. In practice this can be somewhat sensitive to the quality of PCR and the false negative, and it may still be necessary to resort to analysis of individuals. Nevertheless, it may still be useful as an initial pass in analysis via high-throughput marker systems.

For some loci, detailed phenotypic characterizations may yield useful information about the molecular, physiological or biochemical function which can help narrow the range of potential candidate genes to the point that it becomes an efficient strategy to isolate and test linkage to these candidates directly without the need for high-resolution mapping. This is most likely to be successful where the function of the locus can be specifically defined, and pathways known to control the process are well characterized in other species. Examples include the analysis of loci controlling photoperiod sensitivity, for which conserved circadian clock genes were strong candidates [8, 23, 35], and anthocyanin production, where three loci have been associated with mutations in well-known regulatory and biosynthetic genes [34, 38].

3.4 Genetic Analysis of Natural Phenotypic Variation

Genome-wide mapping is also used to define quantitative trait loci (QTL) that contribute to differences in a trait of interest across two or more parent lines. As in other mapping studies, various forms of segregating progeny can be used for QTL mapping, and many of the pea populations listed in Table 2 have been used for this purpose. The main advantage of an inbred population is that each line is effectively homozygous and therefore multiple phenotypic assessments can be conducted. This allows for a wide range of traits to be analyzed, under different conditions and with the potential for replication. In contrast, the effort devoted to genotyping a segregating progeny such as an F_2 is only useful for assessment of traits expressed in single individuals from a single sowing. The process of generating RILs from F_2 is straight-forward but time-consuming, usually involving at least eight generations of inbreeding by single seed descent before effective homozygosity is achieved and genotyping is performed.

3.5 Fine Mapping and Candidate Gene Evaluation

After initial phenotypic characterizations and mapping place a genetic locus within a defined chromosomal region, and potential functional candidates are identified within that region, these candidates can be assessed by expression analysis and sequencing. In a majority of cases, this may result in identification of putative causal mutations that can then be further evaluated. Most of the loci cloned in pea have followed this approach, benefiting from work in other species and a high degree of functional conservation (Table 3). Access to multiple independent mutants for the locus in question is invaluable as the identification of multiple mutations

Table 3
Examples of recently cloned pea loci

Locus	Trait	Approach	Candidate	Species	Reference
DET	Flowering	Candidate	*TFL1*	Arabidopsis/Antirrhinum	[39]
LF	Flowering	Candidate	*TFL1*	Arabidopsis/Antirrhinum	[39]
LATH	Flower development	Candidate	*NAO1*	Lotus	[37]
K	Flower development	Candidate	*CYC*	Arabidopsis/Antirrhinum	[31]
LST	Flower development	Candidate	*CYC*	Arabidopsis/Antirrhinum	[31]
A	Anthocyanin synthesis	Candidate	*TT8*	Arabidopsis/Petunia	[38]
A2	Anthocyanin synthesis	Candidate	*TTG1*	Arabidopsis/Petunia	[38]
B	Anthocyanin synthesis	Candidate	*F3′5′H*	Various	[34]
TL	Leaf development	Deletion scanning			[6]
SYM28	Nodulation	Candidate	*CLV2*	Arabidopsis	[40]
SYM33	Nodulation	Candidate	*IPD3*	Medicago	[41]
NOOT	Nodulation	Candidate	*RDI1*	Medicago	[36]
LATE1	Flowering	Candidate	*GI*	Arabidopsis	[23]
DNE	Flowering	Candidate	*ELF4*	Arabidopsis	[35]
GIGAS	Flowering	Candidate	*FT*	Arabidopsis	[42]
HR	Flowering	Candidate	*ELF3*	Arabidopsis	[8]
ER1	Disease resistance	Candidate	*MLO1*	Various	[33]

in the putative gene significantly strengthens the argument for a causal relationship between gene and mutant phenotype.

However, in other cases there may be no obvious candidates in the defined interval, or candidates may be identified but carry no potentially causal mutations. At this point, a number of possible strategies could be considered. The interval containing the locus of interest can be narrowed by generation of larger progenies and increasing marker density. In parallel, the corresponding genomic region in Medicago and other legumes could be assessed less stringently for potential candidates.

The ability to increase marker density in a given region depends in part on the nature of the original cross. A narrow cross that was initially established for ease of phenotyping and rapid mapping may not provide sufficient polymorphism for higher resolution mapping. For this reason it may also be useful to develop parallel material from a wide cross. As discussed above, it may prove difficult to accurately determine mutant phenotypes in an F_2 or early-generation population of a wide cross. A common approach to get around this problem is instead to use an advanced back-cross population in which F_1 individuals heterozygous at the locus of interest are recurrently backcrossed to the mutant parent, enabling the eventual generation of a large F_2 in which much of the background variation has been eliminated. In this case it may also be necessary to use marker-assisted selection based on linked markers identified in the analysis of a narrow cross to guide the identification of F_1 individuals heterozygous for the locus of interest.

4 Notes

1. How this approach is put into practice depends on the intended purpose. For example, if the aim is to quantify expression of the gene, then it may only be necessary to isolate partial cDNA sequence, ideally spanning a splice junction. If it is only of interest to map the gene, then it may be most useful to isolate cDNA spanning splice junctions, and subsequently to isolate one or more introns from genomic DNA. If the purpose is to express the protein or to conduct transgenic functional analyses, then a cDNA containing the full coding sequence is obviously needed. Finally, to comprehensively evaluate a candidate gene for the presence of putative causal mutations, then it may potentially be necessary to isolate the entire gene, from genomic DNA, including introns and other noncoding regions.
2. One potential complication that can be encountered, particularly where BLAST results from genomic and transcript sequences are combined, may be the presence of multiple very similar sequences from one species. While in some cases this may accurately reflect several highly similar genes, it can also reflect differences in source material or the low technical quality of one or more of the sequences.
3. Some genes may be difficult to amplify from standard cDNA preparations, due to their low, localized, or otherwise specific expression, and it may be necessary to try a wider range of tissue types, developmental stages or environmental conditions. Occasionally, one primer pair may amplify two highly similar or even identical cDNA fragments deriving from distinct genes, and the problem may only become apparent through amplification and sequencing of genomic DNA. Even then it may only be detected following inconsistent amplification and minor ambiguities in sequencing PCR products. Analysis of cloned products from genomic PCR is then essential, following which it may be possible to design specific primers for noncoding regions. Whether this is worth the effort depends on the context of the study.

4. However, it is worth noting that there are still cases where an intron is present in pea but not in other species, and vice versa. It is also worth noting that the sizes of introns are not so well predicted from comparative analysis. It is not uncommon to find, for example, that some pea introns are much larger than corresponding Medicago introns (or vice versa), potentially presenting problems for their isolation using flanking primers designed on cDNA.
5. Some pea and Medicago orthologs apparently map to distinct locations. These situations may reflect inversions, relocations or other rearrangements of genomic regions. However, some of these apparent anomalies between pea linkage maps and Medicago physical maps are more likely to reflect the incomplete nature of the Medicago genome sequence and an incorrect orientation of BAC contigs in Medicago, particularly for some regions where numerous gaps remain in the assembly. One clear example is provided by flowering-related genes *FKF1*, *COLi*, and *MFT*, located on Medicago chromosome 8 (collinear with pea linkage group 4) in MtGenome v3.5 but mapping in pea to linkage group VII (collinear with Medicago chr 7). Also, the many gaps in the Medicago assembly are likely still to reflect significant regions of missing sequence, and there are still genes represented in Medicago transcript databases but not in the latest assembly (e.g., the circadian clock gene *TOC1*).
6. A simple illustration of this is provided by a comparison of cv. Cameor, an increasingly popular line for mutant isolation, and JI1794, a wild line that has been used in wide crosses. These lines differ in at least three major loci controlling flowering time and three major loci controlling stem elongation, and a wide cross between the two would therefore not be practical for mapping additional loci affecting these traits that have been defined by mutant analysis in cv. Cameor. This is a more important consideration where the trait is quantitative rather than qualitative.

References

1. Franssen SU et al (2011) Comprehensive transcriptome analysis of the highly complex *Pisum sativum* genome using next generation sequencing. BMC Genomics 12:227
2. Kaur S et al (2012) Transcriptome sequencing of field pea and faba bean for discovery and validation of SSR genetic markers. BMC Genomics 13:104
3. Rose TM et al (1998) Consensus-degenerate hybrid oligonucleotide primers for amplification of distantly related sequences. Nucleic Acids Res 26:1637–1644
4. Hecht V et al (2005) Conservation of Arabidopsis flowering genes in model legumes. Plant Physiol 137:1420–1434
5. Coyne CJ et al (2007) Construction and characterization of two bacterial artificial chromosome libraries of pea (*Pisum sativum* L.) for the isolation of economically important genes. Genome 50:871–875
6. Hofer J et al (2009) Tendril-less regulates tendril formation in pea leaves. Plant Cell 21:420–428
7. Bordat A et al (2011) Translational genomics in legumes allowed placing *in silico* 5460

unigenes on the pea functional map and identified candidate genes in *Pisum sativum* L. G3 (Bethesda) 1:93–103
8. Weller JL et al (2012) A conserved molecular basis for photoperiod adaptation in two temperate legumes. Proc Natl Acad Sci U S A 109:21158–21163
9. Smykal P et al (2012) Pea (*Pisum sativum* L.) in the genomic era. Agronomy 2:74–115
10. Deulvot C et al (2010) Highly-multiplexed SNP genotyping for genetic mapping and germplasm diversity studies in pea. BMC Genomics 11:468
11. Laucou V et al (1998) Genetic mapping in pea. 1. RAPD-based genetic linkage map of *Pisum sativum*. Theor Appl Genet 97:905–915
12. Gilpin BJ et al (1997) A linkage map of the pea (*Pisum sativum* L.) genome containing cloned sequences of known function and expressed sequence tags (ESTs). Theor Appl Genet 95: 1289–1299
13. McPhee KE et al (2012) Mapping QTL for Fusarium wilt race 2 partial resistance in pea (*Pisum sativum*). Plant Breed 131:300–306
14. Tar'an B et al (2003) Quantitative trait loci for lodging resistance, plant height and partial resistance to mycosphaerella blight in field pea (*Pisum sativum* L.). Theor Appl Genet 107: 1482–1491
15. Bourion V et al (2010) Genetic dissection of nitrogen nutrition in pea through a QTL approach of root, nodule, and shoot variability. Theor Appl Genet 121:71–86
16. Ellis THN et al (1992) Linkage maps in pea. Genetics 130:649–663
17. Lejeune-Hénaut I et al (2008) The flowering locus *Hr* colocalizes with a major QTL affecting winter frost tolerance in *Pisum sativum L.* Theor Appl Genet 116:1105–1116
18. Prioul S et al (2004) Mapping of quantitative trait loci for partial resistance to Mycosphaerella pinodes in pea (*Pisum sativum* L.), at the seedling and adult plant stages. Theor Appl Genet 108:1322–1334
19. Weeden NF et al (1998) A consensus linkage map for *Pisum sativum*. Pisum Genet 30:1–4
20. Davey JW et al (2011) Genome-wide genetic marker discovery and genotyping using next-generation sequencing. Nat Rev Genet 12: 499–510
21. Kilian A et al (2012) Diversity arrays technology: a generic genome profiling technology on open platforms. Methods Mol Biol 888:67–89
22. Sidorova KK, Shumnyi VK (2003) Creation and genetic study of a collection of symbiotic mutants of the pea (*Pisum sativum* L.). Genetika 39:501–509
23. Hecht V et al (2007) Pea *LATE BLOOMER1* is a *GIGANTEA* ortholog with roles in photoperiodic flowering, deetiolation, and transcriptional regulation of circadian clock gene homologs. Plant Physiol 144:648–661
24. Platten JD et al (2005) Cryptochrome 1 contributes to blue-light sensing in pea. Plant Physiol 139:1472–1482
25. Weller JL, Murfet IC, Reid JB (1997) Pea mutants with reduced sensitivity to far-red light define an important role for phytochrome A in day-length detection. Plant Physiol 114: 1225–1236
26. Rameau C et al (1997) New ramosus mutants at loci Rms1, Rms3 and Rms4 resulting from the mutation breeding program at Versailles. Pisum Genet 29:7–12
27. Triques K et al (2008) Mutation detection using ENDO1: application to disease diagnostics in humans and TILLING and Eco-TILLING in plants. BMC Mol Biol 9:42
28. Duc G, Messager A (1989) Mutagenesis of pea (*Pisum sativum* L.) and the isolation of mutants for nodulation and nitrogen fixation. Plant Sci 60:207–213
29. Borisov AY et al (1992) New symbiotic mutants of pea (*Pisum sativum* L.) affecting wither nodule initiation or symbiosome development. Symbiosis 14:297–313
30. Tsyganov VE et al (1994) New symbiotic mutants of pea obtained after mutagenesis of line SGE. Pisum Genet 26:36–37
31. Wang Z et al (2008) Genetic control of floral zygomorphy in pea (*Pisum sativum* L.). Proc Natl Acad Sci U S A 105:10414–10419
32. Koornneef M, Alonso-Blanco C, Stam P (1998) Genetic analysis. Methods Mol Biol 82:105–117
33. Humphry M et al (2011) Durable broad-spectrum powdery mildew resistance in pea er1 plants is conferred by natural loss-of-function mutations in PsMLO1. Mol Plant Pathol 12: 866–878
34. Moreau C et al (2012) The B gene of pea encodes a defective flavonoid 3′,5′-hydroxylase, and confers pink flower color. Plant Physiol 159:759–768
35. Liew LC et al (2009) *DIE NEUTRALIS* and *LATE BLOOMER 1* contribute to regulation of the pea circadian clock. Plant Cell 21: 3198–3211
36. Couzigou JM et al (2012) NODULE ROOT and COCHLEATA maintain nodule development

and are legume orthologs of Arabidopsis BLADE-ON-PETIOLE genes. Plant Cell 24: 4498–44510

37. Zhuang LL et al (2012) LATHYROIDES, encoding a WUSCHEL-related Homeobox1 transcription factor, controls organ lateral growth, and regulates tendril and dorsal petal identities in garden pea (*Pisum sativum* L.). Mol Plant 5:1333–1345
38. Hellens RP et al (2010) Identification of Mendel's white flower character. PLoS One 5:e13230
39. Foucher F et al (2003) *DETERMINATE* and *LATE FLOWERING* are two *TERMINAL FLOWER1/CENTRORADIALIS* homologs that control two distinct phases of flowering initiation and development in pea. Plant Cell 15:2742–2754
40. Krusell L et al (2011) The Clavata2 genes of pea and *Lotus japonicus* affect autoregulation of nodulation. Plant J 65:861–871
41. Ovchinnikova E et al (2011) IPD3 controls the formation of nitrogen-fixing symbiosomes in pea and Medicago Spp. Mol Plant Microbe Interact 24:1333–1344
42. Hecht V et al (2011) The pea *GIGAS* gene is a *FLOWERING LOCUS T* homolog necessary for graft-transmissible specification of flowering but not for responsiveness to photoperiod. Plant Cell 23:147–161

Chapter 12

Gene Silencing in *Medicago truncatula* Roots Using RNAi

Daniela S. Floss, Alexa M. Schmitz, Colby G. Starker, J. Stephen Gantt, and Maria J. Harrison

Abstract

Medicago truncatula is used widely as a model system for studies of root symbioses, interactions with parasitic nematodes and fungal pathogens, as well as studies of development and secondary metabolism. In *Medicago truncatula* as well as other legumes, RNA interference (RNAi) coupled with *Agrobacterium rhizogenes*-mediated root transformation, has been used very successfully for analyses of gene function in roots. One of the major advantages of this approach is the ease and relative speed with which transgenic roots can be generated. There are several methods, both for the generation of the RNAi constructs and the root transformation. Here we provide details of an RNAi and root transformation protocol that has been used successfully in *M. truncatula* and which can be scaled up to enable the analysis of several hundred constructs.

Key words Legume, Gene knockdown, Functional genomics

1 Introduction

Double-stranded RNAs, which have the capacity to "silence" gene expression, have been widely used for studies of gene function in plant, animal and invertebrate systems [1–3]. There is a variety of approaches [4], but RNA interference (RNAi) by inverted repeat or so-called hairpin constructs has proved to be particularly effective in a range of plant species, including legumes [5–9]. Expression of an inverted repeat construct, homologous to the gene of interest, results in the creation of a hairpin (hp) RNA which is recognized and cleaved by an RNAse-III endoribonuclease, Dicer, into 21–22 nucleotides. Generally one strand of these short RNA oligonucleotides is then incorporated into the RNA-induced Silencing Complex (RISC) and this complex facilitates recognition of the homologous mRNA, and subsequent cleavage and degradation, or in some cases inhibition of translation [10]. In plants, RNAi constructs are generally delivered to plant cells in binary vectors.

Ray J. Rose (ed.), *Legume Genomics: Methods and Protocols*, Methods in Molecular Biology, vol. 1069,
DOI 10.1007/978-1-62703-613-9_12, © Springer Science+Business Media, LLC 2013

Efficient cloning of the inverted repeats constructs in binary vectors has been facilitated by the generation of Gateway©(GW) compatible binary vectors such as the pHellsgate vector series [1, 11, 12], which permit the use of the Gateway site-specific recombination cloning technology for simultaneous, directional insertion of inverted repeats [13] (Invitrogen). pHellsgate8, a vector used successfully in many plant species, also contains a spacer that encodes an intron, which has been shown to increase silencing efficiency [5].

The binary vectors containing the RNAi hairpin constructs are readily introduced into plant cells through *Agrobacterium tumefaciens*- or *Agrobacterium rhizogenes*-mediated transformation. In recent years, *A. rhizogenes*-mediated transformation has been used to generate "composite plants," which have genetically transformed roots but wild-type shoots [14, 15]. This approach is particularly useful for studies of many aspects of gene function in roots because of the speed with which composite plants can be generated. This approach has accelerated the pace of root biology research and has been used widely, particularly in legumes [16, 17], where the generation of stable transformants via *A. tumefaciens-mediated* transformation is a lengthy process. Several protocols for efficient *A. rhizogenes*-mediated transformation of *M. truncatula* have been published [14, 18].

Here, we describe a cloning procedure based on Gateway technology to generate RNAi constructs, and a procedure for their introduction into *M. truncatula* roots via *A. rhizogenes*-mediated transformation. The latter is modified from Boisson-Dernier et al. [14]. The resulting composite plants are suitable for studies of gene function in roots, nodules and during AM symbiosis [19].

2 Materials

2.1 General

1. PCR reagents: sterile deionized distilled H_2O (H_2O_{dd}), primers (*see* **steps 1** and **2** in Subheading 3.1.1), template (*M. truncatula* cDNA or PCR product from **step 1** in Methods), dNTPs, Taq polymerase buffer, Taq polymerase.
2. Plasmids: donor vector with *att*P sites; binary destination vector appropriate for RNAi with *att*R sites. For the purpose of this chapter we will present methods concerning the use of pDONR207 (gentamicin selection) and pHellsgate8 [1] (spectinomycin selection) as donor and destination vectors, respectively. The plasmid DNA concentrations should be approximately 150 ng/μL (*see* **Note 1**).
3. Gateway® reagents: BP Clonase™ II enzyme mix, LR Clonase™ II enzyme mix, Proteinase K (Invitrogen).

4. Antibiotics: gentamicin (50 mg/mL stock solution), spectinomycin (50 mg/mL stock solution), streptomycin (100 mg/mL stock solution), and kanamycin (50 mg/mL stock solution).
5. Heat-shock competent DH5α *E. coli* cells.
6. Heat-shock competent *A. rhizogenes* ARqua1 cells.
7. 50 % glycerol in H_2O_{dd}.
8. *M. truncatula* seeds.
9. Concentrated sulfuric acid (H_2SO_4).
10. Bleach solution containing 6 % sodium hypochlorite.
11. 0.1 % Tween 20 in H_2O_{dd}.
12. Petri plates for germinating the seeds (we use glass petri plates as they can be reused).

2.2 Media

1. LB medium: Dissolve 10 g tryptone, 10 g NaCl, and 5 g yeast extract in just under 1 L H_2O_{dd}. Adjust pH to 7.0 with NaOH. Bring volume to exactly 1 L with H_2O_{dd} in a graduated cylinder. Pour solution into a 2-L Erlenmeyer flask. For solid media, add 15 g agar. Autoclave 30 min on a liquid cycle. For solid media, allow solution to cool to 50 °C then add 50 mg/L gentamicin for selection of pDONR207 in DH5α, or 100 mg/L spectinomycin for selection of pHellsgate8 in DH5α and pour plates immediately.
2. YEP media: Dissolve 10 g bacto peptone, 5 g NaCl, and 10 g yeast extract in just under 1 L H_2O_{dd}. Adjust pH to 7.0 with NaOH. Bring volume to exactly 1 L with H_2O_{dd} in a graduated cylinder. Pour solution into a 2-L Erlenmeyer flask. For solid media, add 15 g agar. Autoclave 30 min on a liquid cycle. For solid media, allow solution to cool to 50 °C then add 150 mg/L spectinomycin for selection of pHellsgate8 in *A. rhizogenes* ARqua1 and 100 mg/L streptomycin for selection of ARqua1's tumor-inducing plasmid, and pour plates immediately.
3. TY medium for *A. rhizogenes* ARqua1: Dissolve 5 g tryptone, 880 mg $CaCl_2$, and 3 g yeast extract in just under 1 L H_2O_{dd}. Adjust pH to 7.0 with NaOH. Bring volume to exactly 1 L with H_2O_{dd} in a graduated cylinder. Pour solution into a 2-L Erlynmeyer flask. Add 15 g agar. Autoclave 30 min on a liquid cycle. Allow solution to cool to 50 °C then add 150 mg/L spectinomycin for selection of pHellsgate8 in *A. rhizogenes* ARqua1 and 100 mg/L streptomycin for selection of ARqua1's tumor-inducing plasmid, and pour plates immediately.
4. Modified Fähraeus medium for *M. truncatula*: 0.9 mM $CaCl_2$, 0.5 mM $MgSO_4$, 20 μM KH_2PO_4, 10 μM Na_2HPO_4, 20 μM ferric citrate, 1.0 mM NH_4NO_3, 33 μg/L $MnCl_2$, 33 μg/L $CuSO_4$, 7 μg/L $ZnSO_4 \cdot 7H_2O$, 100 μg/L H_3BO_3, 33 μg/L

Na_2MoO_4, 218 mg/L MES free acid monohydrate, and 2.5 g/L Gelzan™ CM (Sigma-Aldrich®) [20]. This version of Fähraeus [20] was modified from the version described by Boisson-Dernier et al. [14].

Prepare individual stock solutions for the first 11 ingredients. We typically make stocks with the following concentrations: 0.9 M $CaCl_2$, 0.5 M $MgSO_4$, 0.35 M KH_2PO_4, 0.2 M Na_2HPO_4, 20 mM ferric citrate, 0.5 M NH_4NO_3, 1 mg/mL H_3BO_3 and 1 mg/mL Na_2MoO_4. The $MnCL_2$, $CuSO_4$ and $ZnSO_4{\cdot}7H_2O$ can be combined in a micronutrient stock solution of 1 mg/mL $MnCL_2$, 1 mg/mL $CuSO_4$ and 0.42 mg/mL $ZnSO_4{\cdot}7H_2O$.

For 1 L of Fähraeus media add the appropriate amount of each stock solution and 218 mg MES free acid monohydrate to 500 mL H_2O_{dd}. Bring volume very close to 1 L and adjust pH to 7.4 with 1 M NaOH. Carefully add 2.5 g Gelzan™ (*see* **Note 2**). Autoclave for 35 min and allow solution to cool to 50 °C. Add 25 mg/L kanamycin to the media (*see* **Note 3**). Under sterile conditions, pour the media into square plates (100 × 100 × 15 mm, Laboratory Products Sales) on a slant as shown in Fig. 1. 1 L of media is sufficient for the preparation of 25 plates. Allow the plates to dry for 2 h before storage at 4 °C. The plates are normally prepared 1 or 2 days before use.

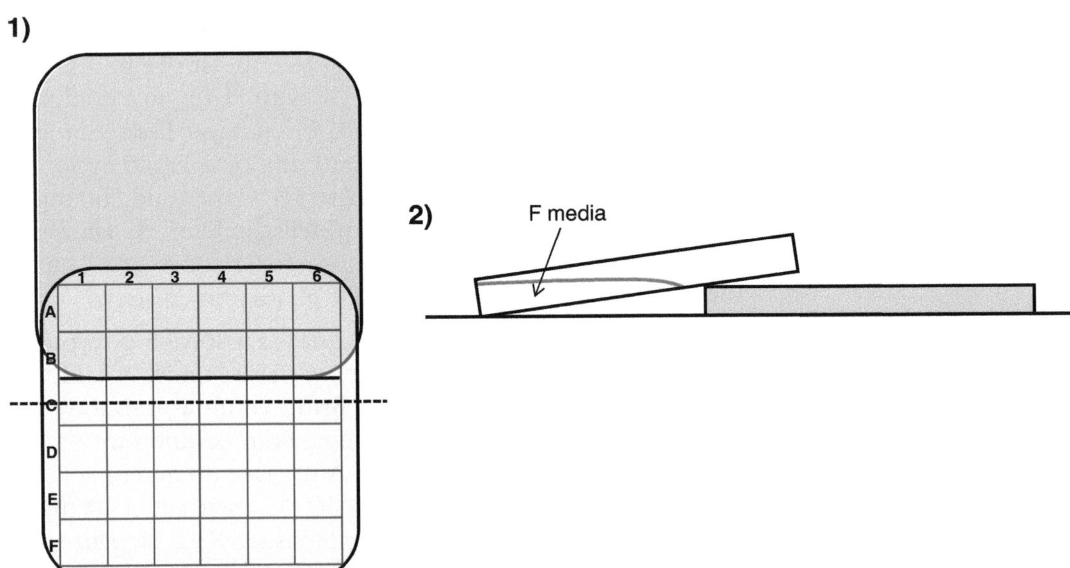

Fig. 1 Preparation of plates containing a slant of F media. 1 Plan view. 2 Lateral view. The *square* petri plates have a grid printed on the *bottom*. This is a convenient marker. To create the appropriate angle, tilt the *bottom* of the plate and rest it on the lid. The point of contact between the *bottom* and the lid (*grey*) should be the line at the lower edge of row B on the grid (see *thick black line*). Add media to the *bottom* plate until it reaches half-way through row C (*dotted line*)

3 Methods

3.1 Generation of RNAi Constructs in pHellsgate8

The Gateway® cloning technology enables easy insertion of a DNA fragment into a destination vector via lambda phage-based site-specific recombination [13] (Invitrogen). This system, which eliminates many of the difficulties involved in restriction enzyme cloning, requires three steps: (1) amplification of a target DNA fragment flanked by *att*B sites via PCR (2) recombination of the DNA fragment with *att*B sites into a donor vector (for example, pDONR207) containing *att*P sites in a BP reaction producing an entry clone with *attL* sites (comprised of parts of the *att*B and *att*P sites) and (3) an LR reaction in which the entry vector containing *attL* sites surrounding the insert recombines with a destination vector (for example, pHellsgate8) containing *attR* sites flanking the ccdB toxin gene to produce an expression plasmid containing the new insert (Gateway Technology Manual, Invitrogen). Transformants containing the empty donor or destination vectors are selected against due to the toxicity of the ccdB gene product (*see* **Note 4**). The destination vector, pHellsgate8, contains two ccdB genes that must each be replaced by the target fragment from pDONR207 for the survival of the bacteria maintaining this plasmid. The site-specific recombination of the Gateway® technology provides the directional insertion of the fragment from pDONR207, creating the inverted repeats necessary for RNAi [5]. Alternate approaches for the initial stages of the procedure are indicated in **Note 5**.

*3.1.1 Amplification of **att**B Flanked Target Fragment*

The target fragment should contain 300–600 base pairs homologous to a region of the targeted mRNA. To avoid cross-silencing the target should have no region of identity longer than 20 nucleotides with any other gene in *M. truncatula* (*see* **Note 6**). Do not choose a region that includes any conserved motifs, and in the case of a highly conserved gene, either the 3′ or 5′ UTR can be used with equal silencing efficiency [11].

1. Design two sets of primers to amplify your target region by PCR. The first set should be complementary to 20–25 nucleotides at the 5′ and 3′ ends of the target region to enable amplification of the region from cDNA. The second set should include approximately 20 nucleotides complementary to the target with the addition of the following *att*B1 and *att*B2 sites to the 5′ end of the forward and reverse primers, respectively [11] (*see* **Note 7**).

   *att*B1: 5′ GGGGACAAGTTTGTACAAAAAAGCAGGCT 3′.

   *att*B2: 5′ GGGGACCACTTTGTACAAGAAAGCTGGGT 3′.

2. With the first set of primers complementary to the target region, PCR amplify the target region from *M. truncatula* cDNA. Purify this product with a method of your choice, and use as template for a second round of PCR with the second

pair of primers. For the second round, program three cycles with a Tm appropriate for the target-homologous sequence, and then all additional cycles with a Tm appropriate for the full primer. This round adds the *att*B sequence a to the target. Gel-purify the PCR product to remove primers, which can interfere with the following clonase reaction (Invitrogen).

3.1.2 Generation of Entry Vector with Target Region

A BP reaction is carried out with the *att*B-flanked PCR product and a donor vector with *att*P recombination sites (pDONR207) to create an entry vector with the target region flanked by *att*L sites, according to Invitrogen's protocol (Invitrogen) modified as shown below (*see* **Note 8**).

1. Add the following components to a 1.5 mL tube at room temperature and mix.

   *att*B-flanked PCR product (=10 ng/μL; final amount ~15–150 ng) 1–7 μL.

   pDONR207 empty vector (150 ng/μL) 1 μL.

   Add TE buffer (pH8) to a total of 8uL.
2. To each sample, add 2 μL of BP Clonase™ II enzyme mix to the reaction resulting in a total volume of 10 μL, and mix well by mix well by flicking the tube. Centrifuge briefly.
3. Incubate reactions at 25 °C for 1–16 h.
4. Add 1 μL of the Proteinase K solution to each sample to terminate the reaction. Vortex briefly. Incubate samples at 37 °C for 10 min.
5. Transform 1–3 μL of reaction mix into heat-shock competent *E. coli* DH5α cells using a standard method. Plate on LB media with 50 mg/L gentamicin to select bacteria with the recombined vector.
6. Select a colony from transformation and inoculate 3 mL LB media containing 50 mg/L gentamicin for plasmid maintenance. Incubate at 37 °C, with shaking, overnight.
7. Extract pDONR207 containing the target fragment from the culture using a miniprep of your choice. Sequence the vector for insertion verification in the forward and reverse directions with the following primers (Invitrogen).

   SeqL-A: 5′ TCGCGTTAACGCTAGCATGGATCTC 3′.

   SeqL-B: 5′ GTAACATCAGAGATTTTGAGACAC 3′.

As indicated earlier, an alternative approach for this initial phase of the cloning is possible (*see* **Note 5**).

3.1.3 Generation of Binary Vector with Target Region as Inverted Repeats

A LR reaction is carried out with the entry clone (pDONR207 containing the target fragment) and a binary destination vector containing *att*R recombination sites (pHellsgate8) to create the final binary vector containing inverted repeats of the target region. The protocol shown below is as described by Invitrogen with minor modifications.

1. Add the following components to a 1.5 mL tube at room temperature and mix.

   pDONR207 containing target fragment (50–150 ng) 1–7 μL
   pHellsgate8 (150 ng/μL) 1 μL.

   Add TE buffer (pH8) to a total of 8uL.

2. To each sample, add 2 μL of LR Clonase ™II enzyme mix to the reaction for a total volume of 10 μL, and mix well by flicking the tube. Centrifuge briefly.
3. Incubate reactions at 25 °C 1–16 h (For large inserts, longer incubation times are better).
4. Add 1 μL of the Proteinase K solution to each sample to terminate the reaction. Vortex briefly. Incubate samples at 37 °C for 10 min.
5. Transform 1–3 μL of reaction mix into heat-shock competent *E. coli* DH5α cells using a standard method. Plate on LB media with 100 mg/L spectinomycin to select bacteria with the recombined vector.
6. Select a colony from the transformation and inoculate 3 mL LB media containing 100 mg/L spectinomycin. Incubate on a 37 °C shaker overnight.
7. Extract pHellsgate8 containing your target inverted repeats from the culture using a miniprep of your choice.
8. Check for proper insertion of inverted repeats into pHellsgate8 with PCR using the reverse primer complimentary to the target region with a primer complimentary to either the promoter region (CaMV35S promoter) or terminator (OCS3′). The following primers are an option [11].

   P27-5: 5′ GGGATGACGCACAATCC 3′.

   P27-3: 5′ GAGCTACACATGCTCAGG 3′.

9. Verify the correct direction of the intron via restriction digest (*see* **Note 9**). Sequences and vector maps can be found online at www.pi.csiro.au/tech_licensing_biol/MapsProtocol.htm.

3.1.4 *A. rhizogenes* Transformation with Binary Vector

For transformation of *M. truncatula*, we routinely use *A. rhizogenes* strain ARqua1 [21] (*see* **Note 10**). Prior to plant transformation, pHellsgate8 containing the target inverted repeats must be transformed into *A. rhizogenes* ARqua1.

1. Add 2 μL (~500 ng) of pHellsgate8 containing target inverted repeats of heat-shock competent *A. rhizogenes* ARqua1 cells.
2. Incubate cells on ice for 30 min.

3. Heat-shock at 37 °C for 2 min.
4. Leave cells on ice for 5 more minutes.
5. Add 500 μL liquid YEP media and shake at 200 rpm for 1–2 h at 30 °C.
6. Plate transformation on YEP plates with 100 mg/L streptomycin and 150 mg/L spectinomycin for selection of pHellsgate8 and the Arqual helper plasmid (*see* **Note 11**).
7. Allow colonies to grow for 2 days at 30 °C.
8. Screen *A. rhizogenes* colonies for pHellsgate8 containing inverted repeats by colony PCR.
9. Grow a positive clone overnight in YEP media containing 100 mg/L streptomycin and 150 mg/L spectinomycin at 30 °C.
10. Create a glycerol stock using 0.4 mL of 50 % glycerol and 0.6 mL of overnight ARqual culture with pHellsgate8 containing inverted repeats. Freeze and store at −80 °C (*see* **Note 12**).

3.2 Generation of M. truncatula Transgenic Roots Expressing the RNAi Construct via A. rhizogenes-Mediated Transformation

3.2.1 A. rhizogenes Cultivation

To enable the investigation of the phenotype in 10–15 independent root systems expressing the RNAi construct, 24 seedlings should be transformed per construct. Start cultivation of ARqual 4 days prior to the day on which seedlings will be inoculated. Streak bacteria carrying the pHellsgate8 construct from glycerol stocks on TY agar plates containing appropriate antibiotics. For ARqual carrying pHellsgate8 include spectinomycin (150 mg/L) and streptomycin (100 mg/L). Grow the bacteria for 48 h at 28 °C. Re-streak bacteria on TY agar plates with the appropriate antibiotics. At this time, it is important to spread bacteria over the whole agar plate so that a thick lawn of *A. rhizogenes* will grow on the plate. This can be done by using a sterile, bent pipet tip or a sterile, flat toothpick. Incubate the plate at 28 °C for 48 h. To calculate the amounts of TY agar plates of *A. rhizogenes* that you will need for one RNAi construct of interest, take in consideration that one TY agar plate is sufficient to inoculate 24 *M. truncatula* seedlings.

3.2.2 Seed Sterilization and Germination

Start seed sterilization procedure 2 days prior to the day on which seedlings will be inoculated. The method described below is suitable for *M. truncatula* ecotypes Jemalong A17 and R108. In our hands, freshly shelled seeds of these genotypes give germination rates ~80 %. The germination rate should be considered for the calculation of the amount of seeds needed for the experiment.

1. Collect unbroken *M. truncatula* seeds and place them into a 50 mL tube. Add a few milliliters of concentrated sulfuric acid (H_2SO_4), just sufficient to cover the seeds (perform this operation in the fume hood and wear gloves, lab coat and safety glasses). Soak the seeds for 10 min to scarify the seed coat. This is required for the subsequent germination steps (*see* **Note 13**).

2. Carefully remove as much H_2SO_4 as possible with a pipette. Quickly add 50 mL of H_2O_{dd} to dilute the H_2SO_4 (*see* **Note 14**). Decant the water and rinse the seeds a further six times with H_2O_{dd}.
3. To surface sterilize the seeds soak them in 10 mL bleach solution (10 % bleach solution [bleach solution contains 6 % sodium hypochlorite], 0.1 % Tween 20) for 10 min with gentle agitation in a laminar flow hood.
4. Remove the bleach solution by decanting. Rinse the seeds at least six times using sterile H_2O_{dd}.
5. Place the seeds in sterile H_2O_{dd} and cover the 50 mL tube with aluminum foil. Leave the tube on a shaker at approximately 60 rpm for 3 h at room temperature. The seeds will imbibe.
6. Decant the water and rinse the seeds once with sterile H_2O_{dd}. Place the seeds in fresh sterile H_2O_{dd} and incubate the seeds at 4 °C for 26 h in the dark. This breaks seed dormancy and allows synchronized germination.
7. Decant the water and transfer the seeds in sterile petri plates of 10 cm diameter (we use glass petri plates that can be washed, sterilized, and reused). Since crowding lowers the germination rate, distribute ~60 seeds uniformly across the plate using sterile forceps or tips. Add some drops of sterile H_2O_{dd} between the seeds. To allow the development of straight radicles, invert the plates. The seeds will stick to the bottom because of surface tension. Add a thin layer of sterile H_2O_{dd} in the lid of the petri plates, which is now on the bottom (Fig. 2). This will help maintain higher humidity. Wrap the inverted plates in aluminum foil and incubate at 28 °C for 18 h. During this 18 h period, the seeds will germinate and the radical will grow straight down (*see* **Note 15**).

3.2.3 Induction of Transgenic Roots

The following steps should be performed in the laminar flow hood to ensure sterile conditions.

1. After 18 h, remove aluminum foil from the plates. The seeds will have germinated and the roots will be approximately 1 cm in length and will have grown straight down. This is helpful for inoculation with *A. rhizogenes*. Add a shallow layer of sterile H_2O_{dd} to the petri dish containing the seedlings to avoid desiccation of their radicles.
2. Select 12 seedlings with straight radicles of 1 cm length and remove their seed coat with a sterile forceps or scalpel while gently holding the hypocotyl with a sterile, flat forceps (*see* **Note 16**). Transfer the seedlings to a sterile petri plate containing a shallow layer of sterile H_2O_{dd}.
3. Using a sharp, sterile scalpel, cut the radicle approximately 3 mm from the root tip (the 3 mm excised root tips are discarded).

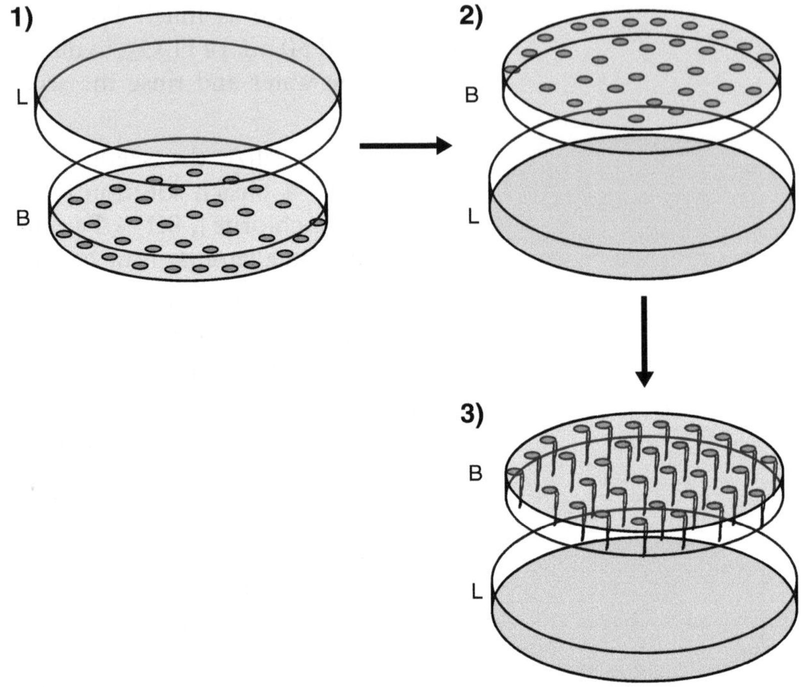

Fig. 2 Plate setup for seed germination. B: *Bottom* of petri plate. L: lid of petri plate. 1 Seeds are spread uniformly in the *bottom* of a petri plate. 2 Invert the plates. Add a thin layer of sterile H_2O_{dd} in the lid. Wrap the inverted plates with aluminum foil. *3* Seedlings after 18 h at 28 °C. The seeds have germinated and their roots have grown *straight down*

4. Using sterile, flat forceps, hold the seedling by the cotyledons the cut end of the radicle with *A. rhizogenes* by lightly scraping the cut end of the radicle on the surface of the TY agar plate with *A. rhizogenes* containing the appropriate construct. This will result in the accumulation of bacteria on the cut end of the radicle.
5. Place the seedlings on a Fähraeus agar plate with the appropriate antibiotics. To keep the radicles in place, shallow grooves can be made with the forceps on the agar. All 12 seedlings can be placed side by side on a single Fähraeus agar plate (*see* **Note 17**).
6. Carefully seal the plate with parafilm. On the top of the plate, make several incisions in the parafilm with a scalpel to allow gas exchange.
7. Place the plates in a slanting position (angle of approx. 45°) in an 18 °C incubator (16 h light/8 h dark) with a light intensity of 6.5 μm/m^{2}s. To create an appropriate slant rest one edge of the plate on top of the lid of a petri dish (*see* **Note 18**).
8. 5 days after inoculation transfer the plates to a growth chamber with a 16 h light (25 °C)/8 h dark (22 °C) regime with a light intensity of 50 μm/m^{2}s. Keep the plates in a slanting position.

Within 7–10 days transgenic roots emerge from the inoculated radical. By using the protocol described above 50–75 % of the plants develop transgenic roots (*see* **Note 19**).

9. 2–3 weeks after inoculation with *A. rhizogenes* the roots are well developed, and if required, the plants can be transplanted in 11 cm pots filled with growth medium such as Turface or soil (10 plants/pot). Since the plants have been growing on plates in an environment with a high humidity, they have to be adapted to a lower humidity. To do this, open the lids of the petri plates, add enough H_2O_{dd} to cover the roots, and leave the lids partially offset for 1 day (the lid covers 1/3 of the plate). The plants should be transplanted the next day. Cover the pots with domes to keep high humidity conditions for a few more days (*see* **Note 20**).

4 Notes

1. pDONR207 and pHellsgate8 have origins of replication with low and high copy numbers, respectively. A standard Qiagen© miniprep (Qiagen) from overnight culture will yield about 50–100 ng/μL of pDONR207 and 200–250 ng/μL of pHellsgate8 if the final elution is 50 μL. This concentration of pDONR207 will suffice, provided the molar concentration of *att*B-flanked PCR product is less than that of the donor vector. To increase the yield from the miniprep, Qiagen© recommends including the PB washing step or eluting in only 30 μL EB buffer (Qiagen protocol). Additionally, one can increase the concentration by doubling the volume of the overnight culture.
2. Gelzan™ tends to agglomerate, and the clumps won't dissolve during the sterilization process. Therefore, slowly add the appropriate amount of Gelzan™ while shaking the flask. Alternatively, Plant Cell Culture certified agar may be used as a gelling agent.
3. In order to increase the percentage of roots cotransformed with the binary T-DNA the use of antibiotic counter selection is recommended. The pHellsgate 8 RNAi constructs carry the neomycin phosphotransferase II gene, which inactivates, by phosphorylation, a range of aminoglycoside antibiotics such as kanamycin. We typically use kanamycin at 25 mg/mL but it may be necessary to determine the best selective concentration experimentally.
4. Due to the toxicity of the ccdB gene product, Gateway®-compatible empty vectors must be maintained in *E. coli* resistant to this toxin, such as DB3.1 (Invitrogen).

5. Alternatively, portions of the cDNA to be targeted can be amplified and cloned into the Gateway-compatible pENTR/D-TOPO or similar vector without the need to add *att*B sites onto PCR products since *att*L sites are already present in this vector. This allows the user to avoid the expense of primers that contain the *att*B sites and BP clonase as well as the associated effort. After this, go to the Subheading 3.1.3 and continue with the protocol as described to recombine to pHellsgate8 through an LR reaction. The approach using pENTR/D-TOPO vectors was used for the generation of approximately 1,500 RNAi constructs by Starker and Gantt.
6. In some instances, silencing multiple genes may be desirable. Up to three transcribed genes can be targeted at the same time by amplifying regions from each and fusing the fragments together via PCR before adding the *att*B sites to the ends of the fused product. Each gene should be represented by at least 200 nucleotides complimentary to its mRNA, and the total fragment should not be much greater than 600 nucleotides [11].
7. Alternatively, one can use only the second set of primers directly, especially when amplifying from a more concentrated source of the targeted gene, such as a cloned vector. The success of this method depends on the target and quality of template.
8. A half reaction (5 μL final volume) is sufficient for both Gateway reactions. For larger inserts, allow reactions to incubate overnight before stopping with Proteinase K. Addition of Proteinase K is essential.
9. In some cases the direction of the intron reverses during the LR reaction, presumably due to the intron's *att*P2 sites on both sides, allowing either side to recombine with either insert's *att*B2. Frequency of intron inversion depends on the insert sequence, but no correlation has been determined [11]. After sequencing >200 pHellsgate8-derived plasmids that were first confirmed by colony PCR for correct recombination of the 5′ portion of the RNAi construct, sequence data confirmed that all clones had the undergone recombination of the 3′ portion of the RNAi construct as well.
10. To generate transgenic roots with growth and morphology comparable to normal roots, low virulence *A. rhizogenes* strains such as ARqua1 are used. These strains elicit a limited number of genetically transformed roots. Roots generated by using high virulence strains may show abnormal root structure and growth, which could negatively impact phenotypic characterization.
11. Up to Spectinomycin 300 mg/L may be used for selection of pHellsgate-derived plasmids. This concentration will help

reduce background, but will also cause Arqual to grow slower than normal. Spectinomycin 150 mg/L is sufficient for maintenance of the plasmid.

12. For larger scale, it is recommended to transform 100–110 pHellsgate-derived plasmids (separately) into Arqual in a single day. This allows more convenient, full 96-well screening of Arqual clones, and the eventual preparation of glycerol stocks in 96 well plates. More than 96 transformations are recommended as a few will fail on the first transformation attempt.
13. The duration of the H_2SO_4 treatment is critical and may vary between seed batches. Therefore, it is necessary to monitor the progress of scarification. If small black dots appear on the seed surface (usually after 5–12 min), immediately remove H_2SO_4 and rinse the seeds with water.
14. Addition of water to sulfuric acid-coated seeds causes an exothermic reaction, with the result that the excessive heat may kill some or all seeds. To mitigate the local heating of seeds, which may damage them, it is extremely important to use a large amount of cold water.
15. Alternatively, a very thin layer of 1 % water-agar can be poured into the bottom of 2 cm deep sterile plastic petri plates. The seeds will stick to the agar and it will help maintain appropriate humidity.
16. The removal of the seed coat helps the cotyledons to open.
17. If antibiotic counter selection is used, include seedlings without *A. rhizogenes* inoculation as a control. This will confirm that growth of non-transformed roots is inhibited in presence of the antibiotic.
18. An experienced researcher can transform 250–450 plants in a single day. However, since planting and analyzing transgenic plants is far more time consuming than transforming them, care should be taken to avoid producing more plants than can be examined.
19. Since non-transformed roots may occasionally escape antibiotic selection, we recommend using a construct which includes a fluorescent marker protein such as DsRED [22], which further enables identification of transgenic roots.
20. These composite plants should express the RNAi hairpin construct and as each independent root arises from an independent transgenic event, the level of gene expression can be expected to vary between independent roots. Silencing of gene expression should be confirmed by RT-PCR. If possible, phenotypic analysis and analysis of transcript levels should be performed on the same roots.

Acknowledgments

The authors thank Armando Bravo and Sergey Ivanov for their reviews and useful comments on the manuscript. Financial support for research was provided by the US National Science Foundation, grants DBI-0421676 and IOS-1127155.

References

1. Helliwell CA, Wesley SV, Wielopolska AJ, Waterhouse PM (2002) High-throughput vectors for efficient gene silencing in plants. Funct Plant Biol 29:1217–1225
2. Pasquinelli AE, Ruvkun G (2002) Control of developmental timing by microRNAs and their targets. Annu Rev Cell Dev Biol 18:495–513
3. Hannon GJ (2002) RNA interference. Nature 418:244–251
4. Ossowski S, Schwab R, Weigel D (2008) Gene silencing in plants using artificial microRNAs and other small RNAs. Plant J 53:674–690
5. Wesley SV, Helliwell CA, Smith NA, Wang MB, Rouse DT, Liu Q, Gooding PS, Singh SP, Abbott D, Stoutjesdijk PA, Robinson SP, Gleave AP, Green AG, Waterhouse PM (2001) Construct design for efficient, effective and high-throughput gene silencing in plants. Plant J 27:581–590
6. Gubler F, Hughes T, Waterhouse P, Jacobsen J (2008) Regulation of dormancy in barley by blue light and after-ripening: effects on abscisic acid and gibberellin metabolism. Plant Physiol 147:886–896
7. Limpens E, Franken C, Smit P, Willemse J, Bisseling T, Geurts R (2003) LysM domain receptor kinases regulating rhizobial Nod factor-induced infection. Science 302:630–633
8. Javot H, Penmetsa RV, Terzaghi N, Cook DR, Harrison MJ (2007) A *Medicago truncatula* phosphate transporter indispensable for the arbuscular mycorrhizal symbiosis. Proc Natl Acad Sci U S A 104:1720–1725
9. Subramanian S, Graham MY, Yu O, Graham TL (2005) RNA interference of soybean isoflavone synthase genes leads to silencing in tissues distal to the transformation site and to enhanced susceptibility to *Phytophthora sojae*. Plant Physiol 137:1345–1353
10. Dunoyer P, Himber C, Ruiz-Ferrer V, Alioua A, Voinnet O (2007) Intra- and intercellular RNA interference in *Arabidopsis thaliana* requires components of the microRNA and heterochromatic silencing pathways. Nat Genet 39: 848–856
11. Helliwell C, Waterhouse P (2003) Constructs and methods for high-throughput gene silencing in plants. Methods 30:289–295
12. Helliwell CA, Waterhouse PM (2005) Constructs and methods for hairpin RNA-mediated gene silencing in plants. In: Engelke D, Rossi J (eds) Methods in enzymology. Academic, San Diego, pp 24–35
13. Hartley J, Temple G, Brasch M (2000) DNA cloning using in vitro site-specific recombination. Genome Res 10:1788–1795
14. Boisson-Dernier A, Chabaud M, Garcia F, Becard G, Rosenberg C, Barker DG (2001) *Agrobacterium* rhizogenes-transformed roots of *Medicago truncatula* for the study of nitrogen-fixing and endomycorrhizal symbiotic associations. Mol Plant Microbe Interact 14:695–700
15. Collier R, Fuchs B, Walter N, Kevin Lutke W, Taylor CG (2005) Ex vitro composite plants: an inexpensive, rapid method for root biology. Plant J 43:449–457
16. Floss DS, Hause B, Lange PR, Kuster H, Strack D, Walter MH (2008) Knock-down of the MEP pathway isogene 1-deoxy-D-xylulose 5-phosphate synthase 2 inhibits formation of arbuscular mycorrhiza-induced apocarotenoids, and abolishes normal expression of mycorrhiza-specific plant marker genes. Plant J 56:86–100
17. Pumplin N, Mondo SJ, Topp S, Starker CG, Gantt JS, Harrison MJ (2010) *Medicago truncatula* vapyrin is a novel protein required for arbuscular mycorrhizal symbiosis. Plant J 61:482–494
18. Vieweg MF, Fruhling M, Quandt HJ, Heim U, Baumlein H, Puhler A, Kuster H, Perlick AM (2004) The promoter of the *Vicia faba L.* leghemoglobin gene VfLb29 is specifically activated in the infected cells of root nodules and in the arbuscule-containing cells of mycorrhizal roots from different legume and nonlegume plants. Mol Plant Microbe Interact 17:62–69
19. Ivashuta S, Liu J, Liu J, Lohar DP, Haridas S, Bucciarelli B, VandenBosch KA, Vance CP, Harrison MJ, Gantt JS (2005) RNA interference

identifies a calcium-dependent protein kinase involved in *Medicago truncatula* root development. Plant Cell 17:2911–2921

20. Liu J, Blaylock L, Endre G, Cho J, Town CD, VandenBosch K, Harrison MJ (2003) Transcript profiling coupled with spatial expression analyses reveals genes involved in distinct developmental stages of the arbuscular mycorrhizal symbiosis. Plant Cell 15: 2106–2123
21. Quandt HJ, Puhler A, Broer I (1993) Transgenic root nodules of *Vicia hirsuta*: a fast and efficient system for the study of gene expression in indeterminate-type nodules. MPMI 6:699–706
22. Limpens E, Ramos J, Franken C, Raz V, Compaan B, Franssen H, Bisseling T, Geurts R (2004) RNA interference in *Agrobacterium rhizogenes*-transformed roots of Arabidopsis and *Medicago truncatula*. J Exp Bot 55:983–992

Chapter 13

Molecular Markers for Genetics and Plant Breeding: The MFLP Marker System and Its Application in Narrow-Leafed Lupin (*Lupinus angustifolius*)

Islam Shahidul, Huaan Yang, and Guijun Yan

Abstract

Since the development of molecular markers to tag genes of agronomic traits of interests, molecular markers have played an increasingly significant role in breeding programs. Molecular markers have been implemented for large-scale marker-assisted selection in the breeding program of many important crops including lupin. So far, more than a dozen molecular markers for disease resistance genes and for other agronomic traits of interest have been developed in lupin. The DNA fingerprinting method, "MFLP" has played a pivotal role in the success of lupin breeding program in Australia. Here, we describe the MFLP technique used in lupin breeding which could be easily transferable to other crop species.

Key words Breeding, Lupin, MFLP, Molecular marker

1 Introduction to Lupin

Lupin has been used as a cultivated crop plant for more than 3,000 years [1, 2]. There is evidence of lupin consumption as human food since early Egyptian times [3]. The Mediterranean region and South America are considered the ancient homes of this crop [4] and it was gradually introduced into other parts of the world [3].

In the family Fabaceae, lupin (*Lupinus* L) is a large genus consisting of herbaceous, soft-woody shrubs and small tree species [5, 6]. Herbaceous lupins include 200–500 annual and perennial species [7] that are mainly used as grain crops. The diversified species of lupin are grouped as Old World and New World species [8]. The New World species comprise >90 % of the total species which are distributed from Alaska to South Argentina and Chile [5, 9]. The Old World species include 13 lupin species from Mediterranean and African regions [8, 10] which are annual, herbaceous and large seeded. Three Old World species *L. albus* L., *L. angustifolius* L., and *L. luteus* L. and one New World species *L. mutabilis* are

Ray J. Rose (ed.), *Legume Genomics: Methods and Protocols*, Methods in Molecular Biology, vol. 1069,
DOI 10.1007/978-1-62703-613-9_13,

cultivated intensively as crops for human consumption [11]. Some other species are used as green manure, forage, ornamentals and for land stabilization [12, 13].

Lupin comprises 2.7 % of world pulse production (tonnes) in 2009 [14]. In the last three decades, Australia has been the largest lupin producer capturing approximately 70 % of world production, which increased from about 0.26 million tonnes in 1979 to more than 2 million tonnes in 1999 [14]. Narrow-leafed lupin (*L. angustifolius*) is mostly cultivated in Australia, while white (*L. albus*) and yellow (*L. luteus*) lupins are traditionally cultivated in Europe and South America [15]. Cultivation of *L. angustifolius* is increasing in Europe after an outbreak of anthracnose in Eastern Europe in the 1990s [3]. In African and Mediterranean countries such as Morocco, South Africa, Syria, and Egypt, *L. albus* and *L. luteus* are mainly grown. Bitter *L. mutabilis* is still cultivated in some parts of Ecuador, Peru and Bolivia. Very small areas of *L. albus* are cultivated in the USA and Canada [3].

1.1 Narrow-Leafed Lupin

Narrow-leafed lupin is one of the most cultivated species among the 13 Old World lupin species mainly cultivated in Australia. For thousands of years different species of Old World lupin have been cultivated in the Mediterranean region. Narrow-leafed lupin has been domesticated to suit modern cropping systems. There are three main differences between wild and domesticated lupin. Wild lupins have bitter alkaloids in the seeds which make them inedible and sometimes toxic, an impermeable seed coat that restricts proper germination; and shattering pods spread out seed at maturity. Domesticated lupin species have no or low alkaloid that are consumable; soft seed coat that facilitate germination while sown into moist soil; and non-shattering pods that retain seed on the plant until it is harvested. Fully domesticated lupin crops were developed in the latter part of twentieth century.

The first domesticated narrow-leafed lupin cultivar Uniwhite was released in 1967 by Dr Gladsones in Western Australia. The cultivar was differentiated by and named for its white color flowers from its wild genotypes that have blue color flowers. However, this cultivar did not adapt to the West-Australian cropping system due to its maturing too late. Release of the first early flowering narrow-leafed lupin cultivar Unicrop in 1973 boosted the commercial lupin production.

Breeding of the narrow-leafed lupin cultivar was initially targeted to earlier flowering and resistance to grey leaf spot disease which threatened the industry in the 1970s. Later on improving adaptation, yield potential and resistance to other diseases became the priority for breeding of that species in the Department of Agriculture and Food Western Australia. Cultivars with resistance to phomopsis, which causes lupinosis in grazing animals, and anthracnose (first released respectively in 1988 and 1996), were important breeding milestones. During the last four decades, the breeding program of narrow-leafed lupin released 26 cultivars with improved yield, disease resistance, and other agronomic traits. The breeding program is currently targeting cost-benefit related issues in particular grain quality attributes.

1.2 The Lupin Genome

Lupinus is the genus of the subfamily Papilionoideae within the family Fabaceae. Molecular phylogenetic studies indicated that the genus is in the Genistoid clade which is believed to have deviated from the other crop legume species about 56.4 million years ago [16]. Based on phylogenetic attributes, the clade Genistoid is different from the sister clades which include most of the important legume crops [17], namely, Phaseoloid-Millettioids (*Glycine max* [soybean] and *Phaseolus vulgaris* [common bean]), Hologalegina clade (*Lotus japonicus*, *Medicago truncatula*, *Pisum sativa* [pea], *Vicia faba* [broad bean], *Lens culinaris* [lentil], *Trifolium* [clover] and *Cicer arietinum* [chickpea]), and the Aeschynomnoid-Dalbergioids (*Arachis* [peanut]). During the last few years, considerable effort has been made towards genomic research of model legume crops with a target to generate resources which are useful for genetic improvement of other legume crops. Particularly, remarkable progress has been made in the two model legumes, *Medicago truncatula* and *Lotus japonicus* and the other two major grain legumes, namely, pea and soybean. Meanwhile, recent breakthroughs in plant genomics, specifically in genome sequencing advanced the genomic research to a number of other legume crops like chickpea, peanut, common bean, lotus and pigeon pea.

In contrast, there is very little genomic information available for lupin species. Cytological studies of lupins suggest significant differences in basic genetic components among the species. The 2C nuclear DNA content of 18 species and botanical forms of the genus *Lupinus*, using propidium iodide as a fluorescent dye, showed distinct infra-generic taxonomic groups with differing somatic chromosome numbers [18]. *L. angustifolius* has an estimated DNA content of 2C = 1.89–2.07 pg, which equals 1.85–2.02 Gb [18, 19]. The New World lupins' predominant chromosome number is $2n = 48$ with a basic chromosome number of $x = 6$ [5]. In contrast, Old World lupins have a series of basic chromosome numbers from $x = 5$, 6 (or 9), 7, 8, 13 at different ploidy levels ranging in chromosome number from 32 to 52 [20]. Relatively recent changes in chromosome number are believed to have occurred as a result of intense selection/evolution in changeable environments [12]. Among the smooth-seeded Old World lupins, *L. angustifolius* has lower chromosome numbers ($2n = 40$) compared to *L. albus* ($2n = 50$), *L. micranthus*, *L. luteus*, and *L. hispanicus* ($2n = 52$) [21].

L. angustifolius has the same number of chromosomes and approximately the same DNA content [22] as soybean. The species is predicted to have undergone polyploidy early in the development of the *Lupinus* genus [23]. Morphologically uniform and small sized chromosomes made this species difficult for the classical cytogenetic analysis. However, fluorescent in situ hybridization has been used to differentiate 5 of 20 chromosome pairs [19]. Research in molecular genetics of *L. angustifolius* has been progressed to produce information to facilitate molecular breeding. Particularly, constructing genetic linkage maps and developing DNA markers are remarkable achievements in molecular research of *L. angustifolius* [23].

1.3 Plant Breeding and Molecular Marker Assisted Breeding

During the last decade, genomic research on crop science has been generating novel tools and methodology that are intensively used to increase the efficiency and precision of crop improvement through breeding. In particular, functional bioinformatics and molecular markers are playing an increasingly significant role in breeding programs. Ultimately, understanding of the virtual values of alleles at all loci segregating in a population can provide the opportunity to design a genotype in silico and to practice whole genome selection. But the high costs currently limit the execution of genomics-assisted crop improvement. However, marker-assisted selection in plant breeding is gradually shaping crop improvement.

1.3.1 What Is a Molecular Marker?

Molecular markers are used to indicate the presence of particular genes and/or gene alleles in a genotype. Generally DNA based molecular markers reveal neutral sites of variation at DNA sequence that do not express a phenotype [24]. DNA markers might be characterized by a few or even a single nucleotide difference. DNA markers are generally classified into two groups (Table 1) depending on the way of detecting polymorphism, namely, PCR-based

Table 1
Basic comparison of different types of markers

Types of marker	Way of detecting polymorphism			
	Hybridization polymorphisms	PCR based polymorphism	Dominant	Co-dominant
RFLPs (restriction fragment length polymorphisms)	✓			✓
AFLPs (amplified fragment length polymorphisms)		✓	✓	
SSR (simple sequence repeats or microsatellites)		✓		✓
RAPDs (randomly amplified polymorphic DNA)		✓	✓	
MFLP (microsatellite-anchored fragment length polymorphism)		✓	✓	✓

polymorphism and hybridization based polymorphism [1]. RFLPs (restriction fragment length polymorphisms) are the most common way of revealing hybridization polymorphisms, while AFLPs (amplified fragment length polymorphisms), SSR (simple sequence repeats or microsatellites), and RAPDs (randomly amplified polymorphic DNA) are the most common forms of markers for PCR based polymorphism. Recently MFLP (microsatellite-anchored fragment length polymorphism) has been developed from a combination of the AFLP concept with SSR-anchor primers [25]. This technique has high efficiency in producing DNA markers with each of the detected polymorphisms including an SSR pattern [25–27].

AFLPs and RAPDs are usually expressed as dominant markers (only a small portion (5 %) of AFLP markers are co-dominant). On the other hand RFLPs and SSRs are considered as co-dominant markers. However, MFLPs can be expressed as either dominant or co-dominant markers. Co-dominant markers are more useful due their ability to differentiate heterozygous and homozygous individuals for a particular gene/marker (Table 1).

1.3.2 MAS: Marker Assisted Selection

Since the development of molecular markers to tag genes of agronomic traits of interests they are commonly used in conventional breeding programs. These markers are used to select individual plants bearing the genes and the process is generally called "marker-assisted selection" or "MAS." MAS can facilitate a number of advantages compared to conventional selection which is pointed out as follows:

- *Early generation selection.* With MAS, targeted populations can be selected at the F2 generation whereas it needs to be done at the F5 generation or later in conventional plant breeding.
- *Co-dominance.* MAS can facilitate selection of only homozygous or heterozygous individuals, which cannot be achieved through conventional selection. For example, In MAS, plants are routinely selected for the homozygous disease resistance genes (*RR*), while those with the heterozygous resistant (*Rr*) genes or homozygous susceptible genes (*rr*) are discarded. On the other hand, in conventional breeding this is not possible because traditional disease testing does not differentiate *RR* from *Rr*.
- *More reliable and more accurate.* In MAS, testing is carried out on the genome and hence the outcome is not influenced by environments. In contrast, tests conducted on the phenotype in conventional breeding are often influenced by environmental factors.
- *Pyramid multiple genes.* Molecular markers can differentiate between plants with one, two or multiple disease resistance genes against the same or different diseases. In contrast, conventional breeding does not normally allow testing differentiation between the numbers of resistant genes.

- MAS can provide the opportunity of screening populations for multiple traits at one go. For example, F2 plants from the lupin breeding program were screened for both anthracnose resistance and phomopsis resistance markers simultaneously using a multiplex marker test.
- MAS is quicker and cheaper than glasshouse/field-based testing used in conventional breeding program if managed well.

1.3.3 What Types of Molecular Markers are Useful for MAS?

It is an obvious fact that, only a small portion of the reported thousands of molecular markers can be used in MAS in plant breeding. Actually a marker has to meet certain requirements before being used in breeding. These requirements include.

- *Link to an agronomic trait of industry significance.* A useful molecular marker must be linked to a gene responsible for a trait having agronomic importance. Unfortunately the majority of molecular markers do not meet this criterion.
- *Reliable and reproducible.* The marker must be reliable and reproducible, and the interpretation of marker score must be unambiguous.
- *Co-dominance.* A co-dominant marker is more useful than a dominant marker, mainly because of two reasons: (a) a co-dominant marker differentiates homozygous individuals (e.g., *RR*) from heterozygous individuals (e.g., *Rr*) that is crucial in the breeding program; (b) Dominant markers usually require internal controls (as absence of a band for a dominant marker in a gel may be due to PCR failure or to the absence of a target genetic allele), which is not the case for co-dominant markers.
- *High-throughput and amenable to large number of samples.* A molecular marker for MAS must be able to work even with high impurity levels in DNA. Practically in breeding programs, MAS involves a large number of samples, where DNA extraction is inevitably crude containing high levels of impurities.
- *Applicable to large number of crosses.* A marker is valuable for MAS if it is applicable to a wide range of crosses in a breeding program. Generally a breeding program involves hundreds, even thousands of crosses each year.
- *Cost efficient.* A marker should be as cost efficient compared to the classical phenotypic selection to warrant the use of MAS.

1.3.4 Technical Requirements of Molecular Marker Development for MAS

There are some technical difficulties in respect to specific molecular markers for MAS. For example:

- RAPD markers and DAF markers are not so reliable and reproducible [28].
- The high cost involvement makes the AFLP and the MFLP markers out of reach for a large numbers of samples which is

very common in breeding programs. However, they can be used after conversion into simple PCR-based markers.

- DNA hybridization-based markers, such as RFLP and microarray markers are not effective with crude DNA extracted from high throughput DNA extraction protocol generally practiced in MAS.
- Using CAPS markers in MAS can also be very expensive and time-consuming as they require restriction enzyme cutting after PCR [29].
- Although SNP markers are more and more used for MAS in plant breeding they require sophisticated equipments that are not always available.

Generally speaking, *sequence-specific simple PCR-based markers* can satisfy most the technical requirements and are implementable in MAS. Because they.

1. Only require one step of PCR, followed by gel separation.
2. Are a single copy in a plant genome; with only one set of DNA bands visualized on the gel, making marker scoring easy and unambiguous.
3. Tolerate high levels of impurity of crude DNA extracted from the high-throughput implementation protocol.
4. Are reliable, reproducible and inexpensive to run.
5. Require unsophisticated equipment and a low level of expertise from technical operators.

However, it is notable, that development of sequence-specific PCR markers is a difficult process. They are obtained either by cloning and sequencing genomic DNA (such as STMS marker and SNP markers) or by converting nonspecific markers obtained from generic DNA fingerprinting methods such RAPD [30], AFLP [31] and MFLP [25–27, 32]. A comparison of the Markers is presented in Table 2.

1.3.5 Genetic Requirements of Molecular Markers for MAS

Apart from the need for markers to be linked to genes of agronomic interest, an important challenge is that, the marker should be applicable to a wide range of crosses in a breeding program to implement in MAS. But the reality is that, most published molecular markers can only be used in a small number of crosses, and are therefore not that useful for breeders. It is now well recognized that, the inability of a marker to be applicable to a wide range of crosses in a breeding program is a key limiting factor for large-scale MAS in molecular plant breeding [33, 34].

In MAS, a molecular marker is used to screen progeny from a cross only if (1) the marker is polymorphic between the parents, and (2) the parent with the target marker band possesses the target

Table 2
A comparison of the markers

	Non site-specific markers			Site specific markers	
	RAPD, DAF, ISSR	AFLP, MFLP	RFLP, Microarray	SNP markers	Sequence-specific PCR markers
Source of marker	RAPD/ISSR gels	AFLP/MFLP gels	Cloning	Sequencing	Converted from nonspecific markers
Effort to get marker	Easy	Easy	Difficult	Difficult	Difficult
Reliability and reproducibility	Low	High	High	High	High
Locus specificity/ cross transferability	No	No	Yes	Yes	Yes
Running cost	Low	High	High	Questionable	Low
Application to MAS	Undesirable	Undesirable	Undesirable	Questionable	Desirable

gene of interest, so that the desirable allele can be distinguished from undesirable ones [35]. Unfortunately, most molecular markers developed for MAS are "imperfect" (i.e., not part of the gene itself) and certain genetic distances exist between markers and genes of interest [36, 37]. During evolution and in plant breeding, genetic recombination may occur in the region between the marker and the gene on the chromosome. As a result, cultivars showing desirable markers may not necessarily possess the targeted genes, and vice versa ("false positive") [38]. Obviously, the more false positives in breeding germplasm, the lower the number of crosses the marker can be used to screen breeding progenies [38]. To deal with the prevalence of false positives in breeding lines, molecular biologists must validate these markers to parental combinations before a marker can be used. For example, in the Australian Winter Cereals Molecular Marker Program (AWCMMP), a large proportion of the resources in marker implementation are directed towards "marker validation" [35, 38].

Marker development for anthracnose resistance in lupin illustrates the importance of obtaining markers with wide range application in plant breeding. In the case of narrow-leafed lupin (*Lupinus angustifolius*), there are two sister cultivars, namely, Tanjil and Wonga, which have high levels of resistance to anthracnose caused by the fungal pathogen *Colletotrichum lupini* [39]. Resistance is controlled by a major dominant gene designated as "*Lanr1*." A molecular marker, "AntjM1," was developed linking the *Lanr1* gene [26]. Unfortunately, a number of cultivars without the *Lanr1* gene showed the resistance allele band for marker

AntjM1. This greatly limited the usefulness of this marker in lupin breeding. For example, if breeders cross Tanjil (resistant) with Quilinock (susceptible to the disease), the F2 plants will all have the same DNA banding pattern for marker AntjM1, although these F2 plants are segregating for disease resistance, and the marker cannot be used to differentiate resistant and susceptible F2 plants. As a result, in 2003, marker AntjM1 could only screen the progenies of 23 of over 400 crosses in the Australian National Lupin Breeding Program (ALBP).

In order to solve this problem, marker development researchers developed a second molecular marker, "AntjM2" [40]. AntjM2 is a co-dominant marker, and its banding pattern is consistent with disease phenotype on all cultivars and breeding lines tested by ALBP, enabling widespread implementation. Since 2005, approximately 10,000–20,000 plants per year have been screened with molecular marker AntjM2 for MAS in ALBP.

1.3.6 Molecular Markers and Genetic Linkage Maps

A genetic map is constructed by putting the molecular markers in order that represents the relative genetic distances between the markers [24]. The aim of creating a genetic map is to improve the understanding of genetic behavior and the efficiency of selecting superior genotypes in breeding programs. In 1980s, the concept of genetic mapping was proposed in the case of the study of the human genome using molecular markers. Consequently the idea has been implemented for a wide range of organisms including commercially important crop species.

Building maps involves the production of suitable markers and then assembling the calculated recombination values from all the pair-wise combinations of those markers to make linkage groups. Traits segregating in that population may be physically scored and mapped in the same way as the markers [24]. The linear array of these linkage groups minimizes the number of recombination events between adjacent markers [24, 37].

The first genetic linkage map of *L. angustifolius* was published in 2005 based on MFLP markers [41]. Nelson et al. [23] published the second map based on AFLP markers and markers from *Medicago truncatula*. These two maps were combined and published recently [42].

1.4 Marker Systems Used and Developed in Lupin

The first attempt to develop a molecular marker in *L. angustifolius* was in Poland using 60 RAPD markers and 4 isozyme markers and mapped into 17 linkage groups [1]. In 1999, Brien et al. generated 705 polymorphic AFLP markers which grouped into 52 linkages covering approximately 1,000 cM. Later, in Australia, the MFLP technique was developed [25]. In essence, MFLP is a combination of the AFLP [43] concept with the SSR-anchor primer technique [44]. Sequentially this technology has been used for marker development by several research groups to assist the breeding of lupin.

Particularly the same research group developed sets of molecular markers using MFLP linked to genes conferring resistance to *Diaporthe toxica* [27] and Anthracnose [26]. These markers were commonly used at the early stage of marker assisted breeding of lupin in Australia.

Following the same methodology of MFLP, Boersma et al. [45–48] developed molecular markers linked to several genes corresponding to important traits that includes *mollis* (soft seed coat), *Ku* (early flowering), and *Lentus* and *tardus* (reducing pod-shattering) genes. However, there were some critical issues for a wider application of these markers in MAS of lupin [49]. For examples, markers developed for the *tardus* gene have a low level of association with phenotypes and is not so cost-effective [49] and markers for the *mollis* gene were not efficient for breeding programs [50]. Likewise markers developed for the *lentus* gene were dominant and thus not able to differentiate homozygous and heterozygous individuals among progenies from wild and domesticated crosses [51]. To enhance the applicability of molecular markers, Yang et al. [32] projected an approach in identification of multiple candidate markers linked to the gene of interest followed by a validation step to select a best candidate marker before converting it into an implementable form.

Sequentially, to increase the efficiency of molecular markers for MAS in breeding programs of *L. angustifolious,* Li et al. [49–52] developed more molecular markers for the *tardus, mollis,* and *lentus* genes following the MFLP technique. They also developed markers for the low alkaloid gene [52]. These markers are applicable in MAS and many of them are being used in current breeding programs of lupin at the Department of Agriculture and Food Western Australia.

1.5 Future Directions

The MFLP method uses the "bulked segregant analysis (BSA)" principle of employing a small number of informative progeny plants, where in DNA fingerprinting all the individual plants are kept separate. Thus this effectively got rid of the problem of "false positive" candidate markers [26, 32]. However, this method is not so time efficient. In contrast, the next-generation sequencing (NGS) technology has the potential to detect large numbers of DNA markers within a short time-frame. Recently it has been reported that several marker development methods successfully utilized NGS platforms to sequence complexity reduced representations. For examples, reduced-representation libraries (RRLs) [53, 54], complexity reduction of polymorphic sequences (CRoPS) [55], restriction-site associated DNA sequencing (RAD-seq) [56], sequence based polymorphic marker technology (SBP) [57], low coverage multiplexed shotgun genotyping (MSG) [58], and genotyping by sequencing (GBS) [59]. Recently, we have devised a methodology which cohesively integrated the RAD-seq method into the marker development protocol, by which we discovered nearly 40 molecular markers linked to an agronomic gene of interest, and the markers were readily converted into cost-effective PCR markers desirable for routine implementation in plant breeding [60]. This example certainly indicates that NGS technology would be useful for molecular marker development of major traits in lupin.

Furthermore, most of the genetic linkage maps of *L. angustifolius* were constructed using AFLP, RAPD or MFLP technology primarily based on anonymous dominant genetic markers. Generally speaking, these maps are cultivar-specific and offer no or little opportunity to compare the genome of *L. angustifolius* with those of other legume species [23]. Thus construction of genetic maps for *L. angustifolius* by using co-dominant markers transferable between lupin species and across the legume family would be a valuable resource for lupin genetic research.

Sequence variations between individuals, cultivars, and subspecies showed that single nucleotide polymorphisms (SNPs) and insertions and deletions (InDels) are highly abundant and distributed throughout the genome in various species including plants [61–63]. The abundance of these polymorphisms in plant genomes makes the SNP marker system an attractive tool for mapping, marker-assisted breeding and map-based cloning. SNP marker is just a single base change in a DNA sequence, with a usual alternative of two possible nucleotides at a given position. SNPs seem very exciting markers but they require extensive investment in equipment and manpower.

DArT (diversity arrays technology) is one of the recently developed molecular techniques and it has only been used in some crops. DArT is a microarray hybridization-based technique that enables the simultaneous typing of several hundred polymorphic loci spread over the genome [64] when sequence information is available for that species. It is a high throughput, quick, and highly reproducible method and cost effective, with an estimated cost per data point tenfold lower than SSR markers [65]. However, it remains to be thoroughly tested in various species for a wider application. This chapter will focus on the MFLP method but there continue to be new developments.

2 Materials for MFLP

2.1 Plant Material

An F_8 recombinant inbred line (RIL) population from a cross between a domesticated breeding line 83A:476 (*lentus/lentus*) and a wild type P27255 (*Lentus/Lentus*) of *L. angustifolius* using a single-seed-descent method was used [48]. This population segregates a number of traits including pod shattering, soft seedness, vernalization, and resistance to several diseases. All 25 commercial cultivars released in Australia and 125 accessions from the

Australian lupin collection hosted at Department of Agriculture and Food Western Australia (DAFWA) were also used to evaluate the applicability of developed markers.

2.2 Chemicals for Isolation of Genomic DNA

1. 6 M Ammonium acetate.
2. DNA extraction buffer (1 % sarkosyl, 100 mM Tris–HCl, 100 mM NaCl, 10 mM EDTA; pH 8.5.).
3. Isopropanol.
4. Ethanol.
5. 96-deep-well plate with mat cover from Life Technologies.
6. 4-mm diameter steel ball bearing.
7. TE 0.1 buffer (10 mM Tris; 0.1 mM EDTA; pH 8.0).
8. QIAGEN tissuelyser II.

2.3 Restriction Ligation

1. Primers, e.g., MSAD1 (GACGATGAG TCCTGAG) and MSAD2 (TACTCAGGACTCAT).
2. 10× DNA ligation buffer from Life Technologies.
3. 0.5 M NaCl.
4. BSA (0.5 mg/mL).
5. *Mse*I and SSR adapters.
6. *Tru*9I (an isoschizomer of *Mse*I; Roche Diagnostics Australia) (10 U/μL).
7. T4 DNA-ligase (Roche Diagnostics Australia) (5 U/μL).
8. *Hae*III solution (*see* **Note 3**).

2.4 Pre-selective MFLP Amplification

1. MS-primer: *Mse*I-A (GAT GAG TCC TGA GTA A–A).
2. SSR-primers: MF01 GTCC $(GA)_6$/MF11 GGAC $(CT)_6$.
3. Thermal cycler (Perkin Elmer).

2.5 Selective MFLP Amplification

1. ^{33}P Primer.
2. 5× Kinase buffer from Life Technologies.
3. [g-^{33}P] ATP.
4. T4 polynucleotide kinase (10 U/μL).
5. Gel running unit [5 % polyacrylamide denaturing sequencing gel (7 M urea) in a Sequi-Gen GT sequencing cell (BioRad)].

2.6 Conversion of MFLP Markers into Simple PCR-Based Markers

1. ^{33}P radioisotope is generally used for labeling the MFLP product.
2. pGEM-T Easy Vector, Promega and JM109 High Efficiency Competent Cells, Promega for cloning the MFLP fragments.
3. UltraClean® 6 min Mini Plasmid Prep kit (MO BIO, USA) High Pure PCR Product Purification Kit (Roche, USA) for DNA extraction and purification.

3 The MFLP Method for MAS

The MFLP method was developed by Yang et al. [25], by combining the concept of AFLP [43] with the SSR-anchor primer technique [44] (*see* **Note 1**). Polymorphisms detected in MFLP are SSR-*Mse*I fragments, which contain a microsatellite motif sequence at one end, with an AFLP primer sequence at the other end [25]. Briefly, in this method, genomic DNA is firstly digested by one restriction enzyme (e.g., *Mse*I). One AFLP adaptor (e.g., *Mse*I adaptor) is ligated onto the restriction fragments. DNA fingerprints are produced by PCR using one microsatellite-anchor primer in combination with one *Mse*I-primer. The method allows co-amplification of over 100 DNA fragments containing microsatellite motifs per PCR. The MFLP technique involves the following steps.

3.1 Selecting RIL (Recombinant Inbred Line) Plants

It is necessary to choose F8 RIL plants for a certain agronomic trait of interest to run the MFLP method and to search for markers. Figure 1 represents the outcome of MFLP fingerprinting in the case of anthracnose and phomopsis resistance genes (*see* **Note 2**).

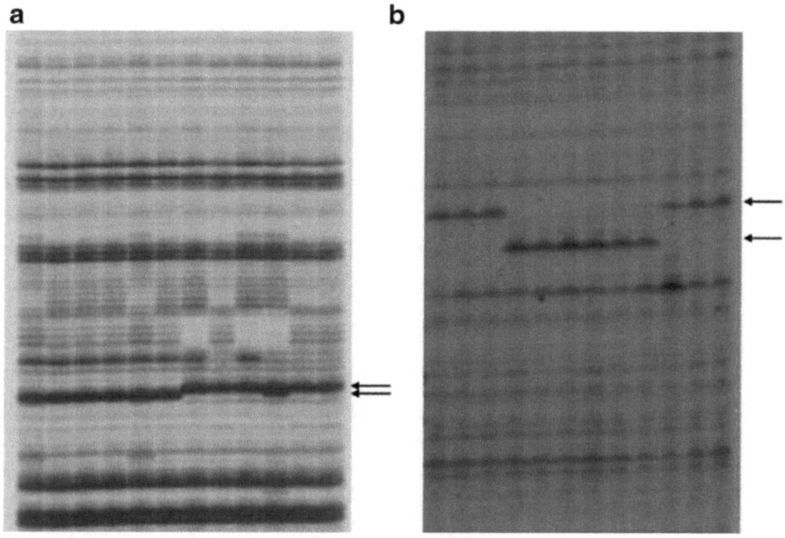

Fig. 1 Design of MFLP fingerprinting by choosing 12 plants so that they represent the extremes of the phenotypes—e.g., the resistant and susceptible ones in this case. *Arrows* in (**a**) indicate a co-dominant marker tagging the anthracnose resistance gene with the first six lanes containing DNA from the resistant genotypes and the second six lanes containing DNA from the susceptible genotypes (note one mismatch in here); *Arrows* in (**b**) indicate a co-dominant marker tagging the phomopsis resistance gene with the first and the last three lanes containing DNA from the resistant genotypes and the middle six lanes containing DNA from the susceptible genotypes (note this marker shows a perfect match)

3.2 Extraction and Estimation of DNA Concentration of Each Plant

To extract high quality DNA from plant materials the following steps need to be followed.

1. Turn on water bath and set to 80 °C.
2. Take 96-deep-well plate. Put three leaflets in each well.
3. Add 600 μL "Plant DNA Extraction Buffer" (*see* Subheading 2.2) using bottle pipetting system. Add one 4-mm diameter steel ball bearing into each well. Put on a mat cover. Smash leaf tissue with the Tissue Lyser at speed "1350" for 5 min. Change the side of plate, run smashing machine again.
4. Spin plates for 5 min to get rid of bubbles using a bench top centrifuge.
5. Put plates (without rubber cover) in water bath to heat at 80 °C for 30 min.
6. Float the plate on cold water for 10 min to cool down. Put in −20 °C freezer for 15–30 min (not too long or it will freeze).
7. Add 300 μL (=twice of 150 μL) 6 M ammonium acetate. Put on a new (or clean) mat cover. Mix by inverting up and down three times. Put in the −20 °C freezer for 8 min. While the samples are in the freezer, take a new 96-deep-well plate. Write sample number on all four sides. Add 300 μL (=twice of 150 μL) extraction buffer in each well.
8. Take samples from freezer and spin at 3,600 rpm (2,200 × *g*) for 15 min. Take 180 μL upper aqueous (do not disturb the bottom rubbish), add to the new deep-well plate, mixing by inverting up and down twice SLOWLY to avoid liquid going into the pipette.
9. Add 400 μL isopropanol. Put on a new (or clean) mat cover. Mixing by inverting up and down twice only. Leave in a −20 °C freezer for 75–90 min.
10. Spin at 3,600 rpm for 15 min.
11. Pour off all the liquid, and leave plate upside down on tissue for 5 s (do NOT try to suck up remaining aqueous!).
12. Add 150 μL 70 % ethanol in each well. Put on the same mat cover. Put the plate on bench for 10 min (do NOT shake the plate).
13. Spin plate at 3,600 rpm for 3 min (so that the pellets stick with the plates).
14. Pour off all the liquidand leave plates upside down on tissue for 10 s (do NOT try to suck up remaining aqueous!).
15. Leave plates on bench for 10 min without mat cover to allow ethanol to evaporate.
16. Add 100 μL TE 0.1 into each tube. Put into refrigerator for 2 h or overnight to dissolve DNA.
17. Mix the DNA by 12-channel pipette setting on 80 μL.

18. DNA concentration can be measured for each individual using a NanoDrop Spectrophotometer following the manufacturer's instructions.

3.3 Setting Up the Restriction-Ligation

1. Prepare an adapter (e.g., "*Mse*I Adapter"): Take the "*Mse*I Adapter," heat in a water bath at 95 °C for 5 min. Cool on bench for 20 min and spin briefly in a microfuge.
2. Prepare the "Master mix" following the table below.

Components	Amount for each sample (μL)
Water	6.6
10× ligase buffer	2
0.5 M NaCl	1
0.5 mg/mL BSA	2
MS adapter	2
*Tru*9I (= *Mse*I) (10 U/μL)	0.6
T4 ligase (5 U/μL)	0.8
Subtotal	15.0

3. R-L reactions are prepared by mixing of DNA (5 μL) and master mix (15 μL), so the final volume is 20 μL. Mix, spin down, run program "R-L" (=37 °C for 2 h, then 65 °C for 20 min to kill the enzyme).
4. Digest with *Hae*III: Take 10 μL of R-L mix in a 0.5-mL tube and add 50 μL of *Hae*III solution (*see* **Note 3**). Incubate at 37 °C water bath for 2 h and add 150 μL TE 0.1 buffer into the mix. This is the DNA template for Pre-selective amplification.

3.4 Pre-selective MFLP Amplification

DNA is amplified using MS-primer (*Mse*I-A; *see* **Note 4**) and SSR primer (MF01; *see* **Note 5**) through the following steps.

1. Add 1.5 μL DNA to each tube.
2. For each sample, add.
   - MS-primer: 0.5 μL.
   - SSR-primer: 0.5 μL.
   - PCR Mix: 7.5 μL.
   - Mix well.
3. Run program "PRE MFLP" (=94 °C for 2 min. 25 cycles of 94 °C 30 s, 52 °C 30 s, 72 °C 1 min, hold at 4 °C).
4. Add 190 μL TE 0.1 buffer into each sample and put the sticky label on the tubes.

   These will be used as DNA template for selective MFLPs.

3.5 Selective MFLP Amplification

1. SSR primers are labelled by using ^{33}P Primer.
   (a) Prepare the master mix (70 μL): 5× Kinase buffer 21.0 μL + [g-^{33}P] ATP 21.0 μL + T4 polynucleotide kinase (10 U/μL) 3.5 μL + Sterile water 24.5 μL.
   (b) Take 35 μL SSR primer stock and add 70 μL of the above master mix.
   (c) Run program "Label" (=37 °C for 60 min; followed by 70 °C to kill the enzyme).
2. Selective MFLP PCR reactions are prepared: 1.5 μL template DNA (from 3.4) + 0.5 μL *Mse*I-primer + 0.5 μL ^{33}P labelled SSR-primer + 7.5 μL PCR mix from FisherBiotech.
3. Program run (PCR) at 94 °C for 30 s, annealing temperature 30 s, 72 °C 1 min, hold at 4 °C. Annealing temperature = starts 60 °C, decrease 0.7 °C each step until 54 °C reached. After 25 cycles annealing at 54 °C.
4. Run the amplified DNA on the sequencing gel at 55 W for 3.5 h.

3.6 MFLP Markers Are Easily Converted into Simple PCR-Based Markers

DNA polymorphisms in MFLP come from three sources.

1. Insertion/deletion derived size difference markers, including number difference in simple sequence repeats (=microsatellite markers) targeted by the SSR-primers used in MFLP.
2. DNA sequence variations between lupin cultivars on the annealing site by the SSR primers.
3. Variations on restriction enzyme sites by the *Mse*I or *Hae*III used in MFLP.

MFLP markers from type (1) and type (2) are easy to convert into simple PCR based markers. After DNA sequencing of marker fragments of SSR-*Mse*I amplified in MFLP, one can simply design a sequence specific primer at the *Mse*I-end. By labeling this primer with radioisotope or fluorescence dye, the labelled primer can be used in PCR in combination with the SSR-primer as a "sequence-specific simple PCR-based marker" for MAS in plant breeding [25–27]. The "conversion" of these two types of MFLP does not need any sequence knowledge beyond the SSR-*Mse*I fragments. MFLP markers from type (3) are similar to AFLP markers which result from restriction sites. The conversion of this type of marker needs DNA sequence knowledge beyond the detected marker fragments, which is difficult to obtain. Fortunately, about one-third of all MFLP markers are from type (1). In lupin marker development, we usually get 6–12 candidate MFLP markers within 3 months, allowing us to focus on co-dominant markers, the most widely used markers in Australian lupin breeding.

4 Notes

1. MFLP has two distinct advantages over other DNA fingerprinting methods (such as RAPD, ISSR, AFLP etc.). These advantages are detailed in [25–27, 32] and You et al. [40], and briefly summarized as follows.
   - MFLP is more efficient at generating markers. Like AFLP, MFLP detects about 100 DNA bands in each reaction, considerably more than that of RAPD, ISSR etc.
   - MFLP targets and detects SSR-related DNA polymorphism. Since SSR is highly polymorphic, the number of polymorphic markers detected by MFLP is much higher than that by AFLP.
   - Several real examples may explain the high efficiency of MFLP in marker development: when developing markers for disease resistance, we use 12 plants (6 *RR*, 6 *rr*) and a 97-well comb to run sequencing gels which allows eight sets of MFLPs to be loaded (12 plants × 8 sets plus + 1 size marker = 97 lanes). Generally, 5–8 polymorphic markers are found in each PCR, or about 50 markers in each gel with lupin. Two sets of gels can be run per day (including PCRs and gels). A 4-week workload produces 40 gels, equivalent to 2,000 markers. The whole genome of *L. angustifolius* ranges from 1,540 to 1,700 centiMorgans [41]. The 2,000 markers give an average of "one marker each cM" in the whole genome, which provide a basis for the discovery of several candidate markers within 5 cM each side of the gene. This explains how so many candidate MFLP markers were found within a few months time [32].
2. In lupin breeding, six undesirable wild major genes must all be eliminated from progenies in wild × domesticated (W × D) crosses, including the high alkaloid gene Iucundis, the later-flowering gene *ku*, the two pod-shattering genes *Tardus* and *Lentus*, the hard-seeded gene *Mollis*, and the blue flower and dark seed color gene *Leucospermus* [66]. Later Boersma et al. [45–48] and Li et al. [49–52] developed molecular markers linked to several genes corresponding to important traits that include *mollis* genes (softening seed coat), *Ku* (early flowering), and *Lentus* and *tardus* gene (reducing pod-shattering) and low alkaloid gene (Table 3).
3. *Hae*III enzyme stock preparation (For 60 μL): Water 38 μL + *Bsu*RI (= *Hae*III; GG/CC) (10 U/μL) 6 μL + Buffer 6 μL.
4. We use *Mse*I-A as an example in this experiment. However, *Mse*I-C, *Mse*I-T, *Mse*I-G all also work well in MFLP.

Table 3
List of candidate markers identified from MFLP fingerprints linked to genes encoding important traits in *Lupinus angustifolius* L

Traits	Name of the marker and author reference	Primer combination in MFLP	Primer pair of converted PCR marker	Amplification size of PCR marker (bp)	Allele linked
Resistance to phomopsis stem blight disease	Ph258M1 [27]	*Mse*I-CAA: GATGAGTCCTGAGTAACAA MF128: AGTAGCTCTCTCTCTCTC	Ph258M1F: TCCAGACTGACTATATTCTTAG Ph258M1R: CAGGCACATATATCTTTATACC	R = 303; S = 253	Co-dominant
Resistance to phomopsis stem blight disease	Ph258M2 [27]	*Mse*I-CAA: GATGAGTCCTGAGTAACAA MF51: GGGAACAACAACAAC	Ph258M2F: TAGAACCATTGTAACTAAATCC Ph258M2R: CTCTTACTGAATCCTATGAAG	R = 265; S1 = 262; S2 = 268	Co-dominant
Resistance to anthracnose disease	AntjM1 [26]	*Mse*I-CAA: GATGAGTCCTGAGTAACAA NF201: CCCATTGTTGTTGTTG	AntjM1F: CATCCTCACATATGAAGC AntjM1R: AAACCCA (TTG)4	R = 303; S = 253	Co-dominant
Resistance to anthracnose disease	AnManM1 [32]	*Mse*I-ACT: GATGAGTCCTGAGTAAACT MF129: HVHTGTGTGTGTGTGTG	AnManF: TTGAGCTTGGTATATAAACG AnManR: TCGAGCAATAAATGATATG	R = 188; S = 164	Co-dominant
Resistance to rust disease	RustM1 (Yang et al. unpublished data)	*Mse*I-CCT: GATGAGTCCTGAGTAACCT MF43: CCTCAAGAAGAAGAAGAAG	RustM1M1: TTAACCTCACCATGGAGG RustM1SR: GAAGGTGGAATTCTAAACG	142	Susceptible allele
Reduced Pod Shattering Gene *tardus*	TaLi [49]	*Mse*I-ACT: GATGAGTCCTGAGTAAACT MF202: GGAATTGTTGTTGTTG	TaLiF: ATCCTACTAAATCCTGGTACAG TaLiR: GATCTGAAAAGGAATATGAAG	400	Wild allele
	DAWA169.305 [45]	*Mse*I-CCT: GATGAGTCCTGAGTAACCT MF128: AGTAGCTCTCTCTCTCTC	TaM1MS: AACAGAGGATTGCAAATC MF128: AGTAGCTCTCTCTCTCTC	306	Domesticated allele

Reduced Pod Shattering Gene *lentus*	LeLi [51]	*Mse*I-ACC: GATGAGTCCTGAGTAAACC MF155: CAACTGTGTGTGTGTG	LeLi F: CCCACAGCTAAAATTATACC LeLi R: GGCAGGATGTAAGTTTAGG	251/247	co-dominant
	DAWA323.150 [1]	*Mse*I-CGA: GATGAGTCCTGAGTAACGA MF51: GGGAACAACAACAAC	LeMS1: (3′-TTAACGAACCTACCATTTG-5′) MF51: GGGAACAACAACAAC	157	Dominant
	DAWA468.290 [1]	*Mse*I-CCA: GATGAGTCCTGAGTAACCA MF42 : (3′-GTCTAACAACAACAACAAC-5′)	LeMS2 (3′-AGAAAAAGATGAATGCACG-5′) MF42 : (3′-GTCTAACAACAACAACAAC-5′)	284	Dominant
Low-alkaloid gene *iucundus*	IucLi [1]	*Mse*I-ATG: GATGAGTCCTGAGTAAACC MF155: CAACTGTGTGTGTGTG	LeLiF: CCCACAGCTAAAATTATACC LeLiR: GGCAGG ATGTAAGTTTAGG	141	Domesticated allele
Soft-seediness gene *mollis*	MoLi [1]	*Mse*I-AGC: GATGAGTCCTGAGTAAAGC MF62: CCCAAACAACAACAAC	MoLiF: TTAAAGCTCCTCCGGGAG MoLiR: CTTATACTATTAGACTAACGCC	314	Domesticated allele
	DAWA561.180 [1]	*Mse*I-CAT: GATGAGTCCTGAGTAACAT MF43: CCTCAAGAAGAAGAAGAAG	MoAMS: TAACATCAACAAGGTGAGAATC MoASR: GAAGCATTCGATGAATTC	179	Dominant
Early-flowering gene *Ku*	[1]	KuHMS1: AGACATACCT TGTATGCGG MF42: GTCTAACAACAACAACAAC	KuHMS1: AGACATACCTTGTATGCGG MF42 (GTCTAACAACAACAACAAC)	298	Domesticated allele

5. We use MF01 and MF11 as examples in this experiment. However, you can find many other SSR-primers for MFLP. Please see MFLP publications [25, 27].

MFLP techniques are being used widely in marker studies of other species. For example, to study genetic diversity of Australian chickpea cultivars [67], chromosome region regulating seed Zn accumulation in barley [68], anthracnose disease resistance in white lupin [69].

References

1. Boersma JG (2007) Contributions to the molecular genetics of the Narrow-leaf lupin (*Lupinus angustifolius* L.)-mapping, marker development and QTL analysis. PhD Thesis, The University of Western Australia, Perth, Australia
2. Gladstones JS (1970) Lupins as crop plants. Field Crop Abstract 2:123–148
3. Lupin.org About Lupins. http://www.lupins.org/lupins/
4. Ainouche A, Bayer RJ, Misset MT (2004) Molecular phylogeny, diversification and character evolution in *Lupinus* (Fabaceae) with special attention to Mediterranean and African lupines. Plant Syst Evol 246:211–222
5. Dunn DB (1984) Cytotaxonomy and distribution of New World lupin species. In: 3rd International Lupin Conference, La Rochelle, France, 1984; International Lupin Association: La Rochelle, France, 1984, pp 67–85
6. Turner BL (1995) A new species of *Lupinus* (Fabaceae) from Oaxaca, Mexico: a shrub or tree mostly three to eight meters high. Phytologia 79:102–107
7. Dunn DB, Gillett J (1966) Lupines of Canada and Alaska. Queen's Press, Ottawa, Canada
8. Gladstones JS (1974) Lupins of the Mediterranean region and Africa. Australian Department of Agriculture, Perth, Australia, pp 1–48
9. Planchuelo-Ravelo AM (1984) Taxonomic studies of Lupinus in South America. In: III International Lupine Congress, La Rochelle, France, 1984, pp 40–54
10. Amaral Franco JD, da Silva AR P (1978) Lupinus L. In: Heywood VH (ed) Flora Europaea, vol 2. Cambridge University Press, London, UK
11. Petterson DS (1998) Composition and food uses of lupins. In: Gladstones JS, Atkins C, Hamblin J (eds) Lupins as crop plants: biology, production, and utilization. CABI, Oxon, UK
12. Gladstones JS (1998) Distribution, origin, taxonomy, history and importance of lupins. CAB International, London, pp 1–39
13. Hoveland CS, Townsend CE (1985) Other legumes. In: Heath ME, Barnes RF, Metcalfe DS (eds) Forages. Iowa State University Press, Ames, IA, pp 146–153
14. FAO (2011) FAOSTAT database. In: FAO in http://faostat.fao.org/site/567/DesktopDefault.aspx: Rome, vol 2011
15. Duranti M, Consonni A, Magni C, Sessa F, Scarafoni A (2008) The major proteins of lupin seed: characterisation and molecular properties for use as functional and nutraceutical ingredients. Trends Food Sci Technol 19:624–633
16. Lavin M, Herendeen PS, Wojciechowski MF (2005) Evolutionary rates analysis of leguminosae implicates a rapid diversification of lineages during the tertiary. Syst Biol 54: 575–594
17. Drummond CS (2008) Diversification of *Lupinus* (Leguminosae) in the western new world: derived evolution of perennial life history and colonization of montane habitats. Mol Phylogenet Evol 48:408–421
18. Naganowska B, Wolko B, Sliwinska E, Kaczmarek Z (2003) Nuclear DNA content variation and species relationships in the genus *Lupinus* (Fabaceae). Ann Bot 92:349–355
19. Hajdera I, Siwinska D, Hasterok R, Maluszynska J (2003) Molecular cytogenetic analysis of genome structure in *Lupinus angustifolius* and *Lupinus cosentinii*. Theor Appl Genet 107:988–996
20. Pazy P, Heyn CC, Herrnstadt I, Plitmann U (1977) Studies in populations of the Old World *Lupinus* species. I. Chromosomes of the east-mediterranean lupines. Israel J Bot 26:115–127
21. Conterato IF, Schifino-Wittmann MT (2006) New chromosome numbers, meiotic behaviour and pollen fertility in American taxa of *Lupinus* (Leguminosae): contributions to taxonomic and evolutionary studies. Bot J Linean Soc 150:229–240
22. Bennett MD, Smith JB, Heslop-Harrison JS (1982) Nuclear DNA amounts in Angiosperms. Proce R Soc Lond B Biol Sci 216:179–199

23. Nelson M, Phan H, Ellwood S, Moolhuijzen P, Hane J, Williams A, O'Lone C, Fosu-Nyarko J, Scobie M, Cakir M, Jones M, Bellgard M, Książkiewicz M, Wolko B, Barker S, Oliver R, Cowling W (2006) The first gene-based map of *Lupinus angustifolius* L.-location of domestication genes and conserved synteny with *Medicago truncatula*. Theor Appl Genet 113:225–238
24. Jones N, Ougham H, Thomas H (1997) Markers and mapping: we are all geneticists now. New Phytol 137:165–177
25. Yang H, Sweetingham MW, Cowling WA, Smith PMC (2001) DNA fingerprinting based on microsatellite-anchored fragment length polymorphisms, and isolation of sequence-specific PCR markers in lupin (*Lupinus angustifolius* L.). Mol Breed 7:203–209
26. Yang H, Boersma JG, You M, Buirchell BJ, Sweetingham MW (2004) Development and implementation of a sequence-specific PCR marker linked to a gene conferring resistance to anthracnose disease in narrow-leafed lupin (*Lupinus angustifolius* L.). Mol Breed 14: 145–151
27. Yang HY, Shankar MS, Buirchell BB, Sweetingham MS, Caminero CC, Smith PS (2002) Development of molecular markers using MFLP linked to a gene conferring resistance to *Diaporthe toxica* in narrow-leafed lupin (*Lupinus angustifolius* L.). Theor Appl Genet 105:265–270
28. Provan J, Thomas W, Forster B, Powell W (1999) Copia-SSR: a simple marker technique which can be used on total genomic DNA. Genome 42:363–366
29. Konieczny A, Ausubel FM (1993) A procedure for mapping Arabidopsis mutations using co-dominant ecotype-specific PCR-based markers. Plant J 4:403–410
30. Paran I, Michelmore RW (1993) Development of reliable PCR-based markers linked to downy mildew resistance genes in lettuce. Theor Appl Genet 85:985–993
31. Shan X, Blake TK, Talbert LE (1999) Conversion of AFLP markers to sequence-specific PCR markers in barley and wheat. Theor Appl Genet 98:1072–1078
32. Yang H, Renshaw D, Thomas G, Buirchell B, Sweetingham M (2008) A strategy to develop molecular markers applicable to a wide range of crosses for marker assisted selection in plant breeding: a case study on anthracnose disease resistance in lupin (*Lupinus angustifolius* L.). Mol Breed 21:473–483
33. Holland J (2004) Implementation of molecular markers for quantitative traits in breeding programs—challenges and opportunities. In: 4th International Crop Science Congress, Brisbane, Australia, 2004. www.cropscience.org.au
34. Snape J (2004) Challenges of integrating conventional breeding and biotechnology: a personal view!. In: 4th International Crop Science Congress, Brisbane, Australia, 2004. www.cropscience.org.au
35. Eagles HA, Bariana HS, Ogbonnaya FC, Rebetzke GJ, Hollamby GJ, Henry RJ, Henschke PH, Carter M (2001) Implementation of markers in Australian wheat breeding. Aust J Agr Res 52:1349–1356
36. Gupta PK, Varshney RK, Sharma PC, Ramesh B (1999) Molecular markers and their applications in wheat breeding. Plant Breed 118: 369–390
37. Staub JE, Serquen FC, Gupta M (1996) Genetic markers, map construction, and their application in plant breeding. HortScience 31:729–741
38. Sharp PJ, Johnston S, Brown G, McIntosh RA, Pallotta M, Carter M, Bariana HS, Khatkar S, Lagudah ES, Singh RP, Khairallah M, Potter R, Jones MGK (2001) Validation of molecular markers for wheat breeding. Aust J Agr Res 52:1357–1366
39. Nirenberg HI, Feiler U, Hagedorn G (2002) Description of Colletotrichum lupini comb. nov. in modern terms. Mycologia 94:307–320
40. You M, Boersma JG, Buirchell BJ, Sweetingham MW, Siddique KHM, Yang H (2005) A PCR-based molecular marker applicable for marker-assisted selection for anthracnose disease resistance in lupin breeding. Cell Mol Biol Lett 10:123–134
41. Boersma JG, Pollotta M, Buirchell BJ, Sivasithamparam K, Yang H (2005) Construction of a genetic linkage map using MFLP, and identification of molecular markers linked to domestication genes in narrow-leafed lupin (*Lupinus angustifolius*). Cell Mol Biol Lett 10:331–344
42. Nelson MN, Moolhuijzen PM, Boersma JG, Chudy M, Lesniewska K, Bellgard M, Oliver RP, Swiecicki W, Wolko B, Cowling WA, Ellwood SR (2010) Aligning a new reference genetic map of *Lupinus angustifolius* with the genome sequence of the model legume *Lotus japonicus*. DNA Res 17:73–83
43. Vos P, Hogers R, Bleeker M, Reijans M, Lee TVD, Hornes M, Friters A, Pot J, Paleman J, Kuiper M, Zabeau M (1995) AFLP: a new technique for DNA fingerprinting. Nucleic Acids Res 23:4407–4414
44. Wu K-S, Jones R, Danneberger L, Scolnik PA (1994) Detection of microsatellite polymorphisms without cloning. Nucleic Acids Res 22:3257–3258

45. Boersma J, Nelson M, Sivasithamparam K, Yang HA (2009) Development of sequence-specific PCR markers linked to the Tardus gene that reduces pod shattering in narrow-leafed lupin (*Lupinus angustifolius* L.). Mol Breed 23:259–267
46. Boersma JG, Buirchell BJ, Sivasithamparam K, Yang H (2007) Development of a PCR marker tightly linked to mollis, the gene that controls seed dormancy in *Lupinus angustifolius* L. Plant Breed 126:612–616
47. Boersma JG, Buirchell BJ, Sivasithamparam K, Yang H (2007) Development of a sequence-specific PCR marker linked to the Ku gene which removes the vernalization requirement in narrow-leafed lupin. Plant Breed 126: 306–309
48. Boersma JG, Buirchell BJ, Sivasithamparam K, Yang H (2007) Development of two sequence-specific PCR markers linked to the le gene that reduces pod shattering in narrow-leafed Lupin (*Lupinus angustifolius* L.). Genet Mol Biol 30:623–629
49. Li X, Renshaw D, Yang H, Yan G (2010) Development of a co-dominant DNA marker tightly linked to gene tardus conferring reduced pod shattering in narrow-leafed lupin (*Lupinus angustifolius* L.). Euphytica 176:49–58
50. Li X, Buirchell B, Yan G, Yang H (2011) A molecular marker linked to the mollis gene conferring soft-seediness for marker-assisted selection applicable to a wide range of crosses in lupin (*Lupinus angustifolius*) breeding. Mol Breed 29:361–370
51. Li X, Yang H, Yan G (2012) Development of a co-dominant DNA marker linked to the gene lentus conferring reduced pod shattering for marker-assisted selection in narrow-leafed lupin (*Lupinus angustifolius*) breeding. Plant Breed 131:540–544
52. Li X, Yang H, Buirchell B, Yan G (2011) Development of a DNA marker tightly linked to low-alkaloid gene iucundus in narrow-leafed lupin (*Lupinus angustifolius* L.) for marker-assisted selection. Crop Pasture Sci 62:218–224
53. Hyten D, Cannon S, Song Q, Weeks N, Fickus E, Shoemaker R, Specht J, Farmer A, May G, Cregan P (2010) High-throughput SNP discovery through deep resequencing of a reduced representation library to anchor and orient scaffolds in the soybean whole genome sequence. BMC Genomics 11:38
54. Gore MA, Chia J-M, Elshire RJ, Sun Q, Ersoz ES, Hurwitz BL, Peiffer JA, McMullen MD, Grills GS, Ross-Ibarra J, Ware DH, Buckler ES (2009) A first-generation haplotype map of maize. Science 326:1115–1117
55. Mammadov J, Chen W, Ren R, Pai R, Marchione W, Yalçin F, Witsenboer H, Greene T, Thompson S, Kumpatla S (2010) Development of highly polymorphic SNP markers from the complexity reduced portion of maize [*Zea mays* L.] genome for use in marker-assisted breeding. Theor Appl Genet 121:577–588
56. Pfender W, Saha M, Johnson E, Slabaugh M (2011) Mapping with RAD (restriction-site associated DNA) markers to rapidly identify QTL for stem rust resistance in *Lolium perenne*. Theor Appl Genet 122:1467–1480
57. Sahu B, Sumit R, Srivastava S, Bhattacharyya M (2012) Sequence based polymorphic (SBP) marker technology for targeted genomic regions: its application in generating a molecular map of the Arabidopsis thaliana genome. BMC Genomics 13:20
58. Andolfatto P, Davison D, Erezyilmaz D, Hu TT, Mast J, Sunayama-Morita T, Stern DL (2011) Multiplexed shotgun genotyping for rapid and efficient genetic mapping. Genome Res 21:610–617
59. Elshire RJ, Glaubitz JC, Sun Q, Poland JA, Kawamoto K, Buckler ES, Mitchell SE (2011) A robust, simple genotyping-by-sequencing (GBS) approach for high diversity species. PLoS One 6:e19379
60. Yang H, Tao Y, Zheng Z, Li C, Sweetingham M, Howieson J (2012) Application of next-generation sequencing for rapid marker development in molecular plant breeding: a case study on anthracnose disease resistance in *Lupinus angustifolius* L. BMC Genomics 13:318
61. Batley J, Barker G, O'Sullivan H, Edwards KJ, Edwards D (2003) Mining for single nucleotide polymorphisms and insertions/deletions in maize expressed sequence tag data. Plant Physiol 132:84–91
62. Garg K, Green P, Nickerson DA (1999) Identification of candidate coding region single nucleotide polymorphisms in 165 human genes using assembled expressed sequence tags. Genome Res 9:1087–1092
63. Nasu S, Suzuki J, Ohta R, Hasegawa K, Yui R, Kitazawa N, Monna L, Minobe Y (2002) Search for and analysis of single nucleotide polymorphisms (SNPs) in rice (*Oryza satva*, *Oryza rufipogon*) and establishment of SNP markers. DNA Res 9:163–171
64. Jaccoud D, Peng K, Feinstein D, Kilian A (2001) Diversity arrays: a solid state technology for sequence information independent genotyping. Nucleic Acids Res 29:e25
65. Xia L, Peng K, Yang S, Wenzl P, Carmen de Vicente M, Fregene M, Kilian A (2005) DArT for high-throughput genotyping of Cassava

(*Manihot esculenta*) and its wild relatives. Theor Appl Genet 110:1092–1098

66. Clements JC, Buirchell BJ, Yang H, Smith PMC, Sweetingham MW, Smith CG (2005) Chapter 9: lupin. In: Singh R, Jauhar P (eds) Genetic resources, chromosome engineering, and crop improvement. Vol. 1, grain legumes. CRC Press, Florida, USA, pp 231–323
67. Lin R, Yang H, Khan TN, Siddique KHM, Yan G (2008) Characterisation of genetic diversity and DNA fingerprinting of Australian chickpea (*Cicer arietinum* L.) cultivars using MFLP markers. Aust J Agr Res 59:707–713
68. Sadeghzadeh B, Rengel Z, Li C, Yang H (2010) Molecular marker linked to a chromosome region regulating seed Zn accumulation in barley. Mol Breed 25:167–177
69. Yang H, Lin R, Renshaw D, Li C, Adhikari K, Thomas G, Buirchell B, Sweetingham M, Yan G (2010) Development of sequence-specific PCR markers associated with a polygenic controlled trait for marker-assisted selection using a modified selective genotyping strategy: a case study on anthracnose disease resistance in white lupin (*Lupinus albus* L.). Mol Breed 25:239–249

Chapter 14

Stable Transformation of *Medicago truncatula* cv. Jemalong for Gene Analysis Using *Agrobacterium tumefaciens*

Youhong Song, Kim E. Nolan, and Ray J. Rose

Abstract

Medicago truncatula is a model legume that has all the genomic resources to be suitable as a model for functional genomics. Transformation to produce transgenic plants is part of this toolkit, enabling a spectrum of approaches to study gene function: including knockdown, overexpression, reporter genes for gene expression, and proteins tagged with fluorescent proteins such as GFP. A special genetic line is necessary for transformation and Jemalong 2HA derived from cv. Jemalong is used in the methods described. Leaf explants can be used for the transformation of the embryonic stem cells to produce the transgenic somatic embryos for regeneration. An auxin and a cytokinin are the key hormone requirements for regeneration by somatic embryogenesis but other hormones such as abscisic acid can be used to augment the system. As the explants used in this system are from leaves, rather than immature embryos or meristematic tissues often used in other species, it is a quite straightforward system. *Agrobacterium tumefaciens* containing a binary vector suitable for the particular objectives is used to deliver the transgene of interest.

Key words *Agrobacterium tumefaciens*, Binary vectors, *Medicago truncatula*, Jemalong 2HA, Regeneration, Somatic embryogenesis, Stable transformation, Tissue culture, Transgenes

1 Introduction

Medicago truncatula or barrel medic is an important legume model [1–3] with substantive genomic resources. There is a draft sequence of the *M. truncatula* euchromatin representing about 94 % of all *M. truncatula* genes [4]. Affymetrix gene chips are available for transcriptomics [5] and Affymetrix gene chip data for *M. truncatula* have been abridged in the *M. truncatula* Gene Expression Atlas (*MtGEA* [6]). There are increasing numbers of insertion mutants [7] and an increasing amount of proteomics [8–11] and *miRNA* data [12–14]. Transformation of *M. truncatula* is a useful tool for overexpressing and knocking down genes [15] and mutant complementation; as well as for the insertion of agronomic genes in this agricultural species used for forage [16]. In addition *M. truncatula*

Ray J. Rose (ed.), *Legume Genomics: Methods and Protocols*, Methods in Molecular Biology, vol. 1069, DOI 10.1007/978-1-62703-613-9_14, © Springer Science+Business Media, LLC 2013

is a close relative of the important fodder crop *M. sativa* with its more complex autotetraploid genetics. Gene expression analysis for all developmental stages in *M. truncatula* is also an advantage for whole plant transformation [17].

M. truncatula transformation was initially dependent on the development of the transformable line Jemalong 2HA (2HA) by a cycle of tissue culture, which greatly enhanced regenerability via somatic embryogenesis, and subsequent selection for regenerability [18, 19]. Similar strategies were used to develop the R108 line using two cycles of tissue culture [20]. Transformation of 2HA uses *Agrobacterium tumefaciens* [21–23] with P4 basal medium [24] and the hormones auxin, cytokinin and abscisic acid [25] for tissue culture. The 2HA transformation uses leaf explants which are easy to prepare. The 2HA line can also be regenerated from leaf protoplasts [26], and protoplasts have the potential for transient expression. In this chapter we present the details for the transformation of *M. truncatula*, Jemalong 2HA.

2 Materials

2.1 Plant Material

The highly regenerable *Medicago truncatula* genotype line Jemalong 2HA is used in this method. The seed is available in the plant community and can also be obtained from the authors' laboratory. We use glasshouse grown plants (*see* **Note 1**) with a 14 h photoperiod and 23 °C/19 °C day/night temperature. Explants are taken from the latest fully expanded trifoliate leaves from individual stems of 2–6 months old plants.

2.2 Equipment and Labware

1. Laminar flow or Biohazard hood.
2. Electroporator—Gene Pulser® II (Bio-Rad).
3. Controlled environment room or cabinet.
4. Orbital mixer incubator (Ratek Instruments) for *A. tumefaciens* growth.
5. Benchtop centrifuge (Eppendorf 5810R).
6. Microfuge.
7. 9 cm petri dishes.
8. Eppendorf tubes (1.5 mL) and Falcon tubes (50 mL).

2.3 Tissue Culture

All instruments and containers are autoclaved (121 °C, 20 min).

2.3.1 Sterilization and Explant Preparation

1. 70 % (v/v) ethanol.
2. 0.5 % (v/v) sodium hypochlorite (White King™ bleach, diluted 1:8).
3. Mesh spring tea infuser to hold leaflets.
4. Screw cap polycarbonate culture pots, 250 mL.

5. Suitable blunt end forceps and scalpel.
6. Chinese take-away lids.
7. Coplin Jar.

2.3.2 Medicago Co-cultivation and Culture Media

Use Milli-Q purified water for solutions.

1. Basal medium (called P4 *see* ref. 24), is made up of major salts (Ca^{2+} is made up separately), minor salts, iron, vitamins, casein hydrolysate, and sucrose.

Major salts: KH_2PO_4 131 mg/L, KCl 225 mg/L, KNO_3 1,875 mg/L, NH_4NO_3 600 mg/L, $MgSO_4{\cdot}7H_2O$ 225 mg/L (make up as a 10× stock). Calcium made up separately (*see* **Note 2**), $CaCl_2{\cdot}2H_2O$, 300 mg/L (make up as 100× stock).

Minor salts: $MnSO_4{\cdot}H_2O$ 10 mg/L, H_3BO_3 3 mg/L, $ZnSO_4{\cdot}7H_2O$ 2 mg/L, KI 0.75 mg/L, $Na_2MoO_4{\cdot}2H_2O$ 0.25 mg/L, $CuSO_4{\cdot}5H_2O$ 0.025 mg/L, $CoCl_2{\cdot}6H_2O$ 0.025 mg/L (made up as 1,000× stock and aliquoted in desired amounts *see* **Note 3**). The $CuSO_4{\cdot}5H_2O$ and $CoCl_2{\cdot}6H_2O$ are at such low concentrations it is easiest to make up a 100,000× stocks to use in preparing the minor salt solution.

Chelated iron: $FeSO_4{\cdot}7H_2O$ 9.267 mg/L, $Na_2EDTA{\cdot}2H_2O$ 37.2 mg/L (made up as 200× stock).

To make a 200× stock dissolve 7.44 g of $Na_2EDTA{\cdot}2H_2O$ in 900 mL of MQ water while stirring, bring the solution to 98–99 °C and slowly add 1.853 g of $FeSO_4{\cdot}7H_2O$, cool and make to 1,000 mL with Milli-Q water. Store in amber bottle at 4 °C (*see* **Note 4**).

Vitamins: Myo-inositol 100 mg/L, Thiamine HCl 10 mg/L, Nicotinic acid 1 mg/L, Pyridoxine HCl 1 mg/L (make up as 1,000× stock and aliquot in desired amounts).

Other: Casamino acids (Casein hydrolysate) 250 mg/L (made up as 100× stock and aliquoted in desired amounts) Sucrose 30 g/L.

Stocks are stored in a –20 °C freezer unless otherwise indicated.

To make 1 L of basal medium

10× Major salts	100 mL
100× Calcium	10 mL
1,000× Minor salts	1 mL
200× Iron	5 mL
100× Casamino Acids	10 mL
1,000× Vitamins	1 mL
Sucrose	30 g
Agar	8 g (Bacto laboratories)

The medium components are made to 1 L with pH adjusted to 5.8 with KOH.

2. P4 10:4 medium, consists of the P4 basal medium, 10 μM Naphthalene Acetic Acid (NAA) and 4 μM 6-Benzylaminopurine (BAP).

NAA is made up as 1,000 μM stock, heat and stir until dissolved (almost to boiling) before making to volume, store at 4 °C. BAP is made up as 1,000 μM stock, add a few drops of 1 N HCl and heat and stir until dissolved (*see* **Note 5**).

3. Cocultivation medium, consists of the P4 10:4 medium with 10 mM glucose and 100 μM acetosyringone. Acetosyringone is made up as a 500 mM stock in DMSO and stored at room temperature (not filter-sterilized). Glucose is made up as a 1 M stock, filter-sterilized and aliquoted into sterile containers and stored at 4 °C. Both acetosyringone and glucose are added after media autoclaving and cooling to about 55 °C.
4. P4 10:4:1 medium, consists of the P4 10:4 medium with 1 μM Abscisic Acid (ABA). ABA is made up as a 1,000 μM stock. Dissolve in a small amount of 1 N NaOH then make to volume in Milli-Q water with stirring and adjust to pH 5.8. Filter-sterilize into sterile containers and store at –20 °C. ABA is added to the medium when the medium is cooled to about 55 °C.
5. The P40 medium consists of the P4 medium without myo-inositol.
6. P40 medium but with 1 % sucrose for plantlet growth and rooting in Magenta pots.
7. Parafilm.

2.4 Agrobacterium Strains, Culture and Storage

2.4.1 A. tumefaciens Strains

The strain we currently use for *Medicago truncatula* is AGL-1 [17, 27, 28] widely available in the plant research community [29]. We have successfully used other strains, e.g., LBA4404 from Clontech Laboratories (*see* **Note 6**).

2.4.2 A. tumefaciens Culture

1. *A. tumefaciens* is cultured in YEP media at 27 °C. YEP liquid media consists of: Tryptone 10 g (Bacto Laboratories), Yeast extract 10 g (Bacto Laboratories), NaCl 5 g made up to 1 L, pH 7 with KOH. For solid agar medium add 15 g agar (Bacto Laboratories).
2. Antibiotics: Antibiotics used to select for AGL-1 are ampicillin 100 μg/mL and rifampicin at 50 μg/mL Stock solutions use ampicillin at100 mg/mL (2.5 g in 25 mL H_2O) and rifampicin at 50 mg/mL (1.25 g in 25 mL DMSO). Filter-sterilize and store at –20 °C. For other strains *see* Hellens et al. [29]. Antibiotics are added to the medium when the medium is cooled to about 55 °C.

2.4.3 A. tumefaciens Storage

Use single colonies grown in YEP medium with ampicillin and rifampicin to OD 0.6–0.8. The culture is stored with glycerol 1:1 in 1.5 mL Eppendorf tubes at −80 °C (*see* **Note 7**).

2.5 Binary Vectors and Electroporation

1. Many binary vectors are available, widely used are the Gateway series ([30] and *see* Life Technologies).
2. "HEPES buffer": 1 mM HEPES buffer in 10 % (v/v) glycerol, pH 7.5 with NaOH, used in preparing electrocompetent cells.

2.6 Transformation

2.6.1 Selection Antibiotics

Selection antibiotics we have used with *Medicago truncatula* are given below and 100× stocks in H_2O can be filter-sterilized and stored at −20 °C (*see* **Notes 8** and **9**). They are added after the temperature of the medium has been cooled to about 55 °C. Suppliers for plant transformation antibiotics and selection agents—include Austratec, KisanBio, Sigma.

1. Hygromycin B 15–25 μg/mL (selectable marker gene *hph*).
2. Kanamycin 50 μg/mL (selectable marker gene *npt*II).
3. Phosphinothricin (active gradient of Basta®) 1 μg/mL (selectable marker gene *bar*).

2.6.2 Antibiotics for Killing Agrobacterium

Timentin or Augmentin (Amoxicillin–Clavulanic acid, 5:1) can be used. Stock solutions at 250 mg/mL. Timentin can be aliquoted and stored at −20 °C. Augmentin is made up fresh. Filter-sterilize and add to medium after cooling to about 55 °C.

2.7 Plant Regeneration

1. 1 Deep (2 cm) 9 cm petri dishes.
2. Magenta pots.
3. Autoclaved Whatman No. 1 filter paper.
4. Plant pots.
5. "Glad® wrap" plastic cling film, small diameter bamboo stakes.

3 Methods

3.1 Transferring the Binary Vector into Agrobacterium

When the transgene of choice has been prepared in the binary vector electroporation is used to transfer the binary vector into *Agrobacterium*. The protocol is as follows.

3.1.1 Preparation of Electrocompetent Cells

1. Streak out AGL-1 cells on to YEP agar plates containing 100 μg/mL ampicillin and 50 μg/mL rifampicin and grow at 27 °C to obtain single colonies.
2. Grow the AGL-1 from single colonies in 20 mL YEP liquid medium with the 100 μg/mL ampicillin and 50 μg/mL rifampicin antibiotics in 50 mL Falcon tubes on an orbital shaker at 27 °C for 2–3 days until OD_{600} of 0.6–0.8.

3. The bacterial cells are pelleted at 1,500 ×*g* in a bench centrifuge and resuspended in 20 mL sterile ice cold "HEPES buffer" and the supernatant removed. This is repeated twice.
4. After the final resuspension 50 μL of cells are aliquoted into sterile 1.5 mL Eppendorf tubes. The cells can be used immediately or stored at −80 °C (*see* Subheading 2.4.3).

3.1.2 Electroporation of Agrobacterium

1. Add 1 μL with about 20 ng of binary vector plasmid DNA to 50 μL competent cells in 1.5 mL Eppendorf tubes.
2. Transfer cells to an ice cold electroporation cuvette (0.2 cm electrode gap, Bio-Rad) making quite sure no ice from the ice-bucket is adhering to the outside of the cuvette.
3. Electroporate cells at 25 μF capacitance, 400 Ω resistance and 2.5 kV using the Bio-Rad Gene Pulser®II. The instrument should be used with due care (see Safety in supplier's instruction manual).
4. Plate the cells on YEP agar plates containing 100 μg/mL ampicillin and 50 μg/mL rifampicin plus the antibiotic selecting for the presence of the transferred binary vector (e.g., 100 μg/mL spectinomycin for pH7GW1WG2(II), 15) and grow at 27 °C to obtain single colonies (*see* **Note 10**).
5. The single colonies can be grown up in liquid medium and stored at −80 °C (*see* Subheading 2.4.3).

3.2 Preparation of A. tumefaciens for Transformation

1. From the prepared and −80 °C stored Agrobacterium (Subheading 3.1.2, **step 5**) use a sterile micropipette tip (attached to the micropipette) to scrape a small quantity of cells and wash into 20 mL of YEP liquid medium containing selection antibiotics in 50 mL Falcon tubes. Incubate 2–3 days at 27 °C on a shaker (250 rpm). Grow the Agrobacterium until OD of 0.6–0.8.
2. Spin tubes at 1,500 ×*g* for 5 min on a bench centrifuge to pellet bacteria. Remove YEP medium and resuspend pellet in 20 mL of sterile P4 10:4 liquid medium. Do shortly before transformation and leave at 27 °C.

3.3 Sterilization of Tissue

Collect leaves from 2HA plants for sterilization. The youngest expanded trifoliate leaves of an elongating stem are used as the explant source. Plants are usually used 2–6 months after sowing and grown under glasshouse conditions. Try to minimize the time from tissue collection until plating. Have everything set up to start the sterilization before collecting the tissue. Plants which are too young may not culture well because the tissue is damaged during the sterilization procedure. It is advisable to collect tissue into an environment where high humidity is maintained, e.g., use a specimen

Fig. 1 (**a**) *M. truncatula* trifoliate leaf showing where explants should be taken from. Bar is 10 mm. (**b**) Plantlet in Magenta pot for root development. Note filter paper arrangement for feeding liquid medium. Bar is 10 mm

jar that has tissue paper soaked in distilled water at the bottom. Keep the lid on the jar to maintain humidity. This prevents the tissue from drying too much.

Working in a UV-sterilized laminar flow or biohazard hood, leaves are sterilized in a sterile mesh spring tea infuser that is moved from solution to solution which are in sterile 250 mL screw cap polycarbonate culture pots.

1. 30 s in 70 % (v/v ethanol).
2. 10 min in 0.5 % (v/v) sodium hypochlorite. Gently swirl the tea infuser, drain excess hypochlorite solution.
3. Move the tea infuser with the leaves to a culture pot with sterile Milli-Q water and gently swirl and drain excess water. Then repeat this in a fresh culture pot of distilled water.
4. Leaves are removed with sterile forceps from the tea infuser to a third culture pot of Milli-Q water using sterile forceps. Screw the sterile cap on, invert and swirl.
5. Keep the leaves floating on the water, until ready to cut up.
6. Cut leaves into rectangular explants (8–10 × 3–5 mm) as shown in Fig. 1a. Black lines represent cuts made to the tissue. Leaves are cut using a sterile scalpel on a sterile support (*see* **Note 11**).

3.4 Co-cultivation with Agrobacterium

1. As leaf explants are cut up, transfer them to a sterile 9 cm petri dish containing about 20 mL of liquid P4 10:4 culture medium. When all of the explants are ready, remove the medium from the dish using a sterile Pasteur pipette and discard. Add bacterial suspension to the petri dish and incubate explants in bacteria for 5–10 min (*see* **Note 12**).

2. Remove explants, blot briefly on sterilized paper towels on top of foil and plate (abaxial side down) onto co-cultivation medium (P4 10:4 + 10 mM glucose + 100 μM acetosyringone). Seal plates with Parafilm and incubate for 2–4 days in the dark at 27 °C. Check the growth of bacteria. They ideally grow 2–3 mm width around the tissue.

3.5 Concluding Co-cultivation and Transfer to Selection Medium

1. When co-cultivation with *Agrobacterium* has been completed, using aseptic techniques the leaf tissue is washed in sterile distilled water for 5 min then 500 μg/mL Timentin for 5 min. Each wash is carried out in 20 mL of solution in a sterile 9 cm petri dish, and then blotted on sterile paper towel.
2. Explants are then plated onto P4 10:4 medium + 750 μg/mL Timentin (or 750 μg/mL Augmentin) + appropriate selection agent. Concentrations for common selection agents are kanamycin 50 μg/mL, hygromycin 15 μg/mL, and the herbicide Basta 1 μg/mL. Incubate plates in the dark at 27 °C for 3 weeks.

3.6 Regenerating Transformed Plants by Somatic Embryogenesis

1. After 3 weeks, explants are transferred to P4 10:4:1 medium + 750 μg/mL Timentin (or 750 μg/mL Augmentin) with appropriate selection agent and incubation continues at 27 °C in the dark. Subculture every 3–4 weeks in the same medium (*see* **Note 13**) Embryos start to appear after 5–6 weeks from initial plating.
2. When robust embryos develop (*see* **Note 14**) they are transferred to P40 medium in a 9 cm petri dish for regeneration and incubated in the light in a culture room or cabinet (light intensity of about 30 μmol/m^2/s, 14 h photoperiod and 27 °C) with the same antibiotics as in P4:10:4:1. Subculture every 3–4 weeks. When shoots or plantlets develop from the embryos they can be transferred to deeper (2 cm) 9 cm petri dishes containing hormone-free P40 agar medium without selection antibiotics and 500 μg/mL Timentin.
3. Plantlets that develop further can be grown in liquid medium in Magenta pots in the same environment. Use a support made from 2 × 9 cm filter papers, folded to form a base sitting on 4 "legs." Use ~8 mL liquid P40 medium with low (1 %) sucrose and 500 μg/mL Timentin (*see* **Note 15**).
4. When plantlets are ready for transfer to soil (this takes at least 3.5 months), i.e., robust shoot and root formation, the roots are gently washed with tap water to remove excess culture medium. The plantlets are placed in a 6 in. pot containing potting mix (coarse sand 1 part, coir peat moss 2 parts and Perlite 1 part). Add nutrient solution to suit experiment requirements. The pot is covered with "Glad® wrap" supported by small bamboo

stakes and the film is removed gradually for 1 week after transfer to potting mix before transfer to the glasshouse. The putative transgenic plants can be analyzed in a number of ways depending on the construct introduced (*see* **Note 16**).

4 Notes

1. For seed germination, seeds are soaked in a small glass beaker containing a small amount of concentrated sulfuric acid for 7 min. Remove acid with a glass Pasteur pipette. Wash thoroughly with tap water. Imbibe until seeds are swollen (can be overnight) and plant. Alternatively other scarifying methods can be used such as piercing seed coat with fine needle, taking care not to damage the embryo.
2. Calcium is made up separately to avoid precipitation with other stocks.
3. With −20 °C storage, aliquoting prevents freeze–thaw cycles, which are best avoided or reduced.
4. This is lower iron than Murashige and Skoog [31] as recommended by Dalton et al. [32] and is made up separately. This prevents precipitation and chelates all the iron. The EDTA must be heated sufficiently to make sure the chelation goes to completion. The color should be straw yellow and pH 2.8.
5. When making up hormone solutions make sure the hormone is dissolved—no very tiny particles at all. NAA and BAP can be autoclaved.
6. Another strain successfully used has been LBA4404 [22] which is less virulent than AGL1 and transformation rates are not as high [27] but is less likely to lead to *Agrobacterium* overgrowth if this becomes a problem. The EHA101 strain has also been successfully used [21].
7. The *A. tumefaciens* can be streaked on to plates on YEP medium with the appropriate antibiotics and stored at 4 °C for 2–3 weeks.
8. For agronomic use of transgenics there are a range of strategies for producing transformed plants without antibiotic or herbicide resistance if that is an issue.
9. Transformed callus grows more slowly on hygromycin and fastest on phosphinothricin with less effect on embryogenesis. If using hygromycin it is best to start with 15 μg/mL and increase to 20 μg/mL once callusing is clearly established.
10. Single colonies can also be checked for the binary vector by colony PCR.
11. The working end of scalpel and forceps are sterilized in 95 % (v:v) ethanol in a suitable container (such as a Coplin Jar)

drained using a suitable stand so the sterile ends are free in the air, and the ethanol completely evaporated off. Flaming after complete ethanol evaporation, then cooling, is carried out to ensure sterility. Extreme care is required given the flammable ethanol. An alternative is glass bead sterilizers, but we have found these less effective. Leaves can be cut up using autoclaved disposable Chinese take-away food container lids as the cutting surface.

12. Various controls can be set up to check the efficacy of the transformation and inserted gene. A positive control to check normal regeneration is untransformed tissue and a negative control is with the selection antibiotic which should stop callussing. An empty vector without the inserted gene of interest is a control for analysis of regenerated plants.
13. If there are any problems with bacterial overgrowth then we recommend rewashing tissue in 750 μg/mL Augmentin and replating.
14. Embryos may not form on every explant as in untransformed tissue. Some explants will do particularly well and may dominate the plate.
15. Robust root development can be a problem in the last phase of regeneration. We have found that liquid medium as illustrated in Fig. 1b is preferable to agar medium at this stage. If rooting does not occur trim off the base of the shoot using sterile technique and return to fresh Magenta pot.
16. The presence of the construct can be verified by PCR and to ensure the absence of fortuitous Agrobacterium contamination one primer from the vector backbone can be used for PCR and no band should be obtained. Southern hybridization can be used to determine copy number by using a restriction digest where the restriction enzyme has a single cut in the transferred DNA [22]. Reporter genes such as GUS or GFP can also be used as GUS or GFP allow direct visualization of gene expression.

References

1. Barker DG, Bianchi S, Blondon F, Datteé Y, Duc G, Essad S et al (1990) Medicago truncatula, a model plant for studying the molecular genetics of the Rhizobium-legume symbiosis. Plant Mol Biol Rep 8:40–49
2. Cook DR (1999) Medicago truncatula—a model in the making! Curr Opin Plant Biol 2:301–304
3. Rose RJ (2008) Medicago truncatula as a model for understanding plant interactions with other organisms, plant development and stress biology: past, present and future. Funct Plant Biol 35:253–264
4. Young ND, Debellé F, Oldroyd GED, Geurts R, Cannon SB, Udvardi MK et al (2011) The Medicago genome provides insight into the evolution of rhizobial symbioses. Nature 480:520–524
5. Tesfaye M, Silverstein KAT, Bucciarelli B, Samac DA, Vance CP (2006) The Affymetrix Medicago GeneChip® array is applicable for transcript analysis of alfalfa (Medicago sativa). Funct Plant Biol 33:783–788
6. Benedito VA, Torres-Jerez I, Murray JD, Andriankaja A, Allen S, Kakar K et al (2008)

A gene expression atlas of the model legume Medicago truncatula. Plant J 55:504–513

7. Cheng X, Wen J, Tadege M, Ratet P, Mysore KS (2011) Reverse genetics in Medicago truncatula using Tnt1 insertion mutants. In Plant reverse genetics: methods and protocols (ed. A. Pereira). Methods Mol Biol 678:179–190
8. Watson BS, Asirvatham VS, Wang L, Sumner LW (2003) Mapping the proteome of barrel medic (Medicago truncatula). Plant Physiol 131:1104–1123
9. Gallardo K, Le Signor C, Vandekerckhove J, Thompson RD, Burstin J (2003) Proteomics of Medicago truncatula seed development establishes the time frame of diverse metabolic processes related to reserve accumulation. Plant Physiol 133:664–682
10. Imin N, de Jong F, Mathesius U, van Noorden G, Saeed NA, Wang X-D, Rose RJ, Rolfe BG (2004) Proteome reference maps of Medicago truncatula embryogenic cell cultures generated from single protoplasts. Proteomics 4: 1883–1896
11. Imin N, Nizamidin M, Daniher D, Nolan KE, Rose RJ, Rolfe BG (2005) Proteomic analysis of somatic embryogenesis in Medicago truncatula. Explant cultures grown under 6-benzylaminopurine and 1-naphthaleneacetic acid treatments. Plant Physiol 137:1250–1260
12. Szittya G, Moxon S, Santos DM, Jing R, Fevereiro MPS, Moulton V, Dalmay T (2008) High- throughput sequencing of Medicago truncatula short RNAs identifies eight new miRNA families. BMC Genomics 9:593
13. Lelandais-Briére C, Naya L, Sallet E, Calenge F, Frugier F, Hartmann C, Gouzy J, Crespi M (2009) Genome-wide Medicago truncatula small RNA analysis revealed novel microRNAs and isoforms differentially regulated in roots and nodules. Plant Cell 21:2780–2796
14. Chen L, Wang T, Zhao M, Zhang W (2012) Ethylene-responsive miRNAs in roots of Medicago truncatula identified by high-throughput sequencing at whole genome level. Plant Sci 184:14–19
15. Mantiri FR, Kurdyukov S, Lohar DP, Sharapova N, Saeed NA, VandenBosch KA, Rose RJ (2008) The transcription factor MtSERF1 of the ERF subfamily identified by transcriptional profiling is required for somatic embryogenesis induced by auxin plus cytokinin in Medicago truncatula. Plant Physiol 146:1622–1636
16. Jayasena KW, Hajimorad MR, Law EG, Rehman A-U, Nolan KE, Zanker T, Rose RJ, Randles JW (2001) Resistance to Alfalfa mosaic virus in transgenic barrel medic lines containing the virus coat protein gene. Aust J Agric Res 52:67–72
17. Nolan KE, Kurdyukov S, Rose RJ (2009) Expression of the SOMATIC EMBRYOGENESIS RECEPTOR-LIKE KINASE 1 (SERK1) gene is associated with developmental change in the life cycle of the model legume Medicago truncatula. J Exp Bot 60:1759–1771
18. Nolan KE, Rose RJ, Gorst JR (1989) Regeneration of Medicago truncatula from tissue culture: increased somatic embryogenesis using explants from regenerated plants. Plant Cell Rep 8:278–281
19. Rose RJ, Nolan KE, Bicego L (1999) The development of the highly regenerable seed line Jemalong 2HA for transformation of Medicago truncatula—implications for regenerability via somatic embryogenesis. J Plant Physiol 155:788–791
20. Hoffmann B, Trinh TH, Leung J, Kondorosi A, Kondorosi E (1997) A new Medicago truncatula line with superior in vitro regeneration, transformation, and symbiotic properties isolated through cell culture selection. Mol Plant-Microbe Interact 10:307–315
21. Thomas MR, Rose RJ, Nolan KE (1992) Genetic transformation of Medicago truncatula using Agrobacterium with genetically modified Ri and disarmed Ti plasmids. Plant Cell Rep 11:113–117
22. Wang JH, Rose RJ, Donaldson BI (1996) Agrobacterium-mediated transformation and expression of foreign genes in Medicago truncatula. Aust J Plant Physiol 23:265–270
23. Chabaud M, Larsonneau C, Marmouget C, Huguet T (1996) Transformation of barrel medic (Medicago truncatula Gaertn.) by Agrobacterium tumefaciens and regeneration via somatic embryogenesis of transgenic plants with the MtENOD12 nodulin promoter fused to the gus reporter gene. Plant Cell Rep 15: 305–310
24. Thomas MR, Johnson LB, White FF (1990) Selection of interspecific somatic hybrids of Medicago by using Agrobacterium transformed tissues. Plant Sci 69:189–198
25. Nolan KE, Rose RJ (1998) Plant regeneration from cultured Medicago truncatula with particular reference to abscisic acid and light treatments. Aust J Bot 46:151–160
26. Rose RJ, Nolan KE (1995) Regeneration of Medicago truncatula from protoplasts isolated from kanamycin-sensitive and kanamycin-resistant plants. Plant Cell Rep 14:349–353
27. Chabaud M, de Carvalho-Niebel F, Barker DG (2003) Efficient transformation of Medicago truncatula cv. Jemalong using the hypervirulent Agrobacterium tumefaciens strain AGL1. Plant Cell Rep 22:46–51

28. Lazo GR, Stein PA, Ludwig RA (1991) A DNA transformation-competent Arabidopsis genomic library in Agrobacterium. Nat Biotechnol 9:963–967
29. Hellens R, Mullineaux P, Klee H (2000) A guide to Agrobacterium binary Ti vectors. Trends Plant Sci 5:446–452
30. Curtis MD, Grossniklaus U (2003) A gateway cloning vector set for high-throughput functional analysis of genes in planta. Plant Physiol 133: 462–469
31. Murashige T, Skoog F (1962) A revised medium for rapid growth and bioassays with tobacco tissue cultures. Physiol Plant 15: 473–497
32. Dalton CC, Iqbal K, Turner DA (1983) Iron phosphate precipitation in Murasgige and Skoog media. Physiol Plant 57:472–476

Chapter 15

Transient Gene Expression in *Medicago truncatula* Leaves via Agroinfiltration

Kelsey Picard, Robyn Lee, Roger Hellens, and Richard Macknight

Abstract

Transient expression is a powerful method for the functional characterization of genes. In this chapter, we outline a protocol for the transient expression of constructs in *Medicago truncatula* leaves using *Agrobacterium tumefaciens* infiltration. Using quantitative real-time PCR we demonstrate that the infiltration of a construct containing the *LEGUME ANTHOCYANIN PRODUCTION 1* (*LAP1*) transcription factor results in the strong upregulation of key biosynthetic genes and the accumulation of anthocyanin pigment in the leaves after just 3 days. Thus, this method provides a rapid and powerful way to the discovery of downstream targets of *M. truncatula* transcription factors.

Key words Agroinfiltration, *Agrobacterium*, Transient expression, *Medicago truncatula*, Legumes

1 Introduction

Plant transformation is a critical tool for understanding plant biology. However, obtaining stable transformants is time-consuming and labor-intensive. *Medicago truncatula* is an important model legume and is being used extensively by researchers throughout the world. Although a number of different *Medicago* transformation protocols have been established, they are all technically demanding and time-consuming. Most laboratories use methods based on *Agrobacterium* transformation of excised tissues followed by plant regeneration in tissue culture [1–4]. Even in *Arabidopsis*, where a reliable and simple transformation protocol is available, it takes several months to obtain transgenic plants for analysis. In contrast to making transgenic plants, the transient expression of genes provides a rapid and simple alternative for analyzing gene function. The most commonly used transient expression method involves the infiltration of *Agrobacterium* into the leaves. This is done by forcing an *Agrobacterium* suspension through the stomata on the underside of the leave using a needless syringe. Copies of the T-DNA are then transferred from *Agrobacterium* into

Ray J. Rose (ed.), *Legume Genomics: Methods and Protocols*, Methods in Molecular Biology, vol. 1069,
DOI 10.1007/978-1-62703-613-9_15, © Springer Science+Business Media, LLC 2013

the leaf parenchyma cells. Although only a tiny proportion of these T-DNA copies integrate into the plant chromosomes, the non-integrated T-DNAs are transiently expressed for several days [5]. The key advantage of transient assays over stable transgenic plants is that experimental results can be generated after a few days, rather than many months.

Agroinfiltration has been used to address a range of biological questions. For example, it provides an excellent way of identifying and/or characterizing transcription factors that regulate specific promoters [6–8]. Other aspects of gene regulation can also be investigated, such as the noncanonical translational initiation [9] or the role of 5′UTR introns on gene expression [10]. These experiments have taken advantage of a binary vector specifically designed for use in transient assays [7]. This vector contains a multiple cloning site into which regulatory sequences of interest can be cloned and thereby fused to the firefly luciferase reporter gene. Within the T-DNA is the CaMV 35S promoter driven renilla luciferase reporter gene that acts as internal control to account for variation in expression between infiltrations [7]. The relative activity of the two luciferases can be determined using commercial dual luciferase assay kits. Agroinfiltration has also been used to identify protein–protein interactions [11], examine protein localization [12], and study host–pathogen interactions [13].

Agroinfiltration of *Nicotiana benthamiana* leaves is widely used by researchers for transient assays, as it provides consistent and high expression of constructs. Other plants have been used with variable success, and often optimization has been needed. For example, agroinfiltration has been demonstrated using leaves of tobacco [6], tomato [14], potato [15], lettuce [14], grapevine [16], grapefruit [17], the medicinal plant *Maesa lanceolata* [18], and *Arabidopsis* [14]. Plant tissue other than leaves has also been used; for example, in strawberry, agroinfiltration of fruit has been proved successful [19, 20], and in *Arabidopsis* a method has been developed using seedlings [21].

In this chapter, we outline a simple protocol for the transient expression of genes in *Medicago truncatula*. *M. truncatula* is commonly used as a model plant species for the study of genes in legumes. Unfortunately, the production of stable *Medicago* transformants is difficult and time consuming. Transient expression of genes in *N. benthamiana* is well-established and adapting the method for use in *Medicago* provides a more relevant tool for researchers of legume species. Although in our experiments we observed an approximately tenfold lower level of expression in this species compared with *N. benthamiana*, a powerful use of the *M. truncatula* transient assay system is to identify the direct targets of a particular transcription factor. Using quantitative real-time PCR of infiltrated leaves, we demonstrate how this method can be used to identify downstream targets of MtLAP1, a MYB transcription factor involved in the regulation of the anthocyanin pathway [22]. Infiltration of *Agrobacterium tumefaciens* containing *35S:LAP1* construct resulted in high *LAP1* expression in the *M. truncatula* leaves 3 days later (Fig. 1). These leaves were red indicating that the entire biosynthetic pathway has been upregulated resulting in

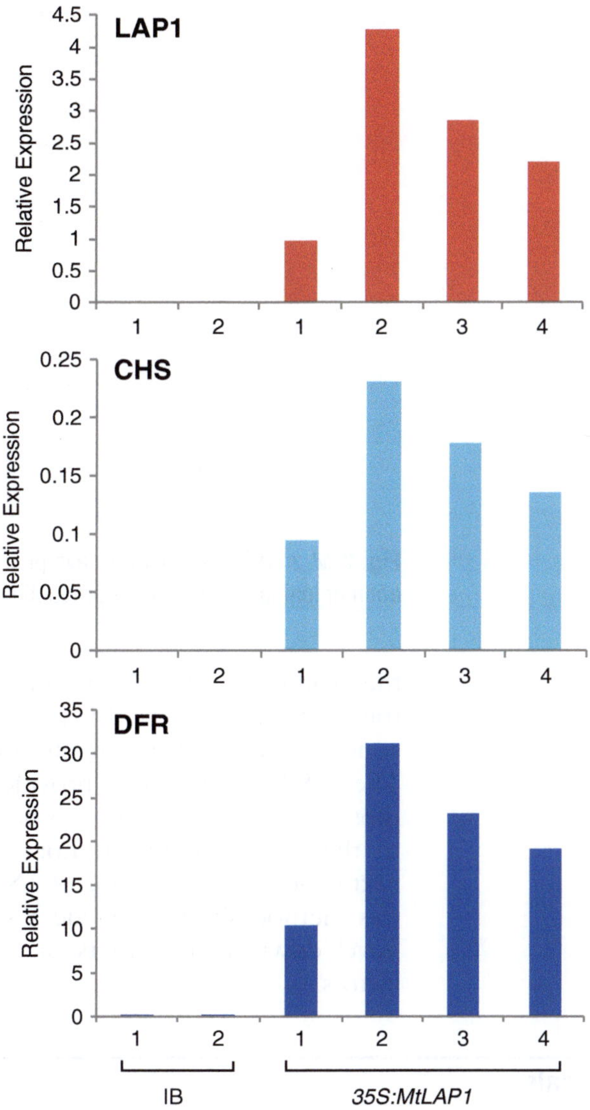

Fig. 1 Quantitative RT-PCR results showing expression of *MtLAP1* and the anthocyanin biosynthetic pathway genes *CHS* (chalcone synthase) and *DFR* (dihydroflavonol reductase). Each sample consisted of three leaflets that were either infiltrated with Infiltration Buffer only (IB, two samples) or Infiltration Buffer containing *Agrobacterium tumefaciens* with the *35S:MtLAP1* construct (*35S:MtLAP1*, four samples). The RNA was isolated 3 days after infiltration. Expression is relative to the housekeeper gene *MtPDF2* (Protodermal Factor 2)

Fig. 2 *M. truncatula* leaflet 3 days post *35S:MtLAP1* agroinfiltration. *Red* coloration of leaf is caused by the production of anthocyanin

the production of anthocyanin pigment (Fig. 2). As expected, key transcriptional targets of LAP1 (chalcone synthase and dihydroflavonol reductase) were highly upregulated in the leaves (Fig. 1). The *35S:LAP1* construct provides an excellent positive control for those wishing to use this protocol and can be obtained by writing to the corresponding author. By combining this method with next-generation sequencing of RNA transcripts or microarray analysis, this method should provide a simple and rapid way to discover novel downstream targets of other *M. truncatula* transcription factors.

2 Materials

2.1 Seeds and Plant Growth Materials

1. *Medicago truncatula* cv R108.
2. Plastic pots (~5 × 5 cm) and labels.
3. Standard seed raising mix.
4. Permanent marker pen.
5. 1,000 grit fine sand paper.

6. Petri dish.
7. Greenhouse or growth room space for plant culture and maintenance.

2.2 Agrobacterium Strains and Plasmids

Agrobacterium tumefaciens GV3101 containing *35S:MtLAP1* in the pB2GW7 binary vector was used as an example in this chapter. This vector provides an excellent positive control and is available from the corresponding author, RCM, on request.

2.3 Media, Buffers and Solutions

1. Luria Broth (LB) medium: 10 g bactotryptone/peptone, 5 g yeast extract, 10 g NaCl into 1 L water. For solid LB agar, add 15 g agar per liter. pH 7.0. Autoclave at 120 °C.
2. Antibiotics (1,000× stock): rifampicin 50 μg/mL, gentamicin 25 μg/mL, spectinomycin 100 μg/mL.
3. 100 mM acetosyringone (3′,5′-Dimethoxy-4′-hydroxy-acetophenone; 1,000× stock): dissolve 196 mg in 10 mL dimethylsulfoxide (DSMO), filter-sterilize.
4. Infiltration Buffer (10× stock): 1 mL of 5 M NaCl, 5 mL 175 mM $CaCl_2$, 125 μL Tween-20, made up to 50 mL with water. Autoclaved at 120 °C.
5. Working solution of Infiltration Buffer: 5 mL 10× Infiltration buffer, 50 μL 100 mM acetosyringone, made up to 50 mL with sterile water (can be stored at 4 °C for ~1 week).
6. Wire loop.
7. Sterile 15 mL screw cap culture tubes.
8. Spectrophotometer and cuvettes.
9. 1 mL tuberculin needleless syringes (Becton, Dickinson and Company).

2.4 RNA Isolation

1. Plant RNA Purification Reagent (Invitrogen).
2. Liquid nitrogen.
3. Small (62 × 75 mm) snap-lock plastic bags.
4. RNase-free microcentrifuge tubes.
5. 5 M NaCl (RNase-free).
6. Chloroform.
7. Isopropyl alcohol.
8. 75 % ethanol.
9. RNase-free Water.

2.5 cDNA Synthesis

1. 1.5 mL RNase-free microcentrifuge tubes.
2. 10× DNaseI Reaction buffer (Invitrogen).
3. DNaseI (Invitrogen).
4. RNase-free water.

Table 1
Primers used for RT-PCR

Primer name	Sequence (5′–3′)
MtLAP1_F	GAAATAGATGGTCATTGATTGCTGG
MtLAP1_R	GGAATACTATTTTTCCCCTTCAACC
MtCHS_F	CCACGACACCATCCTAAATTGTATC
MtCHS_R	TGGTGTGACTAATGCCTTTTTGAC
MtDFR1_F	TGTCCCTATGGAAGGCTGAC
MtDFR1_R	TTCATTCTCAGGGTCCTTGG
MtPDF2_F	GTGTTTTGCTTCCGCCGTT
MtPDF2_R	CCAAATCTTGCTCCCTCATCTG

5. 25 mM EDTA (Invitrogen).
6. 50 μM oligo(dT)$_{20}$ (Invitrogen).
7. First Strand Buffer (Invitrogen).
8. 0.1 M DTT (Invitrogen).
9. 10 mM dNTPs (Invitrogen).
10. RNaseOUT (Invitrogen).
11. Superscript III RT (Invitrogen).

2.6 Quantitative Real-Time-PCR

1. LightCycler® 480 Multiwell Plate 96 (Roche).
2. LightCycler® 480 SYBR Green 1 Master (Roche).
3. Primers specific to the transgene (*see* Table 1).
4. Roche Lightcycler LC480.

3 Methods

3.1 Medicago truncatula Plant Growth

1. Scrape *Medicago truncatula* seeds along sandpaper to scarify the seed coat. A single line of abrasion is all that is required. Germinate seeds by immersing them in water in a petri dish and incubating overnight at room temperature.
2. Once germinated, vernalize seeds in droplets of water at 4 °C for 2 weeks, replenishing with chilled water as required.
3. Plant seedlings in 5 × 5 cm plastic pots containing standard seed raising potting mix and grow at 20 °C, 70 % humidity, under a 16 h light and 8 h dark photoperiod with 74 μmol/m^2 s light intensity.

Fig. 3 Optimal growth stage of *M. truncatula* used for agroinfiltration. Leaves numbered 1–4 are ideal for infiltration. Older leaves (5–8) are more difficult to infiltrate and may senesce before they can be assayed

4. Three weeks after germination the plant will be at the optimal developmental stage. At this stage the plant should have about five fully expanded leaves (Fig. 3). In older plants leaves become toughened and are more difficult to infiltrate and/or demonstrate lower levels of transient expression.

3.2 Preparation of *Agrobacterium tumefaciens* Suspension

1. Transform competent *Agrobacterium* with an expression cassette of your gene of interest in a binary vector. We have used *Agrobacterium tumefaciens* strain GV3101 and the *Medicago* gene LAP1 in the binary vector pB2GW7.
2. Grow the transformed *Agrobacterium* on an LB plate with selective antibiotics (for LAP1-pB2GW7 these are rifampicin, gentamicin and spectinomycin) at 28 °C for 2 days.
3. Suspend a wire loopful (about 1 mm^3) of *Agrobacterium* into ~4 mL of 1× Infiltration Buffer.

4. Incubate at room temperature for 2 h, inverting occasionally to suspend the *Agrobacterium* in the buffer.
5. Measure the OD_{600nm} of 1 mL of sample in a cuvette with a spectrophotometer. The OD_{600nm} should be in the range of 1.0–2.5. Dilute with 1× Infiltration Buffer until the OD_{600nm} is around 1.0–1.2.

3.3 Agroinfiltration

1. For optimal infiltration select healthy trifoliate leaves that are fully expanded and still soft on the underside.
2. Support the leaf with your finger and, using a 1 mL needleless syringe containing your *Agrobacterium* solution, gently apply pressure to the underside of the leaf and slowly depress the syringe to infiltrate the plant tissue (Fig. 4a) (*see* **Notes 1** and **2**). You may need to apply more or less pressure to ensure the liquid is being taken up by the stomata (*see* **Notes 3** and **4**). The leaf will take on a water-soaked appearance as the solution is successfully taken up (Fig. 4b).
3. Ensure all three leaflets are completely infiltrated. This may take a few attempts (*see* **Note 5**).
4. As a control, infiltrate a set of leaves with Infiltration Buffer only.
5. Ensure you label the leaves you are infiltrating so you can easily identify them when sampling. For example, use a small pipette-tip to punch a small hole in the end of the leaflet, make a mark on the stem with a permanent marker pen, or tie tape around the stem.
6. Keep the plants at 20 °C with 70 % humidity, under a 16 h light and 8 h dark photoperiod, and shield them from direct light by placing them under a table, or covering them with a sheet of plastic for 12–24 h to prevent light damage of the infiltrated leaves. Allow the plants to grow for 3 days before assaying (*see* **Note 6**).

3.4 Plant RNA Isolation

This protocol is based on that provided with the Invitrogen Plant RNA Purification Reagent.

1. Combine three leaflets in a microcentrifuge tube and freeze in liquid nitrogen. If you do not wish to isolate RNA straight away, samples can be stored at −80 °C.
2. Add 0.5 mL of cold (4 °C) Plant RNA Reagent into a small snap lock bag.
3. Add the frozen plant tissue and grind by rolling a 30 mL glass test tube over the bag until the tissue is homogenized.
4. Pipette the sample into a fresh 1.5 mL microcentrifuge tube and incubate for 5 min at room temperature.

Fig. 4 Infiltration of *Agrobacterium* suspension into *M. truncatula* leaflets. (**a**) *Agrobacterium* suspension being infiltrated into a leaflet using a needleless syringe. (**b**) The underside of the leaflet turns darker as the *Agrobacterium* is taken up

5. Clarify the solution by centrifuging for 2 min at 12,000 × *g* in a microcentrifuge at room temperature.
6. Transfer the supernatant to an RNase-free tube.
7. Add 0.1 mL of 5 M NaCl to the clarified extract and mix by tapping the tube.
8. In a fume hood, add 0.3 mL of chloroform. Mix thoroughly by inversion or by vortexing for 30 s.
9. Centrifuge the sample at 4 °C for 10 min at 12,000 × *g* to separate the phases. Transfer the upper RNA-containing, aqueous phase into an RNase-free tube.

10. To this, add an equal volume of isopropyl alcohol. Mix by inverting several times and incubate at room temperature for 10 min.
11. Centrifuge the sample at 4 °C for 10 min at 12,000 × *g*.
12. Aspirate the supernatant, taking care not to dislodge the pellet, and add 1 mL of ice cold 75 % ethanol to wash the pellet. Centrifuge for 1 min and then remove the ethanol. Briefly centrifuge to collect residual liquid and remove it with a pipette.
13. Allow any remaining ethanol to evaporate off at room temperature. Any remaining ethanol is likely to interfere with downstream applications.
14. Add 10–30 μL RNase-free water to the RNA pellet and incubate on ice while the RNA dissolves. If any cloudiness is observed, centrifuge the solution at room temperature for 1 min at 12,000 × *g* and transfer the supernatant to a fresh centrifuge tube. Store at −80 °C.

3.5 cDNA Synthesis

1. DNaseI treat the cDNA by adding the following to an RNase-free tube: 1 μg RNA, 1 μL 10× DNaseI Reaction buffer, 1 μL DNaseI (1U/μL), made up to 10 μL using RNase free water.
2. Incubate tubes at room temperature for 15 min. Add 1 μL of 25 mM EDTA to the reaction. Heat at 65 °C for 10 min to inactivate the DNaseI. The RNA is now ready for use in reverse transcription or can be stored at −80 °C until needed.
3. Add 1 μL of 50 μM oligo$(dT)_{20}$ to the tube containing 10 μL of DNaseI-treated RNA, heat at 65 °C for 5 min then incubate on ice for 2 min.
4. Briefly centrifuge contents to bottom of the tube and add: 4 μL First Strand Buffer, 1 μL 0.1 M DTT, 1 μL 10 mM dNTPs, 1 μL RNaseOUT (40 U/μL), 1 μL Superscript III RT (200 U/μL). Mix by pipetting gently up and down.
5. Incubate at 50 °C for 30–60 min.
6. Inactivate the reaction by incubating at 80 °C for 15 min.
7. The cDNA is ready for amplification or can be stored at −20 °C (or −80 °C for long term storage). We routinely dilute cDNA 1/30 with water prior to amplification.

3.6 Quantitative Real-Time PCR

1. Add the following to a multiwell plate; 5 μL 2× SYBR GREEN I master mix, 0.5 μL 10 mM primer 1, 0.5 μL 10 mM primer 2, 1 μL RNase-free water. Spin briefly to collect the contents at the bottom of the plate. The primers used in the example shown in Fig. 1 are provided in Table 1.
2. Add 3 μL of diluted cDNA, seal the plate and spin briefly to collect the contents at the bottom of the plate.

3. We perform quantitative real-time PCR reaction using the Roche Lightcycler 480 (PCR program: 5 min at 95 °C followed by 50 cycles of: 95 °C for 5 s, 58 °C for 5 s and 72 °C for 10 s. A final denaturation cycle of 95 °C for 30 s, 65 °C for 5 s then 95 °C was performed to generate the melt curve).

4 Notes

1. Wear safety glasses and gloves when performing syringe infiltrations. The *Agrobacterium* solution can squirt out sideways if the seal between the syringe and the leaf is not good.
2. Practice infiltrating by using Infiltration Buffer and spare plants to develop your skill and confidence.
3. Infiltration may be improved by the leaf being held taught between the thumb and first finger. It may also help to gently rub your finger across the back of the leaf first to remove any fine hairs to improve infiltration.
4. Be careful not to push too hard as the leaf tissue will become bruised and even tear. This damage can often cause the leaflets to die and fall off and your sample will be lost.
5. Three leaflets are combined for each RNA isolation, and thus, nine leaflets provides three RNA replicates. To reduce variation, it is best to combine leaflets from different leaves for RNA isolation.
6. Expression levels may increase up to about 7 days post infiltration. After this, endogenous silencing mechanisms may reduce expression [11].

Acknowledgments

We thank members of the Macknight laboratory, especially Wen Hann Bong, Jared Fudge, and Manda Safavi for their help in optimizing these protocols and Jane Campbell for technical assistance and plant care. This research was supported by the New Zealand Marsden Fund.

References

1. Trieu AT, Burleigh SH, Kardailsky IV, Maldonado-Mendoza IE, Versaw WK, Blaylock LA et al (2000) Transformation of *Medicago truncatula* via infiltration of seedlings or flowering plants with *Agrobacterium*. Plant J 22:531–541
2. Crane C, Dixon RA, Wang Z-Y (2006) *Medicago truncatula* transformation using root explants. Methods Mol Biol 343: 137–142
3. Wright E, Dixon RA, Wang Z-Y (2006) *Medicago truncatula* transformation using cotyledon explants. Methods Mol Biol 343: 129–135
4. Cosson V, Durand P, d'Erfurth I, Kondorosi A, Ratet P (2006) *Medicago truncatula* transformation using leaf explants. Methods Mol Biol 343:115–127

5. Jones HD, Doherty A, Sparks CA (2009) Transient transformation of plants. Methods Mol Biol 513:131–152
6. Yang Y, Li R, Qi M (2000) *In vivo* analysis of plant promoters and transcription factors by agroinfiltration of tobacco leaves. Plant J 22:543–551
7. Hellens RP, Allan AC, Friel EN, Bolitho K, Grafton K, Templeton MD et al (2005) Transient expression vectors for functional genomics, quantification of promoter activity and RNA silencing in plants. Plant Methods 1:13
8. Dare AP, Schaffer RJ, Lin-Wang K, Allan AC, Hellens RP (2008) Identification of a *cis-regulatory* element by transient analysis of co-ordinately regulated genes. Plant Methods 4:17
9. Simpson GG, Laurie RE, Dijkwel PP, Quesada V, Stockwell PA, Dean C, Macknight RC (2010) Noncanonical translation initiation of the Arabidopsis flowering time and alternative polyadenylation regulator *FCA*. Plant Cell 22:3764–3777
10. Chung BYW, Simons C, Firth AE, Brown CM, Hellens RP (2006) Effect of 5'UTR introns on gene expression in *Arabidopsis thaliana*. BMC Genomics 7:120
11. Walter M, Chaban C, Schütze K, Batistic O, Weckermann K, Näke C et al (2004) Visualization of protein interactions in living plant cells using bimolecular fluorescence complementation. Plant J 40:428–438
12. Goodin MM, Dietzgen RG, Schichnes D, Ruzin S, Jackson AO (2002) pGD vectors: versatile tools for the expression of green and red fluorescent protein fusions in agroinfiltrated plant leaves. Plant J 31:375–383
13. Ma L, Lukasik E, Gawehns F, Takken FLW (2012) The use of agroinfiltration for transient expression of plant resistance and fungal effector proteins in *Nicotiana benthamiana* leaves. Methods Mol Biol 835:61–74
14. Wroblewski T, Tomczak A, Michelmore R (2005) Optimization of *Agrobacterium-mediated* transient assays of gene expression in lettuce, tomato and *Arabidopsis*. Plant Biotechnol J 3:259–273
15. Bhaskar PB, Venkateshwaran M, Wu L, Ane J-M, Jiang J (2009) *Agrobacterium*-mediated transient gene expression and silencing: a rapid tool for functional gene assay in potato. PLoS One 4:e5812
16. Santos-Rosa M, Poutaraud A, Merdinoglu D, Mestre P (2008) Development of a transient expression system in grapevine via agro-infiltration. Plant Cell Rep 27:1053–1063
17. Figueiredo JFL, Romer P, Lahaye T, Graham JH, White FF, Jones JB (2011) *Agrobacterium-mediated* transient expression in citrus leaves: a rapid tool for gene expression and functional gene assay. Plant Cell Rep 30:1339–1345
18. Faizal A, Geelen D (2012) Agroinfiltration of intact leaves as a method for the transient and stable transformation of saponin producing *Maesa lanceolata*. Plant Cell Rep 31: 1517–1526
19. Spolaore S, Trainotti L, Casadoro G (2001) A simple protocol for transient gene expression in ripe fleshy fruit mediated by *Agrobacterium*. J Exp Bot 52:845–850
20. Hoffmann T, Kalinowski G, Schwab W (2006) RNAi-induced silencing of gene expression in strawberry fruit (Fragaria x ananassa) by agroinfiltration: a rapid assay for gene function analysis. Plant J 48:818–826
21. Li J-F, Park E, Arnim AV, Nebenführ A (2009) The FAST technique: a simplified *Agrobacterium*-based transformation method for transient gene expression analysis in seedlings of *Arabidopsis* and other plant species. Plant Methods 5:6
22. Peel GJ, Pang Y, Modolo LV, Dixon RA (2009) The LAP1 MYB transcription factor orchestrates anthocyanidin biosynthesis and glycosylation in *Medicago*. Plant J 59: 136–149

Chapter 16

Subcellular Localization of Transiently Expressed Fluorescent Fusion Proteins

David A. Collings

Abstract

The recent and massive expansion in plant genomics data has generated a large number of gene sequences for which two seemingly simple questions need to be answered: where do the proteins encoded by these genes localize in cells, and what do they do? One widespread approach to answering the localization question has been to use particle bombardment to transiently express unknown proteins tagged with green fluorescent protein (GFP) or its numerous derivatives. Confocal fluorescence microscopy is then used to monitor the localization of the fluorescent protein as it hitches a ride through the cell. The subcellular localization of the fusion protein, if not immediately apparent, can then be determined by comparison to localizations generated by fluorescent protein fusions to known signalling sequences and proteins, or by direct comparison with fluorescent dyes. This review aims to be a tour guide for researchers wanting to travel this hitch-hiker's path, and for reviewers and readers who wish to understand their travel reports. It will describe some of the technology available for visualizing protein localizations, and some of the experimental approaches for optimizing and confirming localizations generated by particle bombardment in onion epidermal cells, the most commonly used experimental system. As the non-conservation of signal sequences in heterologous expression systems such as onion, and consequent mis-targeting of fusion proteins, is always a potential problem, the epidermal cells of the *Argenteum* mutant of pea are proposed as a model system.

Key words *Argenteum* mutant, Confocal microscopy, Fluorescence microscopy, Fluorescent fusion proteins, Gene gun, Green fluorescent protein, Live cell imaging, Onion epidermis, Particle bombardment, Tracer dyes, Transient gene expression

1 Introduction

With the full genomes of *Arabidopsis* and numerous other plants, including the legumes *Medicago*, *Lotus*, *Glycine* (soybean), and *Cajanus* (pigeon pea), now available, the task of understanding the functions of the uncovered genes remains to be completed. Critical to understanding a protein's function is the identification of its localization(s) inside the plant cell. Various annotation routines and prediction algorithms that indicate localizations have been proved and improved through matching the analyses to proteomics

Ray J. Rose (ed.), *Legume Genomics: Methods and Protocols*, Methods in Molecular Biology, vol. 1069,
DOI 10.1007/978-1-62703-613-9_16,

of subcellular fractions, but direct experimental verification of subcellular localization is also critical. As antibody localizations are complicated by issues of isoform specificity, and by the requirement to develop, screen, purify, and test antisera, the last decade has seen the increased use of fluorescent fusion proteins. Green fluorescent protein (GFP) and its many derivatives [1, 2] can be tagged to proteins of interest and their localization inside the cell determined by confocal microscopy following transient expression.

Various approaches exist for the transient expression of fluorescent fusion proteins in plants, including viral expression systems [3], protoplast permeabilization [4], and high-throughput *Agrobacterium*-based systems [5–8]. When analyzed by the number of published studies, however, the primary way in which transient expression has been conducted is through particle bombardment, a process also known as biolistic expression or more colloquially as the gene gun. This technique has mainly transformed the inner epidermal cells of onion bulbs, primarily to analyze protein localizations, with more than a thousand papers published that use the approach. However, microscopic analysis of fluorescent proteins with known localization sequences has generated a greater understanding of the organelles and structures in plant cells. For example, tagging GFP with an N-terminal endomembrane targeting sequence along with the C-terminal endoplasmic reticulum (ER) retention tetrapeptide-HDEL marks the ER, demonstrating its dynamics [9, 10] but also shows the complexity of the organization of the nuclear envelope [11]. Such baseline data on cell organization is not only important for understanding cell physiology, but becomes critical for understanding the localization of unknown proteins. More recently, particle bombardment has been used to introduce fluorescent fusion proteins into cells in order to investigate protein–protein interactions using methods such as FRET (fluorescence resonance energy transfer) and BiFC (bimolecular fluorescence complementation). These processes will be mentioned briefly under Subheading 1.2, but a detailed discussion of them lies beyond the scope of this review.

1.1 Microscopy Hardware

The fundamental importance of microscopic technique cannot be overemphasized in the experimental localization of fusion proteins. Fluorescence microscopy, although conceptually simple, can generate artifacts if not conducted correctly [12, 13]. Optimal imaging requires an understanding of the microscopy hardware being used, of the fluorescent protein being imaged [12, 14–16], and knowledge of the cell biology of the plant being viewed.

Several of the different ways in which fluorescence microscopes are used to view fluorescent fusion proteins, and how this can affect the resulting images, are demonstrated in the hypocotyls of *Arabidopsis* seedlings stably expressing mitochondrial-targeted

GFP and YFP, and photographed using conventional fluorescence microscopy in which the colors of the excitation and emission light are determined by coated filters (Fig. 1d–g). With blue excitation, both GFP and YFP appear green, although of different shades (Fig. 1d, f) with the major difference in appearance depending on whether longpass (LP) or bandpass (BP) emission filters are used. As the name implies, an LP filter transmits light at wavelengths above a set value. For the blue excitation used with GFP and YFP, a typical LP emission filter would transmit light above 500 nm. However, as blue excitation also excites chlorophyll, the LP filter also transmits red chlorophyll autofluorescence at wavelengths greater than 640 nm (Fig. 1d, f). BP filters transmit only a set window of light (e.g., between 500 and 550 nm) and as they eliminate chlorophyll fluorescence, they are often described as a "chlorophyll-blocking" filters (Fig. 1e, g). While the ability to block chlorophyll autofluorescence might suggest that BP filters are superior for observing GFP and YFP, other factors also need to be considered. When epidermal cells die, either from natural causes or associated with gene gun bombardment and transient expression, their cytoplasm becomes condensed and granular, and they fluoresce yellow [16]. This color difference is immediately evident by eye when using an LP filter (Fig. 1h, arrow) but can be hidden when using the BP emission filter because everything appears a similar shade of green (Fig. 1i). Transmitted light images are also important for observing transformed cells, as dead cells will have a grainy appearance (Fig. 1j, arrow).

Confocal fluorescence microscopy is an optical technique that eliminates the out-of-focus light present in conventional microscope images by focussing light through a pinhole (Fig. 1b, c). Although the actual resolution achievable in confocal microscopy is not significantly improved compared to conventional microscopy, the elimination of out-of-focus light results in a large increase in the amount of fine detail that can be observed in cells expressing fluorescent proteins [14]. More importantly, multiple fluorescent images can be recorded concurrently at different emission wavelengths or colors, along with transmitted light images that are non-confocal and which contain out-of-focus light (Fig. 1b). A further, more recent advance is the Leica SP5 confocal system in which a prism separates the light collected by the objective lens and a series of calibrated mirrors and stops allows for specific tuning of emission wavelengths (Fig. 1c). This system is particularly useful when assaying fluorescent fusion proteins because of its ability to record fluorescence emission spectra at different locations within images. The flexibility of modern confocal systems can, however, be daunting for new users, and care needs to be taken that appropriate excitation and emission wavelengths are selected, and that all images are optimized.

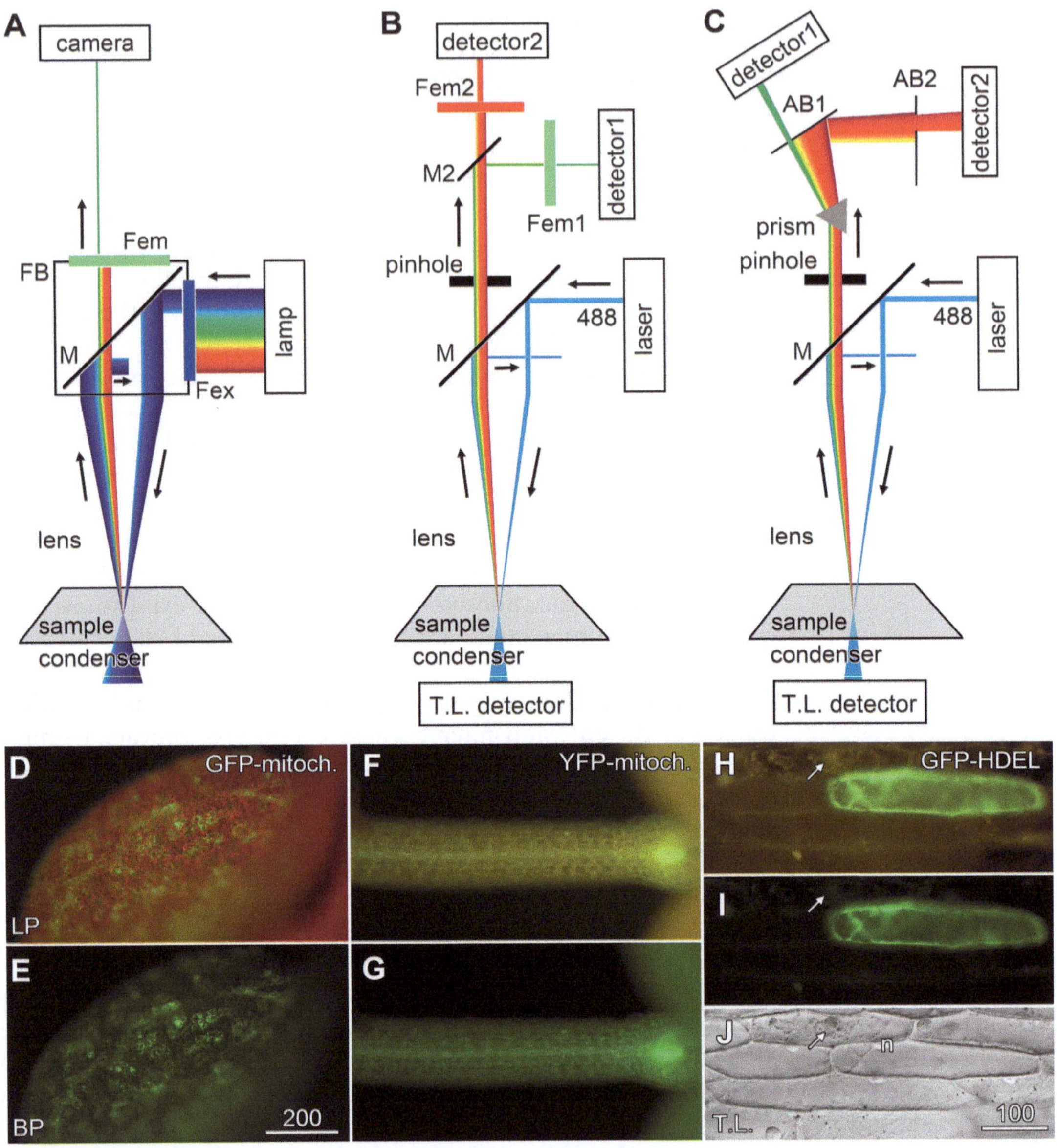

Fig. 1 (**a**–**c**) Light path diagrams for imaging GFP and excluding red chlorophyll fluorescence with conventional fluorescence (**a**) and confocal microscopes (**b**, **c**). (**a**) Conventional fluorescence uses a filter block (FB) composed of excitation and emission filters (Fex and Fem) and a dichroic mirror (M) that reflects at specific wavelengths but which transmits lights of other colors. For GFP imaging, Fex selects blue light from a white light lamp which is directed through the lens onto the sample by M. The lens collects both reflected light and fluorescence, but only the fluorescence passes through M and this is further cleaned of any excitation light by Fem. Single images are collected by a camera: to record different colors, a different filter block would be required. In the example shown, Fem is a bandpass (BP) filter which only transmits green light with chlorophyll fluorescence being filtered out. Many fluorescence microscopes, however, contain longpass (LP) emission filters which would transmit green and red fluorescence. (**b**) As lasers with set wavelengths (e.g., the 488 nm blue line from an argon laser) replace the lamp as a light source in confocal microscopy, Fex is no longer required. Fluorescence passes through a pinhole, removing out-of-focus light and generating optical sections. A combination of further dichroic mirrors (M2) and multiple emission filters (Fem1, Fem2) means

1.2 Fluorescent Proteins

The diversity of fluorescent proteins now available to researchers often appears daunting even to those familiar with conventional GFPs. This diversity reflects not only the basic spectral properties of the original fluorescent proteins, and the ways in which excitation and emission spectra have been modified for laboratory use, but also the various ways in which the chemistry of the different fluorescent proteins has been tinkered with. A guide to selecting fluorescent proteins has been published [1], and descriptions of some of the more common fluorescent proteins are included in Table 1.

GFP was initially isolated from the jellyfish *Aequoria victoria*, although related proteins with diverse spectral characteristics occur in a broad range of marine organisms [2, 17]. Early, conventional GFPs were based on the wild-type protein, and were excitable with both ultraviolet and blue light, although the UV peak was predominant. As UV excitation wavelengths can be more harmful for cells at the light intensities used for fluorescence microscopy than lower energy blue light, one way in which GFP was optimized for laboratory use was through the elimination of the UV excitation peak, and the promotion of blue excitation. The resultant protein contained numerous different point mutations, and was considerably brighter than the wild-type protein. Furthermore, its fluorescence was optimized for standard blue excitation/green emission filter sets on fluorescence microscopes that were then available. Several of the resulting fluorescent proteins, including EGFP (for "enhanced") and EYFP (yellow fluorescent protein), are still commonly used although the fluorescence from YFP is not especially yellow (Fig. 1f). Superior fluorescent proteins such as Emerald (EmGFP), mCitrine and Venus have now become available. Although conventional

Fig. 1 (continued) that multiple fluorescence images can be recorded concurrently by different detectors. Laser light passing through the sample can also be collected by the condenser and imaged with a transmitted light detector (T.L. detector) to give a non-confocal transmitted light image that retains out-of-focus light. (**c**) In Leica spectral confocal microscopes, Fem is replaced by a prism that separates the light collected by the lens into its component wavelengths. A series of adjustable and calibrated barrier mirrors (AB) allow precise wavelengths of light to be collected by detectors, giving more flexibility in imaging and allowing for spectral scanning. (**d–j**) Observations of plant material expressing fluorescent proteins observed by conventional (non-confocal) microscopy through LP (**d**, **f**, **h**) and BP filters (**e**, **g**, **i**). *Arabidopsis* plants stably expressing GFP targeted to mitochondria (imaged in cotyledons) (**d**, **e**) and YFP targeted to mitochondria (in the hypocotyl) (**f**, **g**). The LP filter allowed chlorophyll fluorescence to pass and discrimination between GFP and YFP (**d**, **f**) whereas the BP filter blocked chlorophyll but made which GFP and YFP appear similar (**e**, **g**). (**h–j**) Onion epidermal cell expressing ER-targeted GFP (GFP-HDEL) adjacent to a dead cell fluorescing yellow (*arrow*). With the LP filter (**h**), the colors were different but this was not the case with the BP filter where the weak fluorescence from the dead cell was also green (**i**). Transmitted light images (T.L.) demonstrated that the cytoplasm of dead cells appear granular, and that their nuclei were different to those of living cells (n) (**j**). Bar in **e** = 200 μm for **d–g**; bar in **j** = 100 μm for **h–j**

Table 1
Common fluorescent proteins for use as fusion proteins in plants

Name	Usage notes	Source organism	Excitation (nm)[a]	Emission (nm)[a]	Color
GFP—wild-type green fluorescent protein	Dual excitation peaks present	*Aequoria victoria*	396, 470	504	Green
S65T GFP	A single point mutation eliminates the dual excitation peaks	*A. victoria*	489	510	Green
eGFP—enhanced green fluorescent protein	Early commercial fluorescent protein from Clontech	*A. victoria*	488	507	Green
eYFP—enhanced yellow fluorescent protein	Early commercial fluorescent protein from Clontech	*A. victoria*	514	527	Yellow
eCFP—enhanced cyan fluorescent protein	Early commercial fluorescent protein from Clontech	*A. victoria*	433	475	Cyan
Cerulean	Significantly brighter than eCFP	*A. victoria*	433	475	Cyan
Emerald (EmGFP)	Brighter and more stable than EGFP	*A. victoria*	487	509	Green
mCitrine	More stable than EYFP	*A. victoria*	516	529	Yellow
Venus	More stable than EYFP	*A. victoria*	515	528	Yellow
T-sapphire	Large Stokes shift; violet excitation gives green fluorescence	*A. victoria*	399	511	Green
DsRed	Wild-type protein forming tetramers	*Dictyosoma* sp.	558	583	Red
RFP—red fluorescent protein	Monomeric version of DsRed	*D.* sp.	584	607	Red
mOrange	A common, modified version of RFP	*D.* sp.	548	562	Orange
mCherry	A common, modified version of RFP	*D.* sp.	587	610	Red
mPlum	A common, modified version of RFP	*D.* sp.	590	649	Far red

[a]Emission and excitation wavelengths based on published values in review articles [1, 2, 17]

Aequoria GFP has not been mutated to produce red-fluorescent variants, DsRed, the second fluorescent protein to be discovered that was isolated from the soft coral *Dictyosoma* [18], fluoresces an orange to red color when excited with green light (Table 1). The protein, however, forms tetramers making it a poor choice for use as the basis of fusion protein constructs because the resulting proteins aggregate. DsRed was modified to make a monomeric RFP (red fluorescent protein) [19], and a palette of different fluorescent fusion proteins was derived from RFP that bear the names of fruits [20]. One of these, mCherry, where the "m" indicates that the fluorescent protein is a monomer, has been commonly used in plant transformations.

The biochemical and physical properties of fluorescent proteins have also been modified with the development of variants that can be more or less stable, have faster maturation times or respond to their environment [2]. The most important variations for understanding protein localizations and functions are, however, the development of photoactivatable GFP and fluorescence complementation. A range of different photoactivatable fluorescent proteins that become fluorescent or change their color following irradiation with UV/short wavelength light (Table 2) has been identified. These include a variant of the original GFP, isolated from the jellyfish *Aequoria*, and a range of proteins isolated from soft and hard corals. In some cases, as with proteins such as with photoactivatable GFP, Dendra2 [21, 22] and mEos [23], this UV-triggered conversion from a green to a red fluorescent protein is irreversible. However, Dronpa is a fluorescent protein that will reversibly convert between a nonfluorescent and fluorescent form. These photoactivatable fluorescent proteins are ideal for investigating dynamic processes inside the cell. Bimolecular fluorescence complementation (BiFC) is the most dramatic of the modified fluorescent proteins. In this co-expression technique, two proteins are fused with the nonfluorescent N- and C-terminal halves of a fluorescent protein such as GFP. If the two proteins bind to each other, then the two halves of the fluorescent protein can bind and generate a fluorescent complex [24–26]. BiFC analysis shows considerable promise for investigating protein-protein interactions in plant cells, and as a complement for yeast 2-hybrid studies, and has been extensively used in recent years in both *Arabidopsis* protoplasts and bombarded onion epidermal cells [27–29].

1.3 The Gene Gun

Biolistic transformation methods (the "gene gun") [30] were used for the first transient expression of GFP [31, 32]. The ease with which biolistic systems can be used has made the gene gun increasingly popular. Three different approaches exist for biolistic transformation and in each, plasmid DNA-coated gold or tungsten particles (diameter 1.0–1.6 μm) are shot into epidermal cells by a high-pressure pulse of helium (Fig. 2). For transient expression to

Table 2
Common photoactivatable fluorescent proteins for use in plants

Name	Usage notes	Source organism	Initial excitation (nm)[a]	Initial emission (nm)[a]	Switching color	Induced excitation (nm)[a]	Induced emission (nm)[a]
Photoactivatable GFP (PA-GFP)	Irreversible photoconversion. Monomeric.	*Aequoria victoria*	400	515	UV to violet	504	517
Kaede	Irreversible photoconversion. Tetrameric. Replaced by other proteins.	*Trachyphyllia geoffroyi*	508	518	UV to violet	572	580
Dendra2	Irreversible photoconversion. Monomeric.	*Dendronephthya* sp.	490	507	UV to blue	553	573
mEos	Irreversible photoconversion. Monomeric.	*Lobophyllia hemprichii*	505	519	UV to violet	573	584
Dronpa	Reversible photoconversion. Monomeric.	*Pectiniidae* sp.	–	–	UV induces fluorescence while strong blue light quenches fluorescence	503	518

[a]Emission, excitation and photoconversion wavelengths based on published values in review articles [1, 2, 17]

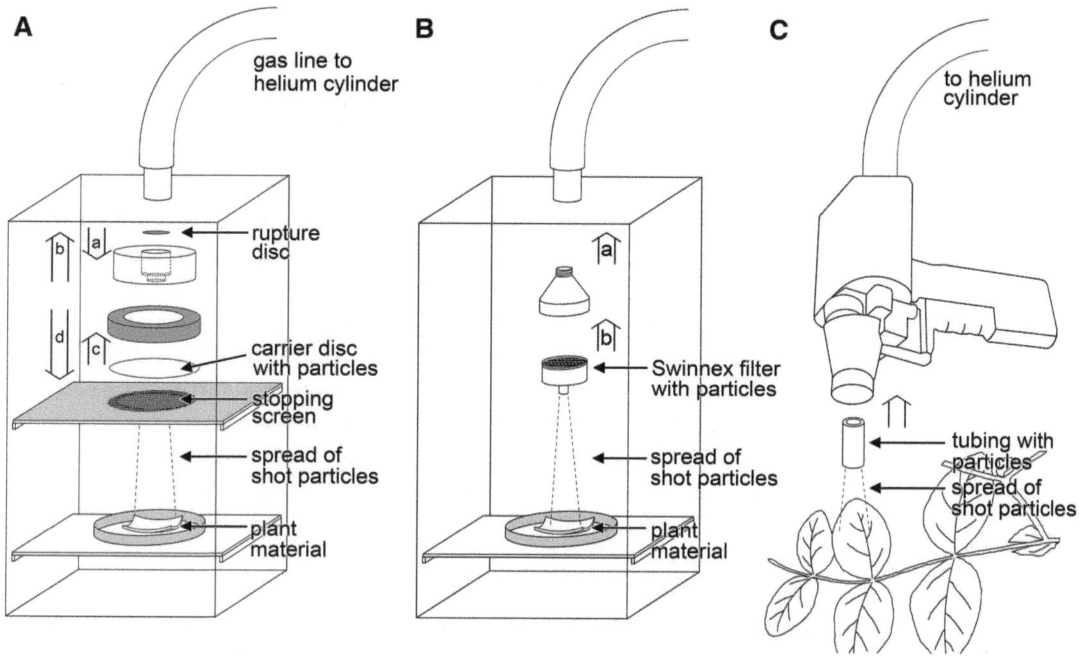

Fig. 2 There are three different types of gene gun used for transformation experiments. The direction for connecting the different components is indicated with *arrows*, and the order with *letters*. (**a**) The Bio-Rad gene gun uses an evacuated chamber. A burst of helium gas caused by the failure of the rupture disk at a predetermined pressure drives a carrier disk with DNA-coated particles against a stopping screen. The particles are released and travel towards the plant tissue. (**b**) Filter-based gene guns also use an evacuated chamber, but the DNA-coated particles are loaded onto a plastic filter and a brief (30–50 ms) pulse of helium gas drives the particles towards the plant tissue. (**c**) The Bio-Rad Helios handheld gene gun is portable and does not require a vacuum, so it can be used to transform whole plants. A pulse of gas drives DNA-coated particles that have been coated onto the inside of a plastic tube towards the plant tissue

occur, gold particles need to lodge within the nucleus. Biolistic transformations are commonly used for transient expression analysis but can also generate stable transformants if the exogenous plasmid DNA is incorporated into the plant genome. As such, it is often used in plant species in which *Agrobacterium* transformations are difficult [33].

The commercially available, vacuum chamber-based gene gun from Bio-Rad (model PDS-1000) (Fig. 2a) is based on an initial design in which a gunpowder charge [30] has been replaced with a gas pulse that is controlled by rupture disks that fail at specific gas pressures between 450 and 2,200 psi. The gas pulse fires a thin sheet of plastic onto which the DNA-coated gold particles have been suspended against a perforated stopping screen, and the gold particles are released to travel to the plant. In alternative chamber-based guns that are often constructed relatively cheaply (Fig. 2b), DNA coated particles are coated into a syringe filter and, under a vacuum, a 30–50 ms pulse of helium fires the particles towards the plant [34]. Bio-Rad also supplies the handheld Helios gene gun in which the DNA-coated particles are coated into the insides of plastic tubes, and these shot directly into the surface of plants (Fig. 2c). As this system does not require the plant material to be placed in a vacuum chamber, it is suitable for transformation of leaves that remain attached to whole plants. Although the three systems all have advantages and disadvantages, they are all suitable for the transient expression of fluorescent fusion proteins in plants.

1.4 Plant Material

High quality fluorescence microscopy of bombarded tissues requires an appropriate target plant, and appropriate handling of the plant material. Foremost among targets have been *Allium* (onion and leek) epidermal cells which have long been model systems for cell biology. Their large, vacuolate cells are particularly amenable to microscopic observations, and the ease of making epidermal peels without disrupting cell structure and function makes viewing the cells more reliable. Moreover, onions and leeks are readily available year round in shops. Since the first biolistic transformation of plant cells, conducted in onions [30], and the first transient expression of GFP, again conducted in onions [31, 32], in excess of 1,000 published studies have used these cells and methods for expression of fluorescent fusion proteins. In some of these studies, however, the cells may have contained over-expression artifacts, while in many cases the published images have been poor. Conclusions drawn from such experiments can be unreliable.

Care is required when selecting other plants for transient expression assays. The relative size of the nucleus becomes important, as does the absence of trichomes that can deflect or block particle entry into the epidermis. Ease of observation is also important. As whole leaves can be difficult if they are uneven or have hairy surfaces, making focussing difficult, epidermal peels can have advantages. For instance, peels lie flat, can be imaged from the mesophyll side without distortions from the cuticle, are easier than whole leaves to image with transmitted light, can more readily be probed with drugs that modulate cell organization, and minimize problems caused by chlorophyll autofluorescence from the mesophyll cells.

Another factor to consider when choosing a species for transient expression is the evolutionary similarity of the system. Assessing the subcellular localization of a protein from a dicot genome in the epidermal cells of the monocot *Allium* makes the assumption that targeting mechanisms will be conserved. This is not always the case: some *Arabidopsis* proteins tagged with GFP are mislocalized when expressed in rice protoplasts [35] and tobacco epidermal cells [6]. This problem has lead to the suggestion “that it is always better to use a homologous system to study protein localization since a heterologous system may result in mis-targeting” [35].

In assessing the localization of proteins in legumes, transient expression should ideally be conducted in the epidermal cells of legumes themselves. To date this has been rare. Examples do, however, exist with GUS (β-glucuronidase) being expressed transiently in epidermal cells of soybean and cowpea [34], while the first reported transformation that labelled microtubules with GFP was in the epidermal cells of fava bean (*Vicia faba*) [36]. There is, therefore, a need to develop legume model systems for transient expression that have the simplicity and ease of use of the onion epidermal system, including the ability to readily isolate the epidermis. The *Argenteum* mutant of pea (*Pisum sativum*) may prove to be extremely useful. In this dominant mutant, the epidermal layers of leaves separate from the underlying mesophyll, creating large air pockets which scatter light and generate a silver coloration [37, 38]. Because the epidermal layers separate so readily from underlying tissue, this mutant has been used for biochemical studies of epidermal cells [39] and, notably, to study stomatal responses [40] but has not yet, apparently, been used for transformations.

2 Materials

2.1 Microscopy Hardware

1. For high quality imaging of subcellular localizations, a confocal microscope is required. Blue excitation (typically 488 nm) is required for excitation of GFP and YFP while green excitation (typically 543 or 561 nm) works for DsRed and RFP. These wavelengths are usually available on most confocal systems. Short wavelength excitation (either ultraviolet at 351 and 361 nm, or more commonly violet at 405 nm) is required for most photoactivatable fluorescent proteins.
2. Where possible, water- or glycerol-immersion lenses should be used to visualize cells. Oil-immersion lenses, while optically good, will often have a working distance that is too short to focus into plant epidermal cells whereas non-immersion objectives will have a small aperture and low image resolution.
3. A fluorescence dissecting microscope with blue and green excitation is often ideal for prescreening transformed material.
4. A color CCD camera attached to both the fluorescence dissecting microscope and to the confocal microscope system is also ideal for recording overall images and, in particular, the color of the fluorescent proteins.

2.2 Fluorescent Proteins and Constructs

1. Plasmid DNA encoding fluorescent proteins fused to characterized targeting sequences and to unknown proteins are purified from *E. coli* liquid cultures in minipreps using the appropriate manufacturers' instructions.

2.3 The Gene Gun

1. A gene gun and vacuum pump (Fig. 2): Gene guns are available commercially from Bio-Rad (Hercules, CA, USA), and include a vacuum-chamber based system (model PDS-1000) and the handheld Helios gene gun. Chamber-based gene guns that use gold particles coated into filters can also be built relatively inexpensively.
2. Gold particles: 1.0 and 1.6 μm diameter particles are available from Bio-Rad.
3. 100 % ethanol and 2.5 M $CaCl_2$ are used in gold particles preparation.
4. Spermidine (Sigma, St Louis, MO, USA) is used to stabilize DNA on gold particles. A 1 M stock solution in water should be stored at −80 °C, and 0.1 M working solutions in water (50 μl aliquots) should also be stored at −80 °C.
5. Peripheral materials for using the PDS-1000 gene gun are available from Bio-Rad and include rupture disks (generally 900 and 1,100 psi disks), stopping screens and DNA carrier disks. 100 % isopropanol is used to wet the rupture disks.
6. Peripherals for using the Helios gene gun are available from Bio-Rad and include Tefzel tubing.
7. Peripherals for using generic gene guns include Swinnex filters (13 mm diameter, Millipore), and 5 % bleach and 100 % ethanol for decontamination.

2.4 Plant Material

1. Onion bulbs and leeks can be purchased from local markets. When used, leeks should stand in water overnight to ensure turgidity.
2. The *Argenteum* mutant of pea, in which the epidermis can be readily pealed, should be available through the USDA's *Pisum* Genetic Stock Collection in Pullman, Washington. Other pea varieties can also be readily peeled, although with more difficulty than this mutant.

2.5 Fluorescent Dyes and Drugs

Drugs and fluorescent dyes that target different organelles and components of the plant cell (Tables 3 and 4) can generally be prepared as dimethyl sulfoxide (DMSO) stock solutions, and stored frozen in the dark until required for experiments.

1. Many dyes used to localize organelles in plant cells are available from Molecular Probes, now part of Life Technologies (Carlsbad, CA, USA). These dyes include acridine orange, BODIPY TR ceramide, CellMask orange, $DIOC_6(3)$, ER tracker blue white, FM4-64, Lysotracker red DND-99, Mitotracker orange CMTMRos, MDY-64, SNARF-1 AM ester, and Syto-13.

Table 3
Organelle-specific fluorescent dyes

Localization	Dye	Staining[a]	Fluorescence Ex (nm)[b]	Em (nm)[b]	Comments	Refs
Cytoplasm	Carboxyfluorescein diacetate (CFDA)	10 μM, 60 min	475	517	Nonfluorescent membrane-permeant molecule that is cleaved by cytoplasmic esterases into a membrane-impermeant molecule that fluoresces green. The fluorescent molecule is gradually pumped into the vacuole, but not as quickly as fluorescein diacetate (FDA).	
	SNARF-AM ester	10 μM, 60 min	576	635	The red-fluorescent membrane-impermeant product produced by esterases is normally used as a pH sensor, but also forms a useful cytoplasmic counter stain to GFP.	
Nucleus	DAPI	10 μM, 60 min	358	461	A DNA stain that is not always plasma membrane permeable in plant cells.	
	Syto-13[c]	1 μM, 10 min	488	509	Labels nucleus and strongly labels mitochondrial nucleoid. Other colors available.	[50]
Mitochondria	Mitotracker orange	0.5 μM, 2 min	551	576	A specific mitochondrial stain. A range of other Mitotracker colors are also available.	
	Syto-13	1 μM, 10 min	488	509	A DNA stain that labels the mitochondrial nucleoid as well as nucleus. Does not visibly label the DNA present in plastids.	[50]
	Rhodamine 6G	2 μM, 10 min	528	551	Fluoresces green through to orange, and can be distinguished from GFP. At higher concentrations, may also label the endoplasmic reticulum.	
	Rhodamine 123	2 μM, 10 min[d]	507	529	Fluoresces green. Specific to mitochondria unlike rhodamine 6G.	[55]
Plastids	(Chlorophyll autofluorescence)	–	440, 630	685	Present in chloroplasts but not in undifferentiated plastids and chromoplasts.	

(continued)

Table 3 (continued)

Localization	Dye	Staining[a]	Fluorescence Ex (nm)[b]	Fluorescence Em (nm)[b]	Comments	Refs
Peroxisomes	8-(4-Nitrophenyl)-BODIPY[d]	5 μM, 5 min[d]	502	~520	Initially nonfluorescent, but converted into a fluorescent form inside cells.	[56, 57]
Endoplasmic reticulum	$DIOC_6(3)$	10 μM, 5 min	484	501	Initially labels mitochondria, and then labels endoplasmic reticulum. Fluoresces green.	[58]
	ER tracker white[c]	1 μM, 60 min[d]	374	575	Questionable localizations and reliability. Other colors available.	[59]
Golgi apparatus	BODIPY TR ceramide[c]	5 μM, 30 min[d]	589	616	Questionable localizations and reliability. Other colors available.	[60, 61]
Endosomes	FM4-64	10 μM, 30 min at 0 °C	558	734	Cold inhibits endocytosis.	[52]
Lipid droplets/oleosomes	Nile red	2 μM, 15 min	~500	~550, 630	Stains lipid droplets/oleosomes yellow but membranes red.	[62]
Plasma membrane	FM4-64	10 μM, 30 min at 0 °C	558	734	Will define plasma membrane as well as label endosomes.	[52]
	CellMask orange	5 μg/ml, 60 min	554	574	Labels plasma membrane and gives prominent labelling of pit-fields. Is weakly taken up into organelles. Other colors available.	
Vacuole	Fluorescein diacetate (FDA)	10 μM, 6 h with lengthy washing	475	517	Initially labels cytoplasm but pumped into vacuole. Longer washes allow label in the cytoplasm to be pumped into the vacuole.	

	Monochlorobimane	1 mM, 6 h	394	490	Initially nonfluorescent, but reacts with thiols such as glutathione to form a fluorescent molecule. Is pumped into the vacuole.	
	Acridine orange	20 μM, 6 h	505	526	Fluoresces green. Slowly accumulates in the vacuole due to loading of weak acids.	[61]
	(Anthocyanin)	–	400–600	~630	Found in the vacuole of red onion inner epidermal cells, and cells of some other plants. Weakly autofluorescent.	[46]
Tonoplast	MDY-64	10 μM, 10 min	456	505	A membrane-permeable label for the tonoplast. Fluoresces green but might be distinguished from GFP on the basis of a slightly shorter excitation wavelength.	[46]
	Lysotracker red	0.2 μM, 10 min	577	590	Supposed to load into acidic organelles (e.g., vacuole) by a weak acid-based effect, although in *Allium* labelling seems to be preferentially to the tonoplast membrane. Fixable. Other colors available.	
Cell wall	Propidium iodide	20 μM, 10 min	493	636	Membrane impermeable DNA stain used to test cell viability. Reported to label pectin.	[63]
	Calcofluor white	0.25 μM, 5 min[d]	365	435	Stains both cellulose and hemicelluloses.	
Plasmodesmata/ pit fields	Aniline blue	50 μM[e], 30 min	370	509	Stains callose associated with pit-fields.	
	CellMask orange	5 μg/ml, 5 min	554	574	Labels plasma membrane and gives prominent labelling of pit-fields. Is weakly taken up into organelles. Other colors available.	

[a]Where possible, washing steps should be followed by brief rinsing of tissue with water
[b]Published values for excitation and emission peaks
[c]One of a series of dyes showing similar properties. Other colors also available
[d]Suggested labelling conditions based on literature reports, and not confirmed by the author
[e]Aniline blue should be prepared as a 50 μM solution in 100 mM K_2HPO_4 buffer at pH 8.2

Table 4
Organelle-specific tests

Localization	Test	Conditions	Comments	Refs
Microtubules	Oryzalin	20 μM, 15 min	Reversibly depolymerizes microtubules. Microtubule recovery is generally slow and irregular.	
Microfilaments	Latrunculin B	1 μM, 15 min	Inhibits cytoplasmic streaming and is not generally reversible even after extensive washing.	
	Cytochalasin D	2 μM, 15 min	Reversibly inhibits cytoplasmic streaming.	
Golgi apparatus	Brefeldin A	50 μM, 40 min	Reversibly degrades the Golgi back into the ER	[47]
Endosomes	Brefeldin A	50 μM, 40 min	Reversibly causes accumulation of endosomes	[47]
Prevacuolar compartments	Wortmannin	20 μM, 2 h[a]	Reversibly causes dilation or vacuolation of prevacuolar compartments.	[48, 49]

[a]Suggested conditions based on literature reports, and not confirmed by the author

2. Fluorescent stains that can be purchased from Sigma include calcofluor white, carboxyfluorescein diacetate, DAPI (4′,6-diamidino-2-phenylindole), fluorescein diacetate, monochlorobimane, Nile red, propidium iodide, rhodamine 123, and rhodamine 6G.
3. Fluorescent stains from other suppliers include aniline blue (BioSupplies, Clayton, VIC, Australia) and 8-(4-nitrophenyl)-BODIPY (Toronto Research Chemicals, Toronto, ON, Canada).
4. Several drugs that target the cytoskeleton and cell organelles (Table 4) are available from Sigma and include brefeldin A, cytochalasin D, latrunculin B, and wortmannin. Oryzalin is available from ChemService (West Chester, PA, USA).

3 Methods

3.1 Preparation of Gold Particles

1. Gold particles (50 mg) are resuspended in distilled water (1 ml), vortexed (20 s) and pelleted by brief centrifugation in a microfuge (15 s, full speed).
2. After supernatant removal, particles are resuspended in 100 % ethanol (1 ml), vortexed and again pelleted by centrifugation.
3. The particles are washed a further four times with 100 % ethanol (1 ml) with vortexing and centrifugation.
4. The particles are resuspended in sterile distilled water (1 ml) and aliquoted into 20 μl samples that can be stored for later use at −20 °C (*see* **Notes 1** and **2**). This means that each aliquot contains about 1 mg of gold.

5. To coat an aliquot of gold, resuspend particles by vortexing and add 5 μl of plasmid DNA (*see* **Note 3**). Suck the solution up and down several times with the pipette tip and vortex (20 s). To transform cells with multiple constructs simultaneously, two (or more) types of DNA can be coated onto particles at the same time.
6. Precipitate the DNA with 25 μl of 2.5 M $CaCl_2$ and vortex briefly, then stabilize the DNA with 10 μl of 0.1 M spermidine (*see* **Note 4**). Vortex for 30 s, pellet briefly, and discard the supernatant.
7. Wash particles by resuspension in 100 % ethanol (200 μl), vortexing briefly, pelleting the particles, and then discarding the supernatant. Resuspend gold particles in 100 % ethanol (100 μl) (*see* **Note 5**).

3.2 Preparation and Handing of Plant Material

It should be noted that plant material from different species may require different handling procedures to those given here, and that these instructions are only intended as a guide.

1. For transformation of onion epidermal cells, the outer, senescing layers of onions are removed, and squares of inner leaf tissue (~20 by 20 mm) prepared. These are stored between moist tissues prior to transformations when the epidermal cells of the inner, concave surface are transformed.
2. For transformation of leek epidermal cells, a system where the transformed cells are actively elongating, mature leeks with fleshy leaves and a root base are left to stand in water overnight to ensure turgidity. Outer leaves are peeled away to reveal the inner leaves that are still elongating and from which squares of tissue (~20 mm^2) are removed (*see* **Note 6**). These are stored between moist tissues prior to transformations when the inner epidermal surface is transformed.
3. For transformation of the epidermal cells of pea, young and expanded leaves are collected and stored between moist tissue paper prior to transformation. The lower epidermis of whole leaves is transformed (*see* **Note 7**).
4. For bombardments, plant material is positioned in the gene gun on moistened tissues (*see* **Note 8**).
5. After bombardment, transformed leaves and leaf segments are stored overnight in the dark between moistened tissues. Alternatively, epidermal peels can be prepared and incubated overnight on agarized nutrient media (*see* **Note 9**).

3.3 Operating the Bio-Rad Gene Gun

1. Turn on the gene gun, vacuum pump, and gas cylinder.
2. Briefly wash DNA carrier disks in 100 % ethanol and stand on their edge to dry.
3. Briefly wash carrier rings in 100 % ethanol and allow to dry.

4. Place dried carrier disks into the dried carrier rings and ensure that these sit flat.
5. Resuspend DNA-coated gold particles by vortexing and dispense ~15 μl evenly across the DNA carrier disks. Allow to fully dry (*see* **Note 10**).
6. Rinse a rupture disk (*see* **Note 11**) with 100 % isopropanol (10 s) and place the moist disk into the top of the round, gun-nozzle ring. This step should only be done when the system is ready to assemble. This should be screwed onto the gun nozzle, and tightened with the tightening tool.
7. Place a piece of stopper mesh and then an upside-down carrier disk into the ring assembly, and tighten the holder ring over this. Place this assembly on the top shelf of the gene gun chamber (Fig. 2a).
8. Place plant material on a lower shelf of the gene gun. Close and lock the door of the chamber, turn the vacuum on and let the pressure drop to about 26 on the scale (*see* **Note 7**).
9. Hold the fire button down until the rupture disk fails. An audible "pop" sound should be accompanied by some movement in the plant tissue associated with the pulse of gas entering the chamber.
10. Release the vacuum.
11. To turn the system off, first turn off the gas cylinder and then bleed the gas lines by closing the gene gun chamber and drawing a vacuum. Hold down the fire button until pressure in the gas lines returns to zero. Vent the chamber and then turn off the vacuum pump. Then turn off the chamber.

3.4 Operating Filter-Based Gene Guns

These instructions refer to non-commercial chamber-based gene guns [34].

1. Turn on the gene gun and vacuum pump. Open the gas flow out of the gas cylinder, and adjust the regulator to give a value of 60 psi (*see* **Note 12**).
2. Ensure that gold particles are well suspended. Add ~15 μl of suspension to the filter side of a Swinnex filter and allow to dry briefly (20 s). Screw filter together and install into the upper port inside the gene gun chamber (Fig. 2b).
3. Place plant material on a lower shelf within the chamber of the gun.
4. Close valves to evacuate the chamber. The pressure should drop to near zero (*see* **Note 7**).
5. Fire!
6. Fire a second time to remove any gold particles still attached (*see* **Note 13**).

7. Release the vacuum.
8. To turn the system off, first turn off the regulator and the gas cylinder. Bleed gas in the line by pulling a vacuum on the gene gun chamber, and press the fire button until both gauges read zero. Turn off the vacuum pump and gene gun.

3.5 Confocal Microscopy

Determining the correct localization of fluorescent proteins requires that confocal microscopes be operated correctly. As the protocols for correct usage vary between different confocal microscopes, the following notes are limited to approaches for maximizing the output of useful information in protein localization experiments. Once an apparent localization has been determined, it is important to identify whether the localization is valid or an artifact. It is important to realize that 35S-dependent over-expression of fusion protein constructs can induce significant changes to cellular organization. Such artifacts are most common in the endomembrane system [5], but seem less common with organelles such as peroxisomes, mitochondria, and plastids that represent the end destination of fusion proteins. Other artifacts are, of course, also possible with fusion proteins as these can compromise targeting information and modify protein turnover [41]. The addition of an intrinsically fluorescent protein can compromise targeting information and result in protein misdirection, and/or modify protein turnover and cell structure [41]. When Golgi apparatus-targeted GFP is transiently expressed in living tobacco epidermal cells, low expression reveals Golgi dynamics whereas over-expression results in labelling of the ER and structural changes [5]. Analysis of random cDNA fusions with GFP also demonstrated that localizations to different organelles were possible even with sequences that contained frame-shifts from their normal frame of reference [42] demonstrating that targeting sequences are comparatively flexible.

1. Cells should initially be checked between 8 and 16 h after transformation (*see* **Note 14**). Epidermal peels, if they can be prepared, should be mounted on water. Epidermal peels provide an excellent way of optimizing imaging for both transmitted and fluorescence light, and for getting dyes and drugs into cells. Otherwise, small leaf fragments can be mounted and viewed (*see* **Note 15**).
2. Most confocal microscopes have transmitted light detectors to record a concurrent non-confocal image (Fig. 1b, c). The transmitted light image should be optimized (Köhler illumination) and should always be recorded.
3. Samples should be screened with conventional fluorescence and by eye before switching to the confocal system. This is because the eye is the best way of determining whether the appropriate color of fluorescence is present within a sample (*see* **Note 16**).

4. Transformation of fluorescent cells should be confirmed by observing one or more gold particles lodged within the nuclei (*see* **Notes 17** and **18**).
5. Although a viewer's eye is naturally drawn to the brightest cells, these cells may contain over-expression artifacts (Fig. 3d, f, h, j). An observer should determine whether weakly expressing

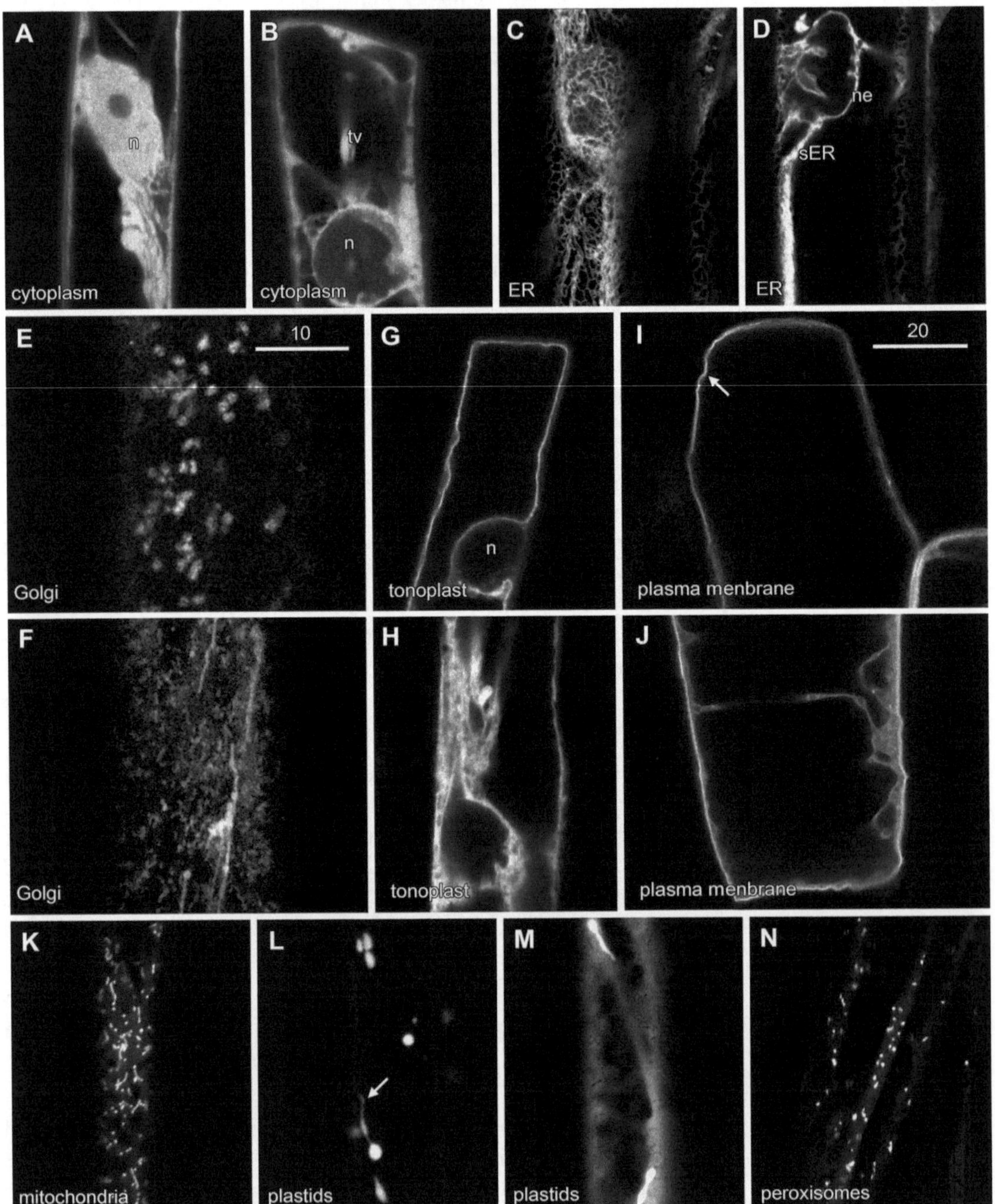

cells show the same patterns of localization as more strongly expressing cells. If variations are apparent, then the stronger expression patterns may be an artifact.

6. Cells should be confirmed as alive. Dead cells will fluoresce with a weak yellow color and will show characteristic changes in cell organization (Fig. 1h). Cell viability can be demonstrated with time-lapse imaging. A series of ten images over 1 min recorded with both fluorescence and transmitted light will normally demonstrate actin microfilament-based cytoplasmic streaming in healthy cells. If the transformed cell shows different cellular dynamics to the adjacent non-transformed cells, there is a strong possibility that the cells are unhealthy.
7. Images should be optimized using the confocal collection software, rather than subsequently using programs such as Photoshop [43].
8. Doubtful localizations should be confirmed with fluorescent dyes or with drugs that demonstrate known functionalities within cells (*see* Subheading 3.6).
9. Examples of typical labelling patterns and some of the artifacts that transformation can generate are shown in Fig. 3. The first three examples typically do not generate artifacts. Fluorescent proteins localized to the cytoplasm fluoresce from throughout the cytoplasm (and sometimes the nucleus), and in high resolution images, dark regions are present in the cytoplasm which are organelles which exclude the fluorescent proteins (Fig. 3a). Cytoplasmic fluorescence is only excluded from the nucleus if the fusion protein is too large to diffuse through the nuclear pores (generally above ~50 kDa) and if no nuclear localization signal (NLS) is present (Fig. 3b). If an NLS is present, then the fluorescence patterns should be limited to the nucleus although

Fig. 3 Typical localization patterns of transiently expressed fluorescent proteins in different organelles of onion epidermal cells viewed by confocal microscopy. (**a**) Cytoplasm, with diffusion into the nucleus (n), but with protein excluded from organelles in the cytoplasm. (**b**) Cytoplasm with protein excluded from the nucleus, but present in transvacuolar strands (tv). (**c**) ER, showing stable cortical arrays adjacent to the plasma membrane. (**d**) ER, showing the nuclear envelope (ne) and subcortical strands (sER). (**e**) Golgi apparatus, showing normal patterns with discrete stacks visible. (**f**) Over-expression in the Golgi apparatus, with ER-like patterns forming in the cell cortex. (**g**) Tonoplast, a single well-defined line inside the nucleus (n). (**h**) Over-expression of tonoplast-directed proteins resulting in fluorescent proteins associated with other membrane throughout the cytoplasm. (**i**) Plasma membrane, a single well-defined line around the cell, often with some indentations (*arrow*) which are pit-fields. (**j**) Over-expression of plasma membrane-directed proteins resulted in fluorescent proteins associated with other membrane throughout the cytoplasm. (**k**) Mitochondria. (**l**) Plastids with crawling stromules (*arrows*) extending from the plastid surfaces. (**m**) Over-expression of plastid-targeted GFP with some fluorescent protein remaining in the cytoplasm. (**n**) Peroxisomes. Bar in **i** = 20 μm for all images except (**e**) where bar = 10 μm

this may be highly convoluted and distorted in shape [11]. Fluorescent proteins that are targeted to and retained by the ER are also typically artifact-free. The ER has three characteristic domains which are a delicate cortical network adjacent to the plasma membrane (Fig. 3c), the nuclear envelope, and large, dynamic subcortical strands (Fig. 3d) [44].

Locations in cells to which proteins are delivered via the endomembrane system are prone to generating artifacts on the over-expression of fluorescent proteins [5]. Paired images are shown (Fig. 3d–j) in which images show how fluorescent proteins should appear, and how they are observed upon over-expression. Golgi-targeted expression, which should label only the individual Golgi stacks that stream rapidly through the cell (Fig. 3e), becomes localized into static, ER-like sheets and tubules on over-expression (Fig. 3f). Tonoplast labelling, which should show a crisp line around the vacuole and with the nucleus outside (Fig. 3g), shows extensive labelling of other membrane components scattered through the cytoplasm when over-expressed (Fig. 3h). A similar over-expression pattern occurs with plasma membrane-targeted fluorescent proteins (Fig. 3i, j). However, fluorescent proteins targeted to organelles directly, and not through the endomembrane system, have fewer possibilities for generating artifacts. Thus for mitochondria (Fig. 3k), plastids (Fig. 3l), and peroxisomes (Fig. 3n), fewer over-expression problems are apparent. However, even with these, it is still possible to overload the system so that some cytoplasmic labelling can occur, as in the case of plastids (Fig. 3m). These organelles show characteristic differences in their organization and dynamics. For example, the chlorophyll-free, undifferentiated plastids typically found in epidermal cells do not stream rapidly but often show long projections referred to as stromules [45], whereas mitochondria and peroxisomes are smaller and stream faster. Mitochondria can be distinguished from peroxisomes because they readily stain with a range of different fluorescent dyes (Table 3) (Fig. 4a, c, d).

3.6 Live Cell Imaging with Dyes

To confirm the localization of problematic fusion proteins, further experimental verification might be required. Fluorescent stains are now available that label most of the different organelles of a plant cell (Table 3). The exceptions to this are plastids for which no dyes are available, although green chloroplasts can be imaged by chlorophyll autofluorescence. Some compartments of the endomembrane system, notably transfer and secretory vesicles also lack specific fluorescent dyes. However, as most fluorescent dyes have been developed for experiments in animal and tissue culture cells, it is important to realize that they might behave differently in plant cells. Further important considerations include that organelle-specific fluorescent dyes vary in their specificity, and can cause changes to organelle and cell structure if used at too high a

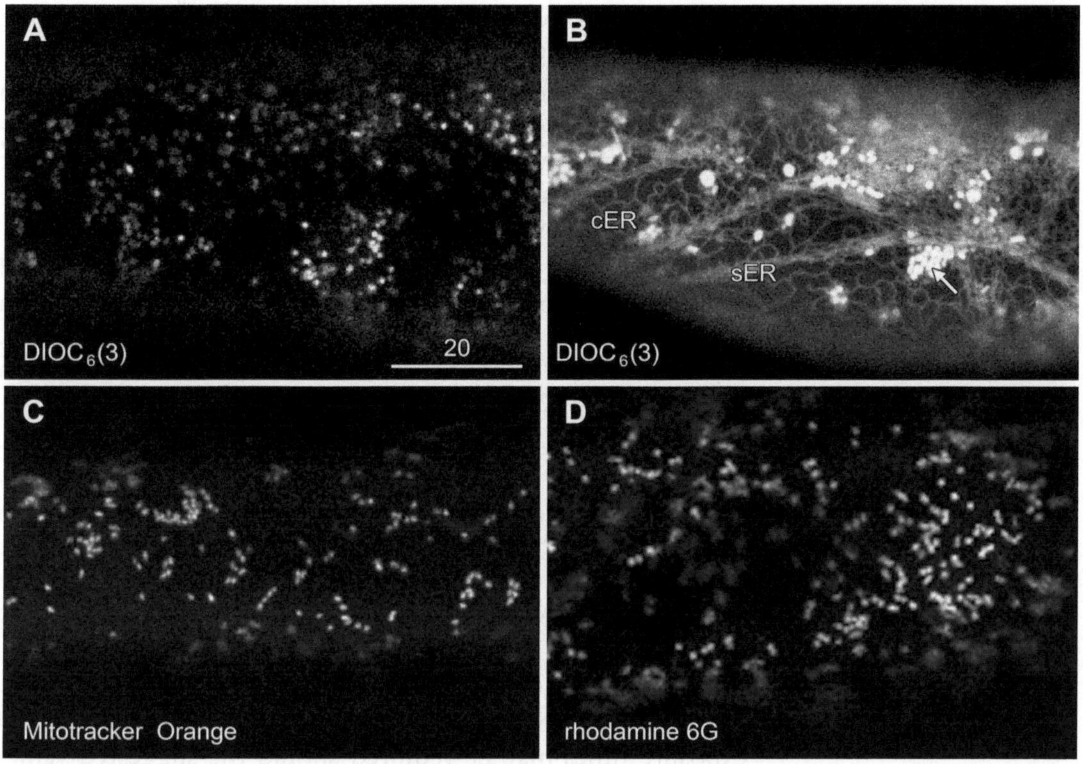

Fig. 4 Typical localization patterns of fluorescent dyes in mitochondria and the endomembrane network of onion epidermal cells viewed by confocal microscopy. (**a**) $DIOC_6(3)$ initially labelled mitochondria but at higher concentrations labelled the ER, including both the cortical (cER) and subcortical (sER) networks. Under these conditions, the bright mitochondria often appeared swollen (*arrow*) (**b**). (**c**) Mitochondrial labelling was possible with specific dyes such as Mitotracker orange. (**d**) Mitochondrial labelling was also possible with rhodamine 6G, a cheaper but less specific alternative to Mitotracker dyes. Bar in **a** = 20 μm for all images

concentration or for too long (Fig. 4) (*see* **Note 19**). Some of these dyes are relatively specific, including the Mitotracker series of dyes which label mitochondria and MDY-64 which labels the tonoplast [46], whereas labelling patterns from other dyes such as fluorescent ceramide labelling of the Golgi apparatus are less specific and more difficult to interpret. Various different drugs are also available that aid in verifying labelling patterns (Table 4) with cytoskeletal drugs that target actin microfilaments and microtubules foremost among these. Other important drugs include brefeldin which disrupts later components of the secretory pathway, leading to distinct changes in Golgi apparatus, secretory vesicle, endosome and ER organization [47] and wortmannin that is specific for the prevacuolar compartment [48, 49].

1. Dye and drug stock solutions (typically ×100 or ×1,000) can be prepared in DMSO and stored frozen.
2. Dyes and drugs can be diluted in distilled water (except aniline blue—*see* Table 3). In most instances, plant cells do not show

responses to the inclusion of 0.1–1.0 % DMSO in the labelling solutions, although controls should be run to confirm this.

3. Epidermal peels can be floated mesophyll-side downwards on labelling solutions in small plastic Petri dishes. To wash epidermal peels, remove the dye or drug solution with a Pasteur pipette and replace with distilled water, or gently lift the epidermal peel to a wash solution.
4. Labelling whole leaf tissue is more difficult because dyes and drugs need to penetrate through the cuticle layer. In such cases, dye and drug solutions can be vacuum infiltrated. Segments of leaf tissue are floated on the dye solution and a vacuum pulled (2 by 5 min) (*see* **Note 20**).
5. As with analyzing the distribution of transiently expressed fluorescent proteins, cells should be confirmed as alive with demonstrations of sustained cytoplasmic streaming. Cells with weak labelling should be compared with cells with stronger labelling and, in most cases, the weaker labelling should be preferred. An exception to this is $DIOC_6(3)$ labelling of the ER (Fig. 4b).
6. Unlike the analysis of transiently expressed fluorescent proteins, some optimization of dye and drug concentrations and exposure times may be required should the suggested values (Tables 3 and 4) prove inappropriate.
7. Images should be optimized using the confocal collection software.
8. Examples of typical labelling patterns for several common dyes are shown in Fig. 4, with the labelling properties of some newer and less well-characterized dyes shown in Fig. 5. The dye $DIOC_6(3)$ stains mitochondria at low concentrations and on short exposures (Fig. 4a) but will then stain the ER with the mitochondria often becoming more rounded (Fig. 4b). The staining of mitochondria on their own is possible with Mitotracker series of dyes, with Mitotracker orange a useful counter-stain to both GFP and YFP (Fig. 4c). Although rhodamine 6G provides a cheaper alternative for counterstaining mitochondria (Fig. 4d),

Fig. 5 Localization patterns of new fluorescent dyes in onion epidermal cells, viewed by confocal microscopy and with transmitted light. (**a**) MDY-64 stained the vacuolar membrane, including transvacuolar strands (tv), the tonoplast adjacent to the cortical cytoplasm (t) and the plasma membrane. (**b**) Longer staining with MDY-64 resulted in vesiculation of the central vacuole, and labelling of further membranes within the cell. Pit-fields labelled prominently, and were visible as thin areas of the cell walls in transmitted light images (*arrows*). (**c**) Lysotracker red, a stain for acidic organelles, labelled vacuolar membranes. (**d**) CellMask orange is a plasma membrane stain which gave distinct labelling of the membranes of adjacent cells (*asterisks*), and which also labelled pit-fields strongly (*arrows*). Bar in **a** = 20 μm for all images

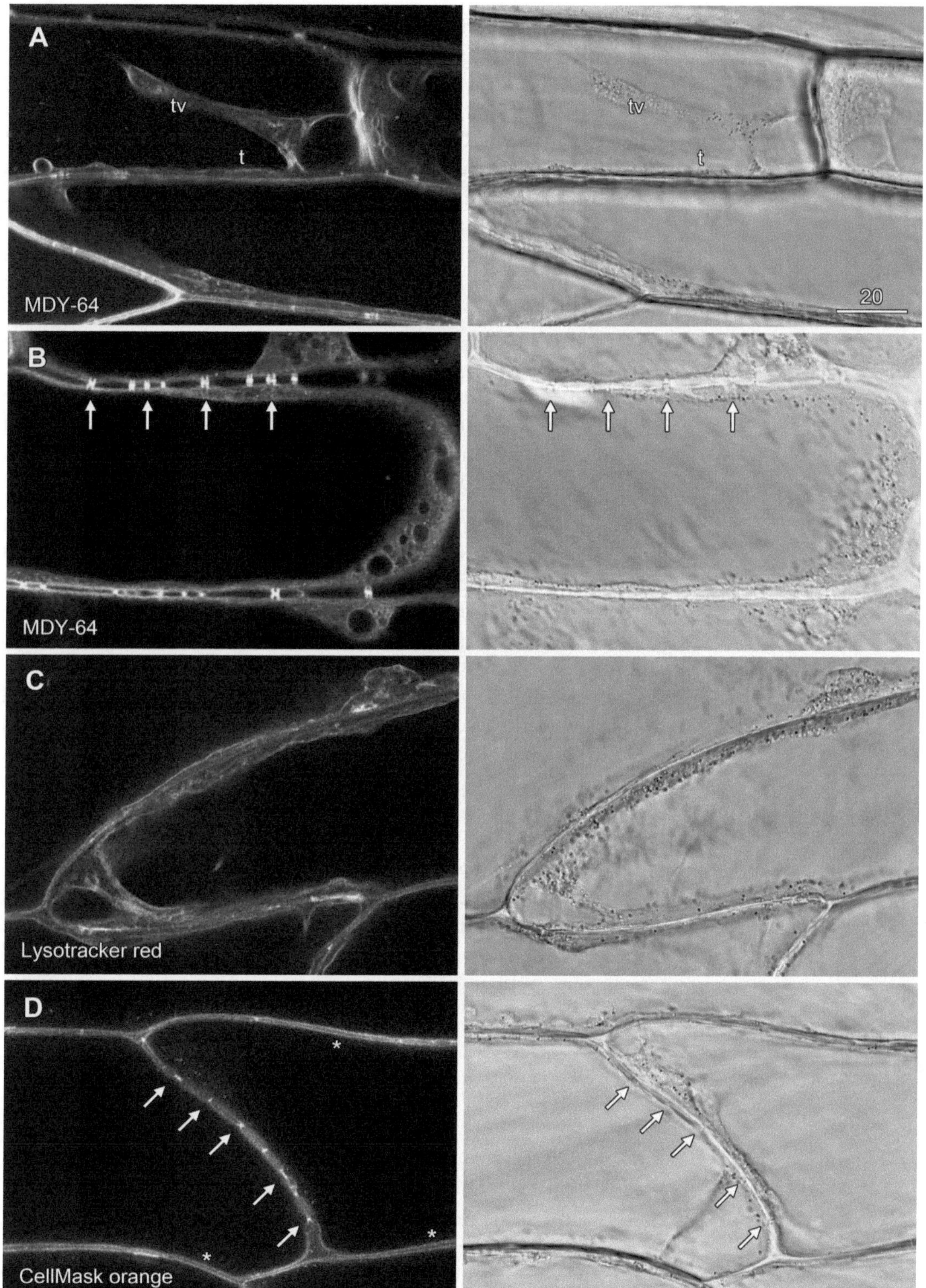
A
tv
t
MDY-64
tv
t
20
B
MDY-64
C
Lysotracker red
D
*
*
*
CellMask orange

at higher concentrations it will also label the endoplasmic reticulum although the related dye, rhodamine 123 is specific for mitochondria [55]. The labelling patterns shown by these dyes match the organization seen with fluorescent proteins targeted to the ER (Fig. 2b, c) and mitochondria (Fig. 2k). ER-specific staining, without the labelling of mitochondria, is also possible using a range of dyes, with the UV-excited ER Tracker white potentially useful as a counter-stain to fluorescent proteins such as GFP. However, observations of the ER in plant cells are difficult, and it has proved difficult to replicate the clear labelling of the different parts of the ER that can be seen with $DIOC_6(3)$. Later stages of the endomembrane network such as the Golgi apparatus and vesicles are also difficult to stain. While probes based on fluorescently labelled ceramide have been suggested to be specific for the Golgi, mainly based on research in animal cells [51], specific labelling the Golgi in plant cells has proved difficult.

Several newer fluorescent probes may also prove useful for cell characterizations, providing confirmation for labelling patterns seen with fluorescent proteins. MDY-64 is a tonoplast-labelling dye that demonstrates the complexity of the vacuolar membrane in the highly dynamic onion epidermal cells (Fig. 5a) [46] but is also prone to generating significant changes in cell structures on over-staining that include vacuole vesiculation (Fig. 5b). Lysotracker red may also label the vacuolar membrane, at least in onion cells and under certain conditions (Fig. 5c). Another novel dye is CellMask orange which is sold as a plasma membrane label. This dye seems to work in onion epidermal cells. The nature of the staining is, however, complex, as the dye seems to preferentially bind to pit-fields, aggregates of plasmodesmata (Fig. 5d). CellMask orange shows relatively little uptake into the cells, although some vesicles can sometimes be seen which are likely endosomes. Endosomal labelling is best seen with the fluorescent dye FM4-64, and has been well characterized in plant cells [52].

4 Notes

1. Gold particles are very dense (70 % denser than lead) and settle extremely rapidly. During aliquoting, the mix should be vortexed between each aliquot.
2. Aliquots can be prepared in 1.5 ml Eppendorf tubes. However, screw-top microcentrifuge tubes work better as these allow better resuspension and mixing of the dense particles. Their screw-top lids are also easier to use than flip-top Eppendorf tubes.
3. Experiments typically use 5 μl of plasmid DNA, although for plasmid preparations with low yields, or for weakly expressing

constructs, this volume can be raised. Alternatively, for strongly expressing constructs, a smaller volume of DNA can be used. The high levels of variability in expression seen in transformed cells seems unlikely to be due to variability between particles. In a typical transformation experiment, where about 1 μg of plasmid DNA is coated onto 1 mg of 1.6 μm diameter gold particles, each particle may carry more than a thousand plasmid copies. Furthermore, when *Allium* epidermal cells were transformed with a combination of cytosolic GFP and RFP constructs coated onto the same gold particles, the ratio of GFP to RFP fluorescence was consistent across a wide range of fluorescence values, and correlated with the relative amounts of the constructs used [53]. This suggests that variations in the cells, or the sites where the gold particle lodge, determine expression levels of fluorescent proteins.

4. A stock solution of 1 M spermidine can be prepared and stored frozen at −80 °C. Aliquots (50 μl) of 0.1 M spermidine can be prepared and should also be stored at −80 °C. A fresh aliquot of spermidine should be used for each day's experiments.
5. Unused, DNA-coated gold particles can be stored for at least several weeks at 4 °C.
6. Leeks that have begun flowering (typically in summer) and which contain a central flower stalk should not be used for transformations as the leaves are no longer elongating rapidly.
7. Under high vacuum, the epidermis of the *Argenteum* pea mutant can lift away from the underlying mesophyll cells. This does not appear to adversely affect transformations.
8. When transforming a different type of tissue, or cells where transformation rates are low, it is sometimes useful to rest this tissue on a piece of onion or leek that will act as a positive control. Particles that miss the first piece of plant tissue can transform the positive control tissue sitting behind.
9. Leek, onion and pea epidermal cells can be incubated overnight or for several days on Hoagland's media containing 3 % (w/v) sucrose and 1.2 % (w/v) agar. This solution contains 2 mM KNO_3, 5 mM $Ca(NO_3)_2$, 2 mM $MgSO_4$, 2 mM KH_2PO_4, 90 μM iron–EDTA complex, 46 μM H_3BO_3, 9.1 μM $MnCl_2$, 0.77 μM $ZnSO_4$, 0.32 μM $CuSO_4$, and 0.11 μM MoO_3 [54], and can be supplemented by drugs to target specific cellular activities that might not be readily added to whole leaves.
10. When completely dried, the gold particles should appear as a smear across the surface of the carrier disks, and not as discrete particles. Discrete particles indicate clumping of the gold particles which will lower transformation efficiency, and is a result of insufficient mixing of the material prior to coating the carrier disks.

11. For most plant tissues, including onion and leek epidermis, rupture disks in the 900–1,100 psi range are appropriate for transformation. For some tissues with thicker epidermal cell walls or tougher cuticles, rupture disks that fail at higher pressures are appropriate as these deliver gold particles that will travel faster and will penetrate better. The pressure reported for transformation of *Vicia* leaves was 1,350 psi [36].
12. For most plant tissues, including onion and leek epidermis, a value of 60 psi gives good transformation rates. For delicate tissues, however, lower pressures might work better and vice versa. Alternatively, the samples can be transformed at different heights within the chamber, with the upper shelves closer to the Swinnex filters allowing more particles to hit a smaller area, and with greater force. The pressure reported for transformation of cowpea leaves was 40–80 psi [34].
13. A major disadvantage of the filter-based gene gun system is that cross contamination occurs easily. This means that loose particles within the chamber can be transformed into plant tissue with subsequent shots which may cause confusion with subsequent protein localizations. Several steps can be taken to avoid cross-contamination:
    (a) A separate Swinnex filter should be used for each construct.
    (b) The gene gun chamber should be wiped with 100 % ethanol prior to the experiment, and between the use of different constructs.
    (c) The Swinnex filters should be cleaned by soaking overnight in 5 % bleach, followed by extensive rinsing with distilled water.
    (d) Cleaned Swinnex filters should be shot blank into plant material before adding DNA-coated particles and, if cleaning protocols are working, no transformation should be seen.
14. Where possible, screening of transformed leaf pieces should be attempted with a fluorescence dissecting microscope because this allows a far faster overview of the entire piece of tissue, and allows hot spots of transformation to be identified. It is important to realize, however, that only brightly transformed cells will be visible with this type of microscope, and that weakly expressing constructs might not be visible at all. Any failure to achieve transformation determined with a fluorescence dissecting microscope requires confirmation with a compound microscope.
15. For upright microscopes, whole leaf samples can be prepared in two ways:
    (a) With a drop of water and a coverslip.
    (b) With a drop of water, no coverslip and a long working distance, water-dipping lens. Although these lenses have a

lower numerical aperture, and therefore a lower resolution, their long working distances means that they can visualize the surface of contoured leaves. Working without a coverslip also allows for experiments with dyes and drugs (assuming that these can penetrate the epidermis and cuticle).

For inverted microscopes, dipping lenses are impractical and samples need to be mounted with a drop of water and a coverslip. For larger leaf fragments, the coverslip will often begin to fall away from the slide. Long coverslips (22 mm wide, and either 40 or 60 mm in length) can be used, and can be fixed to either end of the slide with sticky tape.

16. Some fluorescent microscopes will have both longpass (LP) and bandpass (BP) filter cubes (Fig. 1) for imaging GFP- and YFP-based fusion proteins. Although using chlorophyll-blocking BP filters makes detection of weak signals easier, as the background red fluorescence is eliminated, this means that the subtle color differences often associated with dying and dead cells can be missed. Routinely switching between LP and BP filters when screening transformed cells is, therefore, a useful approach.
17. Gold particles in nuclei can be difficult to see by eye with transmitted light imaging. However, in transmitted light images recorded with the confocal microscope, particles will appear as small black dots within the nuclei. They can also be imaged in reflection mode by the confocal microscope as the particles are considerably more reflective than anything else inside the cell. Fluorescence images of cells that are only weakly fluorescent may sometimes contain a fluorescent dot within the nucleus which is reflected light "contaminating" the fluorescence image.
18. An exception to this occurs when comparatively small, cytosolic proteins transmit through plasmodesmata. In such cases, a brightly fluorescent central cell whose nucleus contains the transforming gold particle will be surrounded by a halo of less fluorescent cells.
19. Confirmation of cell viability is also important with dyes and drugs. This can be confirmed by time lapse imaging, with demonstrations of cytoplasmic streaming being an indication of cell health.
20. Vacuum infiltration can be conducted in the chamber of the gene gun. To reduce the effects of bubbling, dye and drug solutions can be de-gassed for several minutes prior to adding the plant tissue. Experiments can be conducted in Eppendorf tubes or 15 ml Falcon tubes with small holes cut through the lids to limit bubbling from the solution. Increased rates of uptake can be achieved by using small leaf fragments which

allow dye to enter the leaf through the cuts as well as through stomata. During vacuum infiltration, leaf samples floating on the liquid's surface will sink as air is removed from the leaf.

Acknowledgments

This work was supported by funding from the University of Canterbury and the Biomolecular Interaction Centre. The author thanks numerous colleagues for their donations of plasmids, Philippa Barrell and David Goulden (Plant & Food Research, Lincoln, New Zealand) for *Argenteum* pea plants, and Rosemary White (CSIRO Plant Industry, Canberra, Australia), John Harper (Charles Sturt University, Wagga Wagga, Australia), and Krithika Yogeeswaran (University of Canterbury) for comments on the manuscript and discussions about confocal microscopy.

References

1. Shaner NC, Steinbach PA, Tsien RY (2005) A guide to choosing fluorescent proteins. Nat Methods 2:905–909
2. Chudakov DM, Matz MV, Lukyanov S et al (2010) Fluorescent proteins and their applications in imaging living cells and tissues. Physiol Rev 90:1103–1163
3. Baulcombe DC, Chapman S, Santa CS (1995) Jellyfish green fluorescent protein as a reporter for virus infections. Plant J 7:1045–1053
4. Sheen J, Hwang S, Niwa Y et al (1995) Green-fluorescent protein as a new vital marker in plant cells. Plant J 8:777–784
5. Sparkes IA, Runions J, Kearns A et al (2006) Rapid, transient expression of fluorescent fusion proteins in tobacco plants and generation of stably transformed plants. Nat Protoc 1:2019–2025
6. Marion J, Bach L, Bellec Y et al (2008) Systematic analysis of protein subcellular localizations and interaction using high-throughput transient transformation of *Arabidopsis* seedlings. Plant J 56:169–179
7. Li J-F, Park E, von Arnim AG et al (2009) The FAST technique: a simplified *Agrobacterium-based* transformation method for transient gene expression analysis in seedlings of Arabidopsis and other plant species. Plant Methods 5, doi:10.1186/1746-4811-5-6
8. van Loock B, Markakis MN, Verbelen J-P et al (2010) High-throughput transient transformation of *Arabidopsis* roots enables systematic colocalization analysis of GFP-tagged proteins. Plant Signal Behav 5:261–263
9. Haseloff J, Siemering KR, Prasher DC et al (1997) Removal of a cryptic intron and subcellular localization of green fluorescent protein are required to mark transgenic *Arabidopsis* plants brightly. Proc Natl Acad Sci U S A 94:2122–2127
10. Scott A, Wyatt S, Tsou P-L et al (1999) Model system for plant cell biology: GFP imaging in living onion epidermal cells. Biotechniques 26:1125–1132
11. Collings DA, Carter CN, Rink JC et al (2000) Plant nuclei can contain extensive grooves and invaginations. Plant Cell 12:2425–2439
12. North AJ (2006) Seeing is believing? A beginners' guide to practical pitfalls in image acquisition. J Cell Biol 172:9–18
13. Pearson H (2007) The good, the bad and the ugly. Nature 440:138–140
14. Conchello J-A, Lichtman JW (2005) Optical sectioning microscopy. Nat Methods 2: 920–931
15. Lichtman JW, Conchello J-A (2005) Fluorescence microscopy. Nat Methods 2: 910–919
16. Dixit R, Cyr R, Gilroy S (2006) Using intrinsically fluorescent proteins for plant cell imaging. Plant J 45:599–615
17. Chudakov DM, Lukyanov S, Lukyanov KA (2005) Fluorescent proteins as a toolkit for in vivo imaging. Trends Biochem Sci 23:605–613
18. Matz MV, Fradkov AF, Labas YA et al (1999) Fluorescent proteins from nonbioluminescent Anthozoa species. Nat Biotechnol 17: 969–973

19. Campbell RE, Tour O, Palmer AE et al (2002) A monomeric red fluorescent protein. Proc Natl Acad Sci U S A 99:7877–7882
20. Shaner NC, Campbell RC, Steinbach PA et al (2004) Improved monomeric red, orange and yellow fluorescent proteins derived from *Discosoma* sp. red fluorescent protein. Nat Biotechnol 22:1567–1572
21. Chudakov DM, Lukyanov S, Lukyanov KA (2007) Using photoactivatable fluorescent protein Dendra2 to track protein movement. Biotechniques 42:553–558
22. Wu S, Koizumi K, Macrae-Crerar A et al (2011) Assessing the utility of photoswitchable fluorescent proteins for tracking intercellular protein movement in the *Arabidopsis* root. PLoS One 6:e27536
23. Mathur J, Radhamony R, Sinclair AM et al (2010) mEosFP-based green-to-red photoconvertable subcellular probes for plants. Plant Physiol 154:1573–1587
24. Ghosh I, Hamilton AD, Regan L (2000) Antiparallel leucine zipper-directed protein reassembly: application to the green fluorescent protein. J Am Chem Soc 122:5658–5659
25. Hu C-D, Chinenov Y, Kerpppola TK (2002) Visualization of interactions among bZIP and Rel family proteins in living cells using bimolecular fluorescence complementation. Mol Cell 9:789–798
26. Kodama Y, Hu C-D (2012) Bimolecular fluorescence complementation (BiFC): a 5-year update and future persepctives. Biotechniques 53:285–294
27. Diaz I, Martinez M, Isabel-LaMoneda I et al (2005) The DOF protein, SAD, interacts with GAMYB in plant nuclei and activates transcription of endosperm-specific genes during barley seed development. Plant J 42:652–662
28. Ohad N, Shichur K, Yalovsky S (2007) The analysis of protein-protein interactions in plants by bimolecular fluorescence complementation. Plant Physiol 145:1090–1099
29. Kodama Y (2011) A bright green-colored bimolecular fluorescence complementation assay in living plant cells. Plant Biotechnol 28: 95–98
30. Klein TM, Wold ED, Wu R et al (1987) High-velocity microprojectiles for delivering nucleic acids into living cells. Nature 327:70–74
31. Silverstone AL, Ciampaglio CN, Sun T-P (1998) The *Arabidopsis* RGA gene encodes a transcriptional regulator repressing the gibberellin signal transduction pathway. Plant Cell 10:155–169
32. von Arnim AG, Deng X-W, Stacey MG (1998) Cloning vectors for the expression of green fluorescent protein fusion proteins in transgenic plants. Gene 221:35–43
33. Kikkert JR, Vidal JR, Reisch BI (2004) Stable transformation of plant cells by particel bomabrdment/biolistics. Methods Mol Biol 286:61–78
34. Finer JJ, Vain P, Jones MW et al (1992) Development of the particle inflow gun for DNA delivery to plant cells. Plant Cell Rep 11:323–328
35. Zhang Y, Su J, Duan S et al (2011) A highly efficient rice green tissue protoplast system for transient gene expression and studying light/chloroplast-related processes. Plant Methods 7:30
36. Marc J, Granger CL, Brincat J et al (1998) A GFP-MAP4 reporter gene for visualizing cortical microtubule rearrangements in living epidermal cells. Plant Cell 10:1927–1939
37. Hoch HC, Pratt C, Marx GA (1980) Subepidermal air spaces: basis for the phenotypic expression of the *Argenteum* mutant of *Pisum*. Am J Bot 67:905–911
38. Marx GA (1982) *Argenteum* (*Arg*) mutant of *Pisum*. J Hered 73:413–420
39. Elzenga JTM, Staal M, Prins HBA (1997) Calcium-calmodulin signalling is involved in light-induced acidification by epidermal leaf cells of pea, *Pisum sativum* L. J Exp Bot 48:2055–2060
40. Jewer PC, Incoll LD, Shaw J (1982) Stomatal responses of *Argenteum*—a mutant of *Pisum sativum* L. with readily detachable leaf epidermis. Planta 155:146–153
41. Dhanoa PK, Sinclair AM, Mullen RT et al (2006) Illuminating subcellular structures and dynamics in plants: a fluorescent protein toolbox. Can J Bot 84:515–522
42. Cutler SR, Ehrhardt DW, Griffitts JS et al (2000) Random GFP::cDNA fusions enable visualization of subcellular structures in cells of *Arabidopsis* at a high frequency. Proc Natl Acad Sci U S A 97:3718–3723
43. Rossner M, Yamada KM (2004) What's in a picture? The temptation of image manipulation. J Cell Biol 166:11–15
44. Staehelin LA (1997) The plant ER: a dynamic organelle composed of a large number of discrete functional domains. Plant J 11:1151–1165
45. Natesan SK, Sullivan JA, Gray JC (2005) Stromules: a characterisatic cell-specific feature of plastid morphology. J Exp Bot 56:787–797
46. Wiltshire EJ, Collings DA (2009) New dynamics in an old friend: dynamic tubular vacuoles radiate through the cortical cytoplasm of red onion epidermal cells. Plant Cell Physiol 50:1826–1839
47. Nebenführ A, Ritzenthaler C, Robinson DG (2002) Brefeldin A: deciphering an enigmatic

inhibitor of secretion. Plant Physiol 130: 1102–1108
48. Tse YC, Mo B, Hillmer S et al (2004) Identification of multivesicular bodies as prevacuolar compartments in *Nicotiana tabacum* BY-2 cells. Plant Cell 16:672–693
49. Delhaize E, Gruber BD, Pittman JK et al (2007) A role for the *AtMTP11* gene of Arabidopsis in manganese transport and tolerance. Plant J 51:198–210
50. Christensen NM, Nicolaisen M, Hansen M et al (2004) Distribution of phytoplasmas in infected plants as revealed by real-time PCR and bioimaging. Mol Plant-Microbe Int 17: 1175–1184
51. Pagano RE (1989) A fluorescent derivative of ceramide: physical properties and use in studying the Golgi apparatus of animal cells. Methods Cell Biol 29:78–85
52. Bolte S, Talbot C, Boutte T et al (2004) FM-dyes as experimental probes for dissecting vesicle trafficking in living plant cells. J Microsc 214:159–173
53. Collings DA, Collins PP, Chia XR et al (2012) A users guide to transient protein localisations in *Allium* epidermal cells. In preparation
54. Baskin TI, Betzner AS, Hoggart R et al (1992) Root morphology mutants in *Arabidopsis thaliana*. Aust J Plant Physiol 19:427–437
55. Wu F-S (1987) Localization of mitochondria in plant cells by vital staining with rhodamine 123. Planta 171:346–357
56. Landrum M, Smertenko A, Edwards R et al (2010) BODIPY probes to study peroxisome dynamics *in vivo*. Plant J 62:529–538
57. Strader LC, Wheeler DL, Christensen SE et al (2011) Multiple factes of *Arabidopsis* seedling development require indole-3-butyric acid-derived auxin. Plant Cell 23:984–999
58. Quader H, Schnepf E (1986) Endoplasmic reticulum and cytoplasmic streaming: fluorescence microscopical observations in adaxial epidermis cells of onion bulb scales. Protoplasma 131:250–252
59. Villarejo A, Burén S, Larsson S et al (2005) Evidence for a protein transported through the secretory pathway *en route* to the higher plant chloroplast. Nat Cell Biol 7:1224–1231
60. Patel S, Brkljacic J, Gindullis F et al (2005) The plant nuclear envelope protein MAF1 has an additional location at the Golgi and binds to a novel Golgi-associated coiled-coil protein. Planta 222:1028–1040
61. Verbelen J-P, Tao W (1998) Mobile arrays of vacuole ripples are common in plant cells. Plant Cell Rep 17:917–920
62. Greenspan P, Mayer EP, Fowler SD (1985) Nile Red: a selective fluorescent stain for intracellular lipid-droplets. J Cell Biol 100: 965–973
63. Rounds CM, Lubeck E, Winship LJ et al (2011) Propidium iodide competes with Ca^{2+} to label pectin in pollen tubes and arabidopsis root hairs. Plant Physiol 157:175–187

Chapter 17

Proteomics and the Analysis of Nodulation

Nijat Imin

Abstract

Application of proteomics is becoming increasingly important to understand the function of genes and their encoding proteins. This is due to not only the poor correlation between the transcript levels and protein accumulation, but also the critical roles of posttranslational modifications that increase the functional diversity of the proteins and influence almost all aspects of plant growth and its response to the environment. This chapter describes the gel-based quantitative and comparative proteomics that combine two-dimensional gel electrophoresis with mass spectrometry analysis to detect, quantify, and characterize proteins and their post-translational modifications with specific focus on analyzing nodule samples. This method is also applicable for other tissue types.

Key words SDS-PAGE, 2D gel electrophoresis, Mass spectrometry, Symbiosis, Nodules

1 Introduction

To understand the function of all the genes in an organism, one needs to know the expression of both transcripts and the protein products. It is becoming increasingly clear that there is a poor correlation between the transcript levels and protein accumulation. Proteomics aims at describing the whole protein output of the genome and thus complements transcriptomic and metabolomic studies. The proteomes of nodule-forming tissues are complex due to their protein components being derived from both host plant cells and the associating symbiotic bacteria. Two-dimensional gel electrophoresis (2DE) is the basic tool for proteomics studies and it enables the separation of complex proteins by combining isoelectric focusing (IEF) and sodium dodecyl sulfate-polyacrylamide gel electrophoresis (SDS-PAGE). Then, the protein spots can be visualized by staining and proteins and their posttranslational modifications (PTMs) which are critical for protein stability and activity can be identified. This chapter describes the most commonly used steps in high-throughput plant proteomics including gel-based proteomics with specific focus on analyzing nodule samples (*see* Fig. 1).

Ray J. Rose (ed.), *Legume Genomics: Methods and Protocols*, Methods in Molecular Biology, vol. 1069,
DOI 10.1007/978-1-62703-613-9_17, © Springer Science+Business Media, LLC 2013

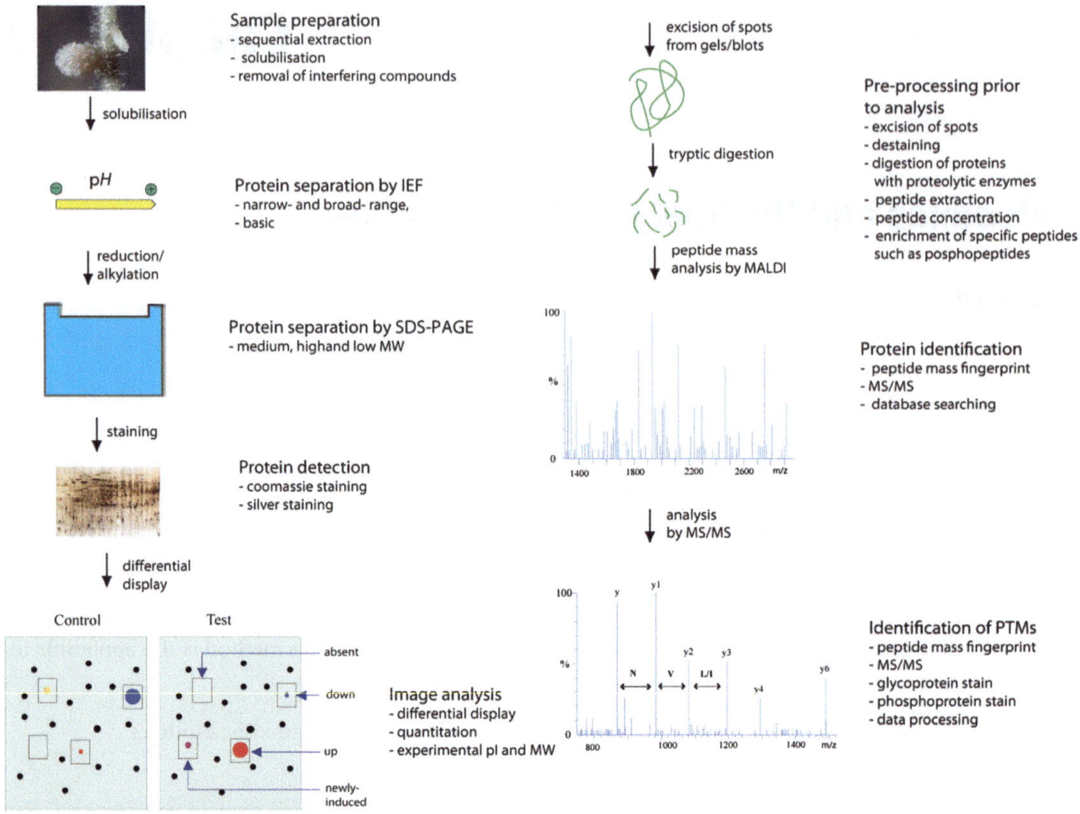

Fig. 1 Two-dimensional gel electrophoresis, quantitation, and peptide analysis

1.1 Protein Extraction

The extraction of proteins from plant tissue requires tissue disruption, separation of proteins from other cell materials and precipitation of proteins with acetone–trichloroacetic acid, resolubilizing proteins in a solution that solubilizes the maximum number of different proteins and deactivation of proteases. The most common method used for plant protein extraction is using trichloroacetic acid (TCA) in acetone and subsequent washings in acetone. This method not only denatures proteins but also removes other interfering compounds such as DNA, RNA, pigments, and phenolic compounds. Nonionic zwitter detergents such as 3-(3-Cholamidopropyl)-dimethylammonio-1-propanesulfonate (CHAPS) in combination with urea can be are used to solubilize and separate proteins in the mixture [1]. Sodium dodecyl sulfate (SDS) is also a strong detergent and used to solubilize membrane proteins. However, it renders a negative charge to proteins and interferes with isoelectric focusing. Reducing agents (e.g., dithiothreitol [DTT] or 2-mercaptoethanol) are needed to disrupt disulfide bonds.

1.2 First-Dimensional Isoelectric Focusing

The isoelectric focusing is a sophisticated and reproducible technique where proteins are immobilized at their isoelectric point or pI. The use of immobilized pH gradients enables clear separation of proteins over different pH ranges [2]. The resolution depends on a clean protein-preparation, a slowly increasing electric field, and a high final voltage (1,000–8,000 V) to enable proteins to move into the gel. Proteins, each with a different pI, will move into the first dimension gel-strip that contains an immobilized pH gradient, until their net charge is zero. High resolution IPG (immobilized pH gradients) strips (pH 3–10 or 4–7) are also a necessity for separating thousands of proteins. Narrow pH range IPG strips (one pH unit) have also proven to be very effective in resolving more proteins (generally about 10 % increase compared to wider IPG strips). The longer the strip, the more protein can be loaded and the better the resolution for total proteins. Typically, 18–24 cm long IPG strips are used.

1.3 SDS

In the second dimension, proteins are separated by their molecular mass. First, proteins are treated with SDS, which introduces negative charges proportional to the size of the protein. Second, proteins are separated by an electric current through a polyacrylamide gel with a certain pore size. Therefore, this technique is referred to as SDS-polyacrylamide gel electrophoresis (SDS-PAGE). Depending on the polyacrylamide concentration, proteins in the range from approximately 5 to 200 kDa can be separated on a gel. For SDS-PAGE runs, large format gels are recommended.

1.4 Protein Detection

For protein detection, either MS (mass spectrometry) compatible silver staining or colloidal Coomassie staining should work. Recently, a 2D fluorescence difference gel electrophoresis (DIGE) system has proven to be a better option due to its sensitivity, better control of gel-to-gel variation and advantages in data analysis [3]. Image analysis with Melanie software (GeneBio, Geneva, Switzerland) has been used successfully.

1.5 Protein Identification

While Matrix assisted laser desorption ionization time-of-flight mass spectrometry (MALDI-TOF-MS) is the first choice for protein identification, tandem MS analysis yields much more valuable data and increases the level of confident matches. Protein identification is typically carried out by MS on Micromass TofSpec 2E-TOF-MS (Manchester, UK) or PE SCIEX (Foster City, CA, USA), QSTAR hybrid LC-MS/MS Q-TOF, or Micromass LC-MS/MS Q-TOF systems. Search engines such as Mascot and ProteinProspect work well both with MS data and tandem MS data. However, a key component of the searches is the databases. Besides general nonredundant database, species-specific EST databases or any other specific databases are needed for high success rates in protein identification. For nodule samples, it is important to include both host plant and the associating symbiotic bacteria sequence databases, so that both plant and bacterial proteins can be identified. Legume specific protein databases are available at semele.anu.edu.au [4–7] and bioinfo.noble.org/manuscript-support [8].

2 Materials

Use analytical grade reagents and prepare all solutions using ultrapure water (prepared by purifying deionized water to attain a sensitivity of 18 MΩ cm at 25 °C).

2.1 Protein Extraction from Nodules or Other Plant Tissues

All solutions should be made fresh. Sample buffer can be frozen at −80 °C for several months. However, the protease inhibitors phenylmethylsulfonyl fluoride (PMSF) and ethylenediaminetetraacetic acid (EDTA) disodium salt have to be added fresh before use.

1. Protein extraction solution TCA/acetone: 10 % trichloroacetic acid in acetone with 0.07 % (0.45 mM) DTT or 0.07 % (9 mM) 2-mercaptoethanol on dry ice.
2. Fine glass powder: Between 0.01 and 0.1 mm grain size (Schott, USA).
3. Probe sonicator: MSE 100 (Thomas Optical and Scientific, Sydney, Australia), or Sonic Dismembrator Model 550 (Fisher Scientific, Pittsburgh, PA, USA) or VC 100 probe sonicator (Sonics and Materials Inc., Danbury, CT, USA) or something similar.
4. Acetone with 0.07 % DTT on dry ice.
5. Sample buffer: 9 M Urea, 4 % (w/v) CHAPS, 1 % (w/v) DTT, 1 % (v/v) BioLyte 3-10 ampholytes (Bio-Rad), 35 mM Tris base, 1 mM PMSF and 5 mM EDTA. Keep at room temperature (20 °C).
6. Sonic bath: Bransonic model 2510 (Branson Ultrasonic Corporation, Danbury, CT USA) or Cole-Parmer model 8845 (Cole-Parmer Instrument Co., Chicago, USA) or something similar.

2.2 2DE

2.2.1 Isoelectric Focusing

Multiphor II horizontal electrophoresis system: GE Healthcare Life Sciences, USA.

Rehydration solution: 8 M urea, 0.5 % (w/v) CHAPS, 0.15 % (w/v) DTT, 0.5 % (v/v) Biolyte 3-10 ampholytes, and a trace of bromophenol blue. Approximately, 360 μL for an 18-cm strip and 480 μL for a 24-cm strip. Longer the strip, better the resolution depending on the complexity of the samples analyzed.

2.2.2 SDS-PAGE

1. Equilibration solution (1): 40 % (v/v) glycerol, 0.05 M Tris–HCI, pH 6.8, 6 M urea, 2 % (w/v) SDS, and 2 % (w/v) DTT.

2. Equilibration solution (2): 40 % glycerol, 0.05 M Tris–HCI, pH 6.8, 6 M Urea, 2 % (w/v) SDS, 2 % (w/v) iodoacetamide, and 0.005 % (w/v) bromophenol blue.
3. Molecular weight markers.

2.3 Protein Detection

1. Photographic tray: can be purchased from Genomic Solutions, Ann Arbor, MI, USA.
2. Orbital shaker: DS-500 (VWR, West Chester, PA, USA) or KJ-201BD Oscillator (Chang Bioscience Inc., Castro Valley, CA, USA) or something similar.

2.3.1 Silver Staining

Prepare all solutions fresh before use and with highest grade chemicals.

1. Fixation solution: 10 % (v/v) acetic acid, 40 % (v/v) ethanol, 50 % ultrapure water.
2. Sensitizer: 30 % (v/v) ethanol, 4.1 % (w/v) sodium acetate, 0.275 % (w/v) potassium tetrathionate, and 0.5 % (v/v) glutaraldehyde.
3. Silver stain: 0.2 % (w/v) silver nitrate, 0.062 % (w/v) HEPES, 0.07 % (v/v) formaldehyde.
4. Developer: 3 % (w/v) potassium carbonate, 0.0012 % (w/v) sodium thiosulfate, 0.025 % (v/v) formaldehyde.
5. Stop solution: 5 % (w/v) Tris-base, 2 % (v/v) acetic acid.

2.3.2 Colloidal Coomassie Staining

1. Colloidal Coomassie staining solution: 5 % (w/v) aluminum sulfate-(14–18)-hydrate, 2 % (v/v) phosphoric acid, 10 % ethanol, 0.02 % (w/v) Coomassie G-250 (Bio-Rad). Briefly: Dissolve aluminum sulfate in Milli-Q water, add ethanol, homogenize, and mix Coomassie G-250 to the solution as soon as the aluminum sulfate is completely dissolved, add phosphoric acid and finally make to volume with Milli-Q water. Do not filter this solution.
2. Destaining solution: 10 % (v/v) ethanol and 2 % (v/v) phosphoric acid.

2.4 Protein Identification by MS Analysis

1. Destain: 100 mM ammonium bicarbonate, pH 7.8.
2. Trypsin solution: 8 μL of 15 ng/mL sequencing-grade modified trypsin (Promega) in 25 mM ammonium bicarbonate, pH 7.8.
3. 50 % (v/v) Acetonitrile and 0.5 % (v/v) trifluoroacetic acid.
4. Matrix: a-cyano-4-hydroxcinnamic acid, 10 mg/mL in 70 % (v/v) acetonitrile, 1 % (v/v) trifluoroacetic acid.
5. Mass spectrometry: Micromass TofSpec 2E-TOF-MS (Manchester, UK) or PE SCIEX (Foster City, CA, USA), QSTAR hybrid LC-MS/MS Q-TOF or Micromass LC-MS/MS Q-TOF systems.

6. Mascot: www.matrixscience.com.
7. Protein Prospector: www.prospector.ucsf.edu.

3 Methods

3.1 Protein Extraction from Nodules and Other Samples

1. Grind tissue in liquid nitrogen after addition of fine glass powder (approximately 10 % of the tissue volume) in a mortar and pestle and suspend in −20 °C cold TCA/Acetone in an acetone-resistant centrifuge tube on dry ice. It is important to grind tissue thoroughly in liquid nitrogen, the finer the powder, the better the yield.
2. Sonicate the suspension on dry ice with a probe sonicator at approximately 20 MHz 6× for 10 s each with intermittent 1 min breaks to avoid overheating (optional).
3. Leave samples for 1 h at −20 °C and centrifuge samples at 35,000 × g for 15 min at 4 °C.
4. Discard the supernatant and resuspend the pellet in cold (−20 °C) acetone containing 0.07 % (w/v) DTT.
5. Place samples at 20 °C for 30 min and then centrifuge at 12,000 × g for 15 min at 4 °C.
6. Repeat the last washing step.
7. Briefly lyophilise the pellet (3–5 min) to evaporate any acetone, and suspend the dry pellet in sample buffer (approx 500 μL for every gram fresh weight) by sonication in ice cold water in a sonic bath (1–3 min; add some ice to the water) and vortex mixing for several minutes.
8. Centrifuge sample at 12,000 × g for 15 min at 20 °C and collect the supernatant that should contain the solubilized proteins.
9. Measure the protein concentration of the sample and keep at −80 °C until used for isoelectric focusing (*see* **Note 1**).

3.2 2DE

3.2.1 Isoelectric Focusing

Depending on the electrophoresis system used, variations of this protocol can be used. We describe here the use of a Multiphor II horizontal electrophoresis system for isoelectric focusing and SDS-PAGE because of its high reproducibility and resolution [2]. Immobilized pH gradient (IPG) strips are highly recommended.

1. Load the samples either on a cup (can be obtained from GE Healthcare) or by rehydrating into the strip with the rehydration solution. The latter is particularly useful for diluted samples.
2. For loading an 18–24 cm strip, use approximately 50–200 μg of protein for silver staining and 500–10,00 μg for Coomassie staining.

3. Cover strips with paraffin oil to avoid drying out and crystallizing of the sample.
4. Focus the rehydrated strips at 20 °C, 1 mA and 5 W for a total of 25–200 kV h with the following voltage gradients: 5–0 min at 150 V, 5 min on a linear gradient from 150 to 300 V, 30 min to 6 h at 300 V, 1–5 h on a linear gradient from 300 to 3,500 V and 7–54 h at 3,500 V (*see* **Note 2**).
5. After isoelectric focusing, strips can either be wrapped in plastic foil and frozen at −80 °C or directly equilibrated for SDS-PAGE.

3.2.2 SDS-PAGE

1. Equilibrate the IPG strips for 10 min in equilibration solution 1 and for another 10 min in equilibration solution 2.
2. Gently blot strips dry on filter paper to remove excess equilibration solution without damaging the gel.
3. Load the IPG strip onto the second dimension gel together with molecular weight markers at one or both ends of the strip. Ensure good contact with the gel and avoid air bubbles. Also avoid water drops on horizontal gels as these will distort.
4. Carry out electrophoresis at 4–15 °C at 300 V for 1 h and subsequently at 600 V for 4–5 h, or until the bromophenol blue front reaches the end of the gel. These conditions will depend on the apparatus used for SDS-PAGE (*see* **Note 3**).

3.3 Protein Detection

3.3.1 Silver Staining

Use only high purity chemicals and make fresh solutions, especially for formaldehyde and glutaraldehyde. Staining is easiest done in photographic trays on an orbital shaker.

1. Fix gels 3× for 30 min each, in fixative directly after SDS-PAGE.
2. Change to sensitizer for 16 h.
3. Wash the gel at least 6x in ultrapure water for 20–30 min each.
4. Incubate gels with silver staining solution for 2 h in the dark. To wash off the silver, quickly rinse for 10 s in ultrapure water. If this washing step is done for any longer, the silver will disassociate from the protein spots.
5. Develop gels for 5–7 min and stop the development by replacing the developer with stop solution. The development should be stopped when no more new spots become visible and before the background of the gel becomes dark (*see* **Note 4**).
6. Leave the gels in stop solution for no longer than 20 min to avoid color changes in the silver stain.
7. Wash gels in distilled water several times, scan, and store sealed in plastic pouches in a few milliliters of 1 % methanol to discourage microbial contamination. Gels can be stored at room temperature or at 4 °C.

3.3.2 Colloidal Coomassie Staining

1. After second dimension separation, rinse the gels three times with Milli-Q water for 10 min on a horizontal shaker (*see* **Note 5**).
2. Shake the Coomassie solution before use to disperse the colloidal particles evenly.
3. Incubate the gels covered with the Coomassie staining solution by agitation on a shaker for 2–16 h (*see* **Note 5**).
4. After the staining procedure remove the Coomassie staining solution and rinse the gels twice with Milli-Q water.
5. Destain for 10–60 min.
6. Rinse the gels twice with Milli-Q water.

3.4 Protein Identification by MS Analysis

3.4.1 Preparation of Protein Spots for MS Analysis

1. Excise protein spots manually with a clean scalpel blade from Coomassie-stained gels and store in a drop of 50 % high-grade methanol.
2. Destain each spot and spin-dry with a SpeedVac for several minutes.
3. Digest proteins in gel with trypsin (or other protease) solution for 16 h at 37 °C.
4. After the tryptic (or other protease) digestion, extract the peptides with 50 % acetonitrile and 0.5 % trifluoroacetic acid.
5. Spot a 1 μL aliquot onto a sample plate along with 1 μL of matrix and air-dry.
6. Submit each digested protein for MS analysis.

3.4.2 Peptide Mass Fingerprint and Tandem MS Analysis

The MS generated data can be compared against theoretical spectra. For this, freely available Internet sites can be used for searches against common organisms, such as Mascot (matrixscience.com) and Protein Prospector (prospector.ucsf.edu). Alternatively, you can create your own database of your specific organism for more specific searches using either genome annotation or EST sequences. Many software packages including Mascot from Matrix Science allow you to search custom nucleotide or other databases (*see* **Note 6**).

3.4.3 Identification of Posttranslational Modifications

In combination with 2DE and MS analysis, it is possible to detect posttranslational modifications (PTMs). Shifting in pH indicates modifications (e.g., phosphorylation) which can be detected by visualizing the spots that shifted in pH by a treatment and then identifying the modified peptide(s) by MS analysis (*see* **Note 7**). Phosphoproteins or glycoproteins can be detected in gels or blots using commercially available staining kits/reagents.

3.5 Quantification of Protein Abundance

Scan silver-stained gels on a high resolution scanner equipped with transparency adaptor and Coomassie-stained gels with an opaque white background in reflective mode at a minimum of 600 dots per square inch (dpi) and save them as TIF images for subsequent analysis. Several software packages are available for spot quantification (*see* **Note 8**). After quantifying protein spots across repeat gels and between treatments, the significance of the observed changes in protein abundance can be calculated with analysis of variance (*see* **Note 9**).

4 Notes

1. The suggested extraction protocol has been successfully used for various species and tissues including nodules. For disruption of soft plant material, tissues can be ruptured by vortex mixing or osmotic lysis, however, most plant tissue will not sufficiently rupture until ground in liquid nitrogen. To maximize extraction of hydrophobic proteins from tissue, it is possible to re-extract the insoluble pellet of the last step with organic solvents, e.g., chloroform–methanol. Protein yields should be approximately 0.5–2 mg protein per l g fresh tissue weight. The yield should range from 0.5 to 2 mg of protein from every gram of tissue fresh weight, but might change with tissue type, depending on water, fiber and other contents. If protein concentrations are too low or volume too large to load onto the first dimension, either precipitate proteins again with TCA/acetone and resolubilize in a smaller volume of sample buffer, or reduce the volume with a Centricon® column (Millipore).
2. In our experience, isoelectric focusing improves if focusing times are increased from the usually recommended 25–30 kVh (for 18-cm strips) to about 200 kVh without leading to over-focusing. Focusing times will need to be adjusted for strips from other manufacturers than those mentioned here as specified. Problems with isoelectric focusing are often due to the absence of sharp spots, resulting in horizontal streaks across the 2DE gel, especially in the alkaline pH range.
3. Second dimension can be performed by SDS-PAGE on vertical electrophoresis systems with a total acrylamide concentration of 12 % or other. The equilibrated IPG strips are placed on the top of the separating gels and fixed with hot agarose solution (0.5 % agarose in running buffer containing bromophenol blue).
4. Timing is important because gels of different runs need to be comparable in staining.
5. This colloidal Coomassie staining is modified from Dyballa et al. [9, 10] and Kang et al. [9, 10]. Insufficient washing can cause poor sensitivity because the remaining SDS on the gels disturbs the binding of the dye to the protein. This staining generates marginal background so you can observe the staining progress in between. After 10 min you can see the first

protein spots appearing, within 2 h of incubation about 80 % staining to its maximum level is completed. For best results, overnight incubation is recommended.

6. Peptide mass fingerprinting can he done on a Micromass TofSpec 2E Time of Flight Mass Spectrometer (Waters). Stringency of matching is paramount to avoid selecting the wrong matches. Generally, a minimum of four peptide matches, a maximum of one mis-cleavage per peptide, >20 % sequence coverage, and 100 ppm molecular weight discrepancy are acceptable for an average protein identification. Peptides derived by tryptic digestion of 2DE protein spots can be chemically derivatized and sequenced by MALDI MS/MS or can be submitted to an Ion Trap system. The determined amino acid sequence of peptides is subsequently used for sequence homology searching against protein sequence databases [11]. The MALDI MS/MS approach is sensitive and robust and a peptide sample can be analyzed several times after it has been deposited onto the MALDI MS target. The combination of 2DE based protein separation with tandem mass spectrometry for peptide sequencing is an attractive method for characterization of organisms for which the genome is not yet available.
7. The most studied PTM which is a key player in cell signalling. Phosphorylation occurs at serine, threonine, or tyrosine residues. Glycosylation is a more complex topic as many glycosylation-branched configurations with different sugar units exist. However, a consensus sequence motif Asn-Xaa-Ser/Thr (Xaa is any amino acid except Pro) has been defined as a prerequisite for N-glycosylation [12]. N-terminal protein modifications are diverse (acetylation, palmitoylation, myristoylation, and methionine excision) and some of them are important for protein localization. For instance, myristoylation can be crucial to target some proteins to a membrane system.
8. Software packages for 2DE image analysis and quantification include BioNumerics 2D, Delta2D, ImageMaster, Melanie, PDQuest, Progenesis, and REDFIN—among others.
9. Use a minimum of three repeats for every sample to evaluate variations between gel runs. The horizontal, precast gels should produce a high reproducibility of gels. If variations between gels of the same biological material are large, spot quantitation will not produce meaningful results.

Acknowledgment

This work was supported by Australian Research Council discovery grant DP120101893.

References

1. Chevallet M, Santoni V, Poinas A, Rouquie D, Fuchs A, Kieffer S et al (1998) New zwitterionic detergents improve the analysis of membrane proteins by two-dimensional electrophoresis. Electrophoresis 19:1901–1909
2. Gorg A, Obermaier C, Boguth G, Harder A, Scheibe B, Wildgruber R et al (2000) The current state of two-dimensional electrophoresis with immobilized pH gradients. Electrophoresis 21:1037–1053
3. Unlu M, Morgan ME, Minden JS (1997) Difference gel electrophoresis: a single gel method for detecting changes in protein extracts. Electrophoresis 18:2071–2077
4. de Jong F, Mathesius U, Imin N, Rolfe BG (2007) A proteome study of the proliferation of cultured *Medicago truncatula* protoplasts. Proteomics 7:722–736
5. Imin N, De Jong F, Mathesius U, van Noorden G, Saeed NA, Wang XD et al (2004) Proteome reference maps of *Medicago truncatula* embryogenic cell cultures generated from single protoplasts. Proteomics 4:1883–1896
6. Imin N, Nizamidin M, Daniher D, Nolan KE, Rose RJ, Rolfe BG (2005) Proteomic analysis of somatic embryogenesis in *Medicago truncatula*. Explant cultures grown under 6-benzylaminopurine and 1-naphthaleneacetic acid treatments. Plant Physiol 137: 1250–1260
7. Prayitno J, Imin N, Rolfe BG, Mathesius U (2006) Identification of ethylene-mediated protein changes during nodulation in *Medicago truncatula* using proteome analysis. J Proteome Res 5:3084–3095
8. Lei Z, Dai X, Watson BS, Zhao PX, Sumner LW (2011) A legume specific protein database (LegProt) improves the number of identified peptides, confidence scores and overall protein identification success rates for legume proteomics. Phytochemistry 72:1020–1027
9. Dyballa N, Metzger S (2009) Fast and sensitive colloidal coomassie G-250 staining for proteins in polyacrylamide gels. J Vis Exp 1431
10. Kang D, Gho S, Suh M, Kang C (2002) Highly sensitive and fast protein detection with Coomassie Brilliant Blue in Sodium Dodecyl Sulfate-Polyacrylamide Gel Electrophoresis. Bull Korean Chem Soc 11:1511–1512
11. Shevchenko A, Sunyaev S, Loboda A, Bork P, Ens W, Standing KG (2001) Charting the proteomes of organisms with unsequenced genomes by MALDI-quadrupole time-of-flight mass spectrometry and BLAST homology searching. Anal Chem 73:1917–1926
12. Blom N, Sicheritz-Ponten T, Gupta R, Gammeltoft S, Brunak S (2004) Prediction of post-translational glycosylation and phosphorylation of proteins from the amino acid sequence. Proteomics 4:1633–1649

Chapter 18

Phosphoproteomic Analysis of Peptides

Mark A. Baker

Abstract

Posttranslational modifications represent a major mechanism by which cells control molecular events. Sample preparation is one of the biggest keys in terms of proteomic analysis, and especially when looking at phosphoproteomics. Here we describe two techniques that are used to identify phosphopeptides. These are Titanium Dioxide and Immobilized Metal Affinity Chromatography (IMAC) which enrich for phosphopeptides. These enrichment techniques typically often identify ~35 % of the same and ~65 % unique phosphopeptides. In this chapter, we describe the use of an ion-trap mass spectrometer for the identification of phosphopeptides.

Key words IMAC, Mass spectrometry, Phosphopeptides, Phosphorylation, Proteomics, Titanium dioxide

1 Introduction

In terms of scientific analysis, the study of proteins, often referred to as "proteomics" only entered mainstream biological research in the late 1980s. It was in 1985, that Franz Hillenkamp and Michael Karas, then of the University of Frankfurt, coined the phrase "MALDI" (matrix assisted laser desorption ionization) [1] and demonstrated that mass of compounds could be measured through matrix assisted, laser ionization, into a time of flight mass spectrometer. It was only 3 years later, that John Fenn (2002 Nobel Prize in Chemistry for the development of electrospray ionization; ESI) ionized an amino-acid which allowed it to be "sprayed" into a mass spectrometer to obtain the mass spectra and molecular weight of the residue [2]. A quarter of a decade later, modern proteomic groups are not only able to identify amino-acid sequences, but readily determine posttranslational modifications within proteins, including phosphorylation [3–6]. Although the mass identification of proteins (shot-gun/MudPit analysis) was once the method of choice [7], this last decade has seen an explosion in the use of more sophisticated technology. A thorough comparison

Ray J. Rose (ed.), *Legume Genomics: Methods and Protocols*, Methods in Molecular Biology, vol. 1069,
DOI 10.1007/978-1-62703-613-9_18, © Springer Science+Business Media, LLC 2013

between IMAC (Immobilized Metal Affinity Chromatography) and TiO_2 of global phosphopeptides demonstrated that only 35 % overlap occurs [8], suggesting that both methods are complementary and for a more complete phosphoproteomic analysis, both methods should be included. We describe the use of an ion-trap mass spectrometer for the identification of phosphopeptides. Although original developed for mammalian cells, similar procedures can be applied to plant proteomics.

2 Materials

Prepare all solutions using ultrapure water (prepared by purifying deionized water to attain a sensitivity of 18 MΩ cm at 25 °C) and analytical grade reagents. Prepare and store all reagents at room temperature (unless indicated otherwise).

2.1 Preparation of Protein and Tryptic Digests

1. Liquid nitrogen.
2. Protein estimation kit—BCA Protein Assay Kit from Pierce.
3. Trypsin buffer: 25 mM $NH_3HCO_3^-$, pH 8.0 with 1 M Urea. Weigh 19 mg $NH_3HCO_3^-$ and 0.6 g of Urea in 10 ml of water. Mix and check pH (should be pH 8.0) (*see* **Note 1**).
4. Trypsin working solution. To one 20 mg vial of trypsin, add 500 ml of the Trypsin buffer (*see* **Note 2**).
5. 10 % TA. Weigh out 10 g of TCA and make up to 100 ml using acetone.
6. 80 % MA. Weigh out 770.8 g of ammonium acetate and add 100 ml of water. Take 80 ml of methanol and add 20 ml of the ammonium acetate solution.
7. Acetonitrile (ACN).
8. P-SDS Buffer: Weigh out 30 g of sucrose, 2 g of SDS into a beaker. Add 100 ml of water (SDS Buffer). Take 80 ml of phenol and mix 1:1 with the SDS Buffer.
9. Microfuge.
10. Eppendorf tubes.

2.2 Preparation of TiO_2 Reagents and Buffers

1. Titanium dioxide. TiO_2 can be obtained from a disassembled column brought through GE Healthcare or Inertsil. 3 mg/ml of titanium dioxide is added to 1 ml of 50 % ACN (v/v).
2. DHB buffer. 350 mg/ml of DHB (2,5 Dihydroxybenzoic acid) is added to 80 % acetonitrile (ACN) containing 0.1 % trifluoroacetic acid (TFA: v/v). Ensure that the pH is ~1.0.
3. Wash Buffer. 80 % ACN containing 0.1 % TFA (v/v).
4. Elution Buffer. 25 µl of ammonium solution (ammonia solution with 25 % ammonia) is added to 1 ml water. Check pH to ensure that it is 10.0 (*see* **Note 3**).

2.3 Preparation of IMAC Solutions and Reagents

1. $FeCl_2$ solution: Weigh out 126.75 mg of $FeCl_2$ and add to 5 ml of water to produce 0.2 M $FeCl_2$. Care must be taken to ensure that anhydrous $FeCl_2$ is used. Adjustments to the calculations should be made if the tetrahydrate $FeCl_2$ is used. In this case, 198.8 mg is weighed out and added to 5 ml of water.
2. EDTA-Salt buffer: Weigh out 146.2 mg of EDTA and 265 mg of NaCl to top up to 10 ml with water to make 50 mM EDTA and 1 M NaCl (10 ml).
3. Equilibration buffer: Make up 6 % acetic acid (v/v). Weigh out 997.5 mg of NaOH and add to 5 ml of water to make 5 M stock solution. This should be made fresh on the day. Adjust the pH of the acetic acid solution to 3.5 by slowly adding (dropwise) the 5 M NaOH solution.
4. Wash buffer: Add 25 ml of ACN to 75 ml of water.
5. Elution buffer: Weigh out 23.06 mg of $NH_4H_2PO_4$ and add to 1 ml of water to make 200 mM.
6. Trifluoroacetic acid (TFA).

2.4 Preparation of IMAC Beads

1. Take approximately 200 μl of nickel NTA beads (50 % slurry), and centrifuge at 2,000 × *g* for 2 min. Remove and discard the supernatant.
2. Add 500 μl of EDTA-Salt buffer to the beads and resuspend by inversion. Roll the beads with roller tube mixer (45 rpm) for 5 min. After this centrifuge at 2,000 × *g* for 2 min. Remove and discard the supernatant. Repeat this step until the Nickel (blue tinge) is removed.
3. Add 300 μl of the $FeCl_2$ solution to the beads. Roll the beads with tube mixer (45 rpm) for 5 min. After this centrifuge at 2,000 × *g* for 2 min, remove and discard the supernatant. Repeat this two more times.

2.5 Equipment

1. Roller tube mixer (Ratek, Australia).
2. Vortex.
3. Mass spectrometry—Thermo-Finnigan (Ultimate 3000) liquid chromatography system and ion-trap mass spectrometer.

3 Methods

3.1 Preparation of Protein and Tryptic Digests

Important to carry out all manipulations at the specified temperature.

1. Grind 100 mg of the sample in liquid nitrogen into a fine powder, then transfer the powder into an Eppendorf tube.
2. Mix the freshly ground sample with 1:1 with cold TA solution. And vortex for 30 s. Centrifuge the sample 14,000 × *g*, 15 min (4 °C) and remove supernatant to waste (*see* **Note 4**).

3. Fill tube with MA and vortex until pellet is fully dispersed. Centrifuge the sample at 14,000×g, 15 min (4 °C) and remove supernatant to waste. Allow pellet to dry (see Note 5).
4. Fill tube with 80 % ACN and vortex until fully dispersed. Centrifuge the sample 14,000×g, 15 min (4 °C) and remove supernatant to waste. Allow pellet to dry.
5. Add 0.8 ml of P-SDS, mix and leave for 5 min on ice. Centrifuge the sample 14,000×g, 15 min (4 °C). Transfer the upper phenol phase to a new 2 ml Eppendorf tube. Fill the tube with MA and leave at −20 °C for 1 h.
6. Centrifuge the sample 14,000×g, 15 min (4 °C) and remove supernatant to waste. Allow pellet to dry (see Note 6).
7. Fill tube with methanol, and vortex until dispersed. Centrifuge the sample 14,000×g, 15 min (4 °C) and remove supernatant to waste. Allow pellet to dry.
8. Fill tube with 80 % (v/v) acetonitrile and vortex until fully dispersed. At this stage, remove an aliquot (50 μl) and place into a separate tube for protein estimation. Centrifuge both samples at 14,000×g, 15 min (4 °C) and remove supernatant to waste. Allow pellets to dry. For protein estimation, resuspend the pellet in a 2 % SDS solution. Centrifuge the sample at 14,000×g, 15 min, and transfer the supernatant to a new Eppendorf tube. Protein estimation can then be performed (e.g., BCA kit from Pierce, see Note 7).
9. After drying the sample, it is then digested in trypsin. A 1:50 (w:w) trypsin–protein ratio is recommended. Use the trypsin working solution (which contains 10 μg/ml) and add 200 μl of the buffer. This is left on a rotator (20 rpm) at 37 °C overnight (*see* **Note 8**).

3.2 Enrichment of Phosphopeptides Using TiO_2

1. We use batch-chromatography for the separation of phosphopeptides, therefore 3 mg TiO_2 is weighed out in an Eppendorf tube (*see* **Note 9**). Eppendorf brand "eppendorf" tubes are used, which have little polymer-carry over when running on sample on a mass spectrometer. All procedures are carried out at room temperature.
2. To 100 μl of sample, add 500 μl of DHB buffer. Invert to mix.
3. Centrifuge the TiO_2 beads in order to wash them and remove any "fines" that may be present. Resuspend the beads in 200 ml of 50 % ACN and recentrifuge to ensure that fines are gone.
4. To the TiO_2 beads, add the 600 μl of the sample plus DHB buffer. Invert to mix and allow the peptides to bind for at least 30 min.
5. Gently centrifuge the sample (2,000×*g*, 2 min). Carefully remove the supernatant and discard (or could be kept as back up).

6. Wash the beads using the DHB buffer. Add 500 μl of buffer to the beads, mix by inverting and centrifuge at 2,000 × *g*, 2 min. Carefully remove the supernatant and discard.
7. Was the beads using wash buffer. Add 500 μl of buffer to the beads, mix by inverting and centrifuge at 2,000 × *g*, 2 min. Carefully remove the supernatant and discard. Repeat this step three times.
8. Once the beads have been spun for the final time, remove the supernatant and add 25 μl of the elution buffer. Ensure that the beads are mixed. Centrifuge the sample at 10,000 × *g* for 5 min. Take the supernatant which consists of the phosphopeptides.
9. Acidify the sample by adding 0.3 μl of TFA to the supernatant.
10. Transfer to a vial for mass spectrometry analysis (*see* **Note 10**).

3.3 Enrichment of Phosphopeptides Using IMAC

1. Equilibrate the IMAC beads in equilibration buffer. Wash the beads at least three times in this buffer before proceeding.
2. Repeat steps in Subheading 3.1.
3. To the freshly washed beads, add 100 μg of trypsin digest. Allow the sample to bind for 30 min on the roller tube mixer (30 rpm) at room temperature for 20 min.
4. Wash the IMAC beads by adding 500 μl of wash buffer to the beads. Allow 2 min on the roller tube mixer (30 rpm), centrifuge (2,000 × *g*, 2 min) and discard the supernatant. Repeat this step twice.
5. Elute the peptides using 100 μl of the elution buffer. Allow the sample to rotate at 30 rpm for 5 min (*see* **Note 11**). Ensure that the beads are mixed. Centrifuge the sample at 10,000 × *g* for 5 min. Take the supernatant which consists of the phosphopeptides.
6. Acidify the sample by adding 0.3 μl of TFA to the supernatant.
7. Transfer to a vial for mass spectrometry analysis (*see* **Note 10**).

3.4 Mass Spectrometry

1. The analysis is carried out on a high performance, liquid chromatography system from Thermo-Finnigan (Ultimate 3000).
2. The sample is typically placed into a glass vial, where it is picked up by a liquid chromatography system. The peptides are trapped on C-18 packing material when buffer A (0.1 % formic acid) is applied. The peptides are slowly eluted over time with an increasing amount of acetonitrile (typically 2–40 %, where Buffer B contains 0.1 % formic acid plus 80 % acetonitrile). As the peptides elute from the column they are sprayed into the mass spectrometer for analysis.
3. A typical analysis is shown in Fig. 1.

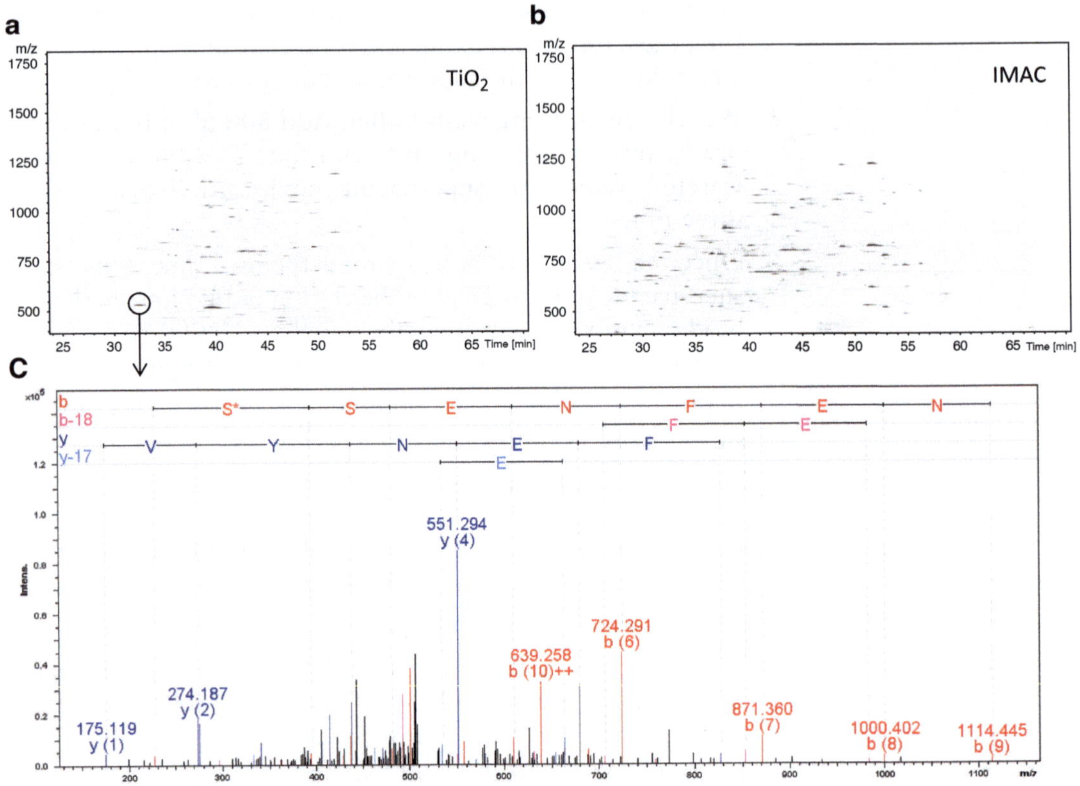

Fig. 1 The different strategies for phosphopeptide enrichment. One sample was prepared in the fashion as described in this chapter and split into two identical fractions. (**a**) Represents phosphopeptides eluting from TiO_2 beads. (**b**) Represents the same sample run and eluted from IMAC. Clearly, major differences in the peptides that are seen can be shown. (**c**) Demonstrates the MS/MS data of one peptide (*circled* in (**a**)) which depicts phosphorylation of the serine residue (S*)

4 Notes

1. Urea should not be heated as this may introduce considerable carbamylation of proteins.
2. This must be made fresh on the day. If wanting to store it can be kept at −20 °C for up to 1 month. One freeze–thaw only is recommended. Trypsin will slowly lose its activity if repetitive freezing–thawing cycles are performed. Do not heat above 37 °C, otherwise trypsin loses activity.
3. We have found this to be the critical step in loss of phosphopeptides. This buffer must be made fresh and ensure that the pH is correct.
4. Care must be taken to use fresh solutions of the highest grade possible. Many modern mass spectrometers will detect plasticizers

in the sample if lower grade "eppendorf" tubes are used. Use Eppendorf brand "eppendorf" tubes.

5. If 14,000 ×*g* cannot be obtained generally the highest setting on a microfuge is adequate.
6. At this stage of the procedure the pellet should be white in color.
7. Ensure that any protein estimation kit is compatible with SDS.
8. Adequate digestion time must be allowed in order for trypsin to work effectively. To test for digestion, a fraction can be run into an SDS-PAGE gel and silver stained. Peptides may appear at the bottom of the gel, however no polypeptide or protein should be present in the sample.
9. TiO_2 can be obtained from a disassembled column or brought through GE Healthcare or Inertsil. Care should be taken since TiO_2 has eight modifications and it is unclear on the specificity of each. Anatase and Rutile may produce non-phosphorylated artifacts (*see* note http://www.inertsil.nl/pdf/Phosphopeptide%20Purifying%20&%20Enrichment%20Products_GLS.pdf).
10. Running the sample fresh is also better than storing and running later. Peptides tend to "stick" to the sides of plastic and glass. As such, storage should only be done at −20 °C or less.
11. As per the TiO_2 protocol ensure that the pH of this elution buffer is correct. This is the major cause for lack of peptide.

References

1. Karas M, Bachmann D, Hillenkamp F (1985) Influence of the wavelength in high-irradiance ultraviolet-laser desorption mass-spectrometry or organic-molecules. Anal Chem 57:2935–2939
2. Griffiths J (2008) A brief history of mass spectrometry. Anal Chem 80:5678–5683
3. Baker MA et al (2010) Use of titanium dioxide to find phosphopeptide and total protein changes during epididymal sperm maturation. J Proteome Res 10:1004–1017
4. Larsen MR et al (2005) Highly selective enrichment of phosphorylated peptides from peptide mixtures using titanium dioxide microcolumns. Mol Cell Proteomics 4:873–886
5. Thingholm TE et al (2006) Highly selective enrichment of phosphorylated peptides using titanium dioxide. Nat Protoc 1: 1929–1935
6. Thingholm TE, Larsen MR (2009) The use of titanium dioxide micro-columns to selectively isolate phosphopeptides from proteolytic digests. Methods Mol Biol 527:57–66
7. Schirmer EC, Yates JR III, Gerace L (2003) MudPIT: a powerful proteomics tool for discovery. Discov Med 3:38–39
8. Bodenmiller B et al (2007) Reproducible isolation of distinct, overlapping segments of the phosphoproteome. Nat Methods 4:231–237

Chapter 19

Plant Metabolomics: From Experimental Design to Knowledge Extraction

Amit Rai, Shivshankar Umashankar, and Sanjay Swarup

Abstract

Metabolomics is one of the most recent additions to the functional genomics approaches. It involves the use of analytical chemistry techniques to provide high-density data of metabolic profiles. Data is then analyzed using advanced statistics and databases to extract biological information, thus providing the metabolic phenotype of an organism. Large variety of metabolites produced by plants through the complex metabolic networks and their dynamic changes in response to various perturbations can be studied using metabolomics. Here, we describe the basic features of plant metabolic diversity and analytical methods to describe this diversity, which includes experimental workflows starting from experimental design, sample preparation, hardware and software choices, combined with knowledge extraction methods. Finally, we describe a scenario for using these workflows to identify differential metabolites and their pathways from complex biological samples.

Key words LCMS, Metabolic profiling, Extraction, Homogenization, Fragmentation of metabolites, Metabolite identification, Metabolomics data mining, *Arabidopsis*

1 Introduction

Metabolites are low molecular weight products of enzyme-catalyzed reactions that occur naturally within cells or are secreted post-synthesis, and perform important structural and functional roles. Primary metabolites such as amino acids, sugars, sugar phosphates, and organic acids, are mostly involved in essential functions of growth and development. Secondary metabolites, on the other hand, are mostly involved in specialized functions, such as environmental response or defense against pathogens. Phenylpropanoids, terpenoids, and alkaloids form some of the major classes of secondary metabolites [1]. Since primary metabolites are directly involved in growth and development, they are present in all living tissues or cells. Secondary metabolites play a somewhat specialized role; hence, they are generally not localized in all cells, but are synthesized and localized in specialized cells, tissues, or plant parts. The basis of

Ray J. Rose (ed.), *Legume Genomics: Methods and Protocols*, Methods in Molecular Biology, vol. 1069, DOI 10.1007/978-1-62703-613-9_19,

classification of primary and secondary metabolites, however, is now considered highly overlapping. For instance, in plants, several isoprenoids, such as chlorophyll and carotenoids act synergistically during photosynthesis, while gibberellic acid and abscisic acid act antagonistically during seed germination [2]. Since many secondary metabolites have essential functions and also contribute towards plant growth and development, many plant scientists consider the classification of primary and secondary metabolites as obsolete [1].

Biosynthetic pathways of metabolites form highly connected metabolic networks. A model plant, such as *Arabidopsis thaliana*, contains at least 3,000 metabolites that are synthesized from nearly the same number of enzyme-catalyzed reactions, and are organized into more than 500 pathways (AraCyc-Arabidopsis metabolism database [3, 4]).

1.1 Metabolome and Metabolomics

Metabolome refers to the complete set of metabolites that arise from cellular activities. *Metabolomics* is the study of metabolites that aims to identify, measure, and interpret complex time-related concentration, activity, and fluxes in cells and tissues. In the context of genes-to-phenotype flow of information, metabolites represent the closest molecular entities to the phenotype of a cell. Such a metabolic phenotype of a cell can, therefore, also be referred to as its "*metabotype*." Study of metabolites through metabolomics approaches offer several unique benefits. Firstly, changes in gene and protein expression can cause amplified changes in the metabolism, thereby making it easier to capture such fluxes. Next, changes in metabolite levels represent the integrated responses at gene and protein expression levels. In most cases, such changes in metabolite levels are less prone to rapid fluctuations, hence making detection of metabolites easier. Lastly, metabolites are not genetically encoded. Metabolite structures are, therefore, conserved in all biological systems. Hence, metabolomics can be readily used with non-model organisms. Developments in high throughput metabolomics techniques now make it possible to measure thousands of metabolites simultaneously, thus offering an insight into the vast diversity of plant metabolites that remain yet to be discovered.

1.2 Diversity and Role of Metabolites

In the plant kingdom, metabolites are extremely diverse, with estimates indicating nearly 200,000 different types [5] (Table 1). This diversity enables them to be metabolically active in a number of biochemical pathways specific to a variety of organs and composed of different cell types. Expression of secondary metabolism pathways is in general, a feature of cell specialization and is integrated into the pattern of differentiation of those cell or tissue types [6]. Specialization of pathways also exists between different compartments within the cell, such as different organelles and the endomembrane system. Diversity of secondary metabolites is characterized by a number of specialized enzymes that provide the

Table 1
Diversity of plant metabolites

Types of plant metabolites	Number
Terpenoids	22,000
Alkaloids	21,000
Flavonoids, tannins	5,000
Phenylpropanoids, lignins, coumarins, lignans	2,000
Polyacetylenes, fatty acids, waxes	1,500
Polyketides	750
Carbohydrates, organic acids	200

Approximate number of known structures of selected classes of metabolites (adapted from Wink et al. [5])

basic chemical skeleton for the metabolites. Such core metabolites can then be further modified by conjugations, hydrolysis, or condensation reactions among others, resulting in new structures.

Plant hormones represent a good example of how metabolite conjugations serve as an important mechanism to regulate different biological processes. Processes such as de novo biosynthesis, import and export rate in cells, interconversion of distinct forms of cytokinins, transient inactivation by conjugation (mainly glucosylation), and catabolic reactions together regulates levels of cytokinin in plants, thereby resulting in complete loss of their biological activity [7]. In case of the hormone auxin, its conjugates are generally considered as temporary reservoirs of inactive indole acetic acid (IAA), which on hydrolysis regenerates the free active IAA [8]. Another good example of structure–function relationships that are affected by conjugations is of flavonoids. Several flavonoids, such as quercetin, are toxic in nature and are thus stored in a conjugated form. Core flavonoid skeletons can undergo various modifications such as glycosylation, malonylation, methylation, hydroxylation, acylation, prenylation, or polymerization, leading to an even larger diversity of end-products [9, 10]. Conjugation of flavonoids, mainly glycosylation, leads to conversion of flavonoids into less toxic molecules, which are then stored in their conjugated forms in cells. While core flavonoid structures are synthesized by similar sets of enzymatic reactions in different plant species, several specialized flavonoid pathways also exist in different species [11]. For example, isoflavone pathways are somewhat restricted to legumes and these metabolites have strong antimicrobial activities which help in defense against bacterial or fungal pathogen attacks. Isoflavones, such as genistein, daidzein, and others in soyabean also make them popular anticancer nutraceuticals for human consumption.

This diversity of metabolites not only provides benefits to plants and humans but also has a strong ecological role. Metabolites have important roles in maintaining metabolic homeostasis and have long been associated with ecology, plant defense, and human health. A comprehensive review on the functional and ecological role of metabolites is provided by Wink et al. [5]. More than a hundred flavonoids and other secondary metabolites secreted by Arabidopsis roots and identified using metabolomics, play a critical role in providing nutritional advantage to root-associated growth promoting bacteria [12]. Such ecological roles have been widely reported to influence plant-microbe interactions [13, 14].

1.3 Plant Metabolic Networks

A large number of metabolic pathways are involved in the synthesis of metabolites. PlantCyc, a metabolic pathway reference database contains more than 800 pathways from over 300 plant species, determined using both experimental procedures and computational predictions [4]. This represents a growing number with the continuous addition of newer pathways as they are discovered using metabolomics. The number of core biosynthetic pathways is restricted and distinct despite an enormous variety of secondary metabolites. These metabolic pathways in a cell are organized in the form of networks. They are characterized by high connectivity, with certain metabolites serving as common substrates at branch points to form a number of different subnetworks. Metabolic networks allow plants to respond efficiently to changes in their environment. Plants are subjected to various abiotic (temperature, light, UV, nutrient availability) and biotic stresses (pathogen attack, herbivores). Being sessile, plant response relies on modulation of biochemical processes, which leads to fluxes in metabolome. Several studies on metabolic pathways regulation in Arabidopsis has shown multiple pathways affected in response to perturbational effect. Studies on salicylic acid (SA) and jasmonic acid (JA) cross-talk have shown mutual antagonistic interaction between SA and JA-dependent signaling, while in a few cases, synergistic interactions have been described as well [15]. Such cross-talk of signaling process can occur through connectivity of metabolic networks. Plant metabolome and the links that connects its members within the metabolic networks are, therefore, characterized by considerable plasticity and modularity.

"*Plasticity*" of a system or a network is its ability to tolerate alterations in its components or structure and it cannot be directly measured [16]. A metabolic network has a fundamental property of being robust that allows it to adapt to the perturbations introduced by environment or genes. Plasticity of metabolic networks is also related to the importance of particular pathways for the biological system. Adaptability of metabolic networks to perform its functions by a number of alternative pathways increases the survivability of the organisms under diverse environmental conditions. Metabolomics approaches, such as those using stable isotopes to study the dynamic nature of metabolic networks are helping in better understanding of plasticity of networks. In general, primary metabolic pathways are more robust to the changes, as they are critical and directly related to growth and survival of the system and hence are under tight regulatory control [17]. Secondary metabolism networks, on the other hand, do not have in relative terms, as tight a regulatory system as primary metabolism networks. This comparison underlines the role of evolutionary processes to the plasticity of metabolic network.

The concept of "*modularity*" of metabolic networks stands for independent and self-contained property of a system. Studies on plant systems have suggested that the modular structure of a network emerges in response to changes in environmental conditions, since a small part of a metabolic network is needed to establish homeostasis against the applied changes [18]. The isoprenoid metabolic network in Arabidopsis presents an excellent example of network modularity, where different branches such as carotenoids and brassinosteroids metabolic pathways exist independently despite being connected at branch points. The isoprenoid metabolic network has two alternative routes for synthesis of metabolites due to subcellular compartmentalization of metabolic pathways [2]. Comparison of the secondary metabolite composition of plant tissues in *Arabidopsis thaliana* [19] and primary metabolites in *Lotus japonicas* [20] revealed that plants have evolved metabolic systems for producing a variety of metabolites in a tissue-dependent manner in order to adapt to various environmental factors. Different alternative modules, therefore, function in different tissues or environmental conditions.

1.4 Qualitative vs. Quantitative Approaches in Metabolomics

Characterization of metabotype provides systems-wide understanding of the plant developmental changes or their interactions with environmental factors [21]. Metabolomics approaches are suited to provide a detailed description of the members of the metabolic networks (i.e., metabolites and their levels). Metabolites from samples can, therefore, be characterized by performing either qualitative or quantitative analyses. Qualitative analysis is performed by comparing metabolic profiles of different samples such as from different legume germplasm members, cultivars, genotypes and others. For example, metabolite profiles of legumes can be compared when grown in a fertile soil with those grown in drought conditions to identify the classes of metabolites enriched in each case. This can lead to identification of biomarkers indicating the physiological state of the plant in response to its environment, such as metal toxicity, herbicide treatments as well as to compare wild-type and transgenic plants [22].

Absolute or relative levels of metabolites can be quantitatively measured from minimal amounts of biological material from samples differing in genotype or treatment conditions. For example, a global approach may be taken to observe system-wide changes in response to an environmental perturbation, in which as many metabolites as possible are measured and compared between samples without bias [23]. Once the biological question has been framed, then either a targeted approach or a non-targeted approach can be used.

Targeted metabolomics: This approach is used for testing a specific biological hypothesis, wherein selective classes of metabolites or pathways are affected. These sets of metabolites are then isolated and quantified from the biological system. As targeted metabolomics involves detecting a small number of metabolites having specific properties, it requires highly sensitive and accurate instruments along with good knowledge of the chemical nature of the metabolites. Biological knowledge is used to choose sampling strategy to identify conditions when differential metabolites of choice are expected to have maximal changes. Extraction chemistry is required for differential extraction wherein metabolites not of interest are eliminated. With innovative techniques, new families of metabolites and metabolic networks can be discovered using this approach. For example, this approach can be effective in characterizing the influence of fertilizers or effects of genetic modifications on a specific enzyme [24].

Non-targeted metabolomics: This approach produces a comprehensive coverage of the metabolome and can identify the metabolite profiles of samples without any bias. Thus it can be used for discovering novel metabolites as biomarkers [25]. By performing global metabolite profiling, new links in cellular pathways to biological mechanisms are being discovered and are aiding in a better understanding of cell biology, physiology and plant biotechnology. However, a number of challenges exist in achieving this goal (described in detail in Subheading 1.8). Firstly, profiling diverse class of metabolites requires multiple extraction strategies to recover different classes of metabolites such as hydrophobic, hydrophilic, neutral or charged [26]. Furthermore, it requires instruments having ultra-high resolution, high scan speeds producing large amounts of data, which further need advanced statistical analysis to obtain meaningful information [27]. Several metabolomics studies have used a combination of targeted and non-targeted approach to identify novel metabolites [28]. For example, targeted analysis can be performed on the biologically interesting pathways, which were initially determined using the non-targeted approach, to identify and quantify putative or novel metabolites.

1.5 Choice of Metabolomics Hardware Based on Experimental Approach

Nucleic acid-based methods can be used for obtaining information about the genetic sequence and expression from the genome. However, unlike transcripts and proteins, the molecular identity of metabolites cannot be extracted from genomic information. Identification and quantification of metabolites requires high-resolution instrumentation. Thus, with recent developments in mass spectrometry (MS) and nuclear magnetic resonance (NMR), metabolomic studies can be performed with high specificity and reproducibility in a qualitative and quantitative manner. The choice of experimental approach, sample type and experimental design are the basis for choosing an analytical instrument. A brief overview of these instruments is described below.

1.5.1 Nuclear Magnetic Resonance

NMR is one of the most widely used techniques for metabolite structural elucidation and quantification because of its highly selective and non-destructive approach [29]. NMR is also the preferred method for confirming chemical structure of an unknown metabolite as techniques such as mass spectrometry can only predict putative structure and chemical formula of an unknown metabolite. However, it suffers from relatively low sensitivity with a lower limit of detection of about 1–5 μM from relatively large sample sizes of 500 μl, thus restricting the metabolic coverage that can be achieved from it. Hence, NMR is not the preferred technology for non-targeted approach of metabolomics. Nevertheless, NMR has widely been used in plant metabolomics for quality control purposes, in chemotaxonomy to classify and characterize plants based on their distinct metabolic signatures and for analyzing the effects of targeted genetic modifications in plants [29, 30].

NMR based methods can be used to analyze both solutions and insoluble or solid-state samples. For example, NMR was used for identifying the chemical forms of aluminum in the xylem sap of plants [31]. One-dimensional and two or multi-dimensional NMR with higher sensitivity can also be used in metabolite identification or flux analysis. However, NMR spectroscopy is significantly less sensitive than MS despite several technological advances such as using higher field-strength superconducting magnets, cryogenically cooled probe technology, and miniaturized radiofrequency coils [32]. A good review on NMR is given by Wishart et al. [33].

1.5.2 Mass Spectrometry

High sensitivity, resolution and high throughput approach for identification and quantification of metabolites make mass spectrometry-based metabolomics one of the widely used technologies in targeted and non-targeted metabolomics. Mass spectrometry is an analytical technique that measures the mass-to-charge ratio of charged particles. Equipped with research and advancement of technology, modern MS provides highly specific chemical information directly related to the chemical structure such as accurate mass, isotope distribution patterns for elemental formula determination, and characteristic daughter ions by fragmentation for structure elucidation or identification via spectral matching to authentic compound data. High sensitivity of MS allows detection and measurement of primary and secondary metabolites even at picomole to femtomole levels [34]. New generation instruments, such as Fourier transform-ion cyclotron resonance MS (FT-ICR-MS), Orbitrap MS and others provide high resolution of 100,000 and offer high mass accuracy with the ability to detect differences less than <2 ppm (parts per million). Hence, for example, a metabolite weighing 80 Da can be measured with an accuracy within 0.16 mDa, whereas, one weighing 10,000 Da with an accuracy of 20 mDa. One of the major weaknesses of MS in metabolomics is quantification as the number of ions detected is not directly proportional to the metabolites in the sample. Furthermore, the MS signal intensity of any compound is also affected by sample preparation and the molecular environment of the sample matrix. Dettmer et al. [35] provide a thorough analysis on the various steps involved in mass spectrometry-based metabolomics.

1.5.3 Hyphenated Mass Spectrometry Techniques

Methods for sample separation, such as gas chromatography (GC), high-performance or ultra-performance liquid chromatography (LC) and capillary electrophoresis (CE), each having its own advantages and limitations, have been used in conjunction with various types of mass spectrometers in plant metabolomics research.

MS-based metabolic profiling can be performed either by directly injecting the complex metabolite mixture, after particulate removal, into MS or separating metabolites first by chromatography before injecting it into MS. Users can also choose an intermediate approach by performing high throughput, solid phase extraction and directly injecting the extracts into MS [36]. Direct injection approaches provide a high throughput screening tool for large number of samples, such as those encountered during screening of large mutant populations or geographic region-wide study. As this approach does not involve time required for separation of complex mixtures in chromatography techniques (which can take up to 20 min per sample), it is relatively faster and efficient for handling large number of samples. Although metabolites can be identified by direct injection at a faster speed, they are susceptible to many problems in obtaining data having good coverage from biologically complex extracts. One of the major problems of injecting complex mixtures is ion suppression or enhancement. This happens when the solution/buffer matrix is co-extracted with the analytes, which leads to inaccurate quantification of the metabolites. This is caused by competition between ions for ejection from the droplet during desolvation. In addition to ion suppression effects, MS data interpretation from direct injections is challenging, as the differences between adducts such as sodium or potassium and product ions after ionization are difficult to distinguish as there is no separation stage.

Chromatography coupled to MS offers an excellent solution for complex mixture analyses as it enables the separation of ions (based on size, charge, and such) and hence has been extensively used in metabolomics. This helps in removing matrix effects and provides additional orthogonal data (i.e., retention time/factor/index) valuable for accurate metabolite identification. GC, LC, and CE are three predominant chromatographic techniques that have been incorporated in MS-based metabolomics. These coupled methods are also known as hyphenated MS techniques.

The GC-MS based approaches are suitable for the analyses of both volatile and nonvolatile compounds following derivatization. Since GC based chromatography requires metabolites to be in gas phase, metabolites are required to be thermolabile. However, relatively few compounds in plant metabolome meet this requirement in their native state (e.g., short chain alcohols, acid, esters, and hydrocarbons) [37]. Plants are a major source of specialized class of volatile metabolites that contribute to vegetable and fruit aromas and plant defense responses. This has made GC-MS as a popular choice for certain users in plant metabolomics [38–40]. A major advantage of GC-MS is the availability of comprehensive robust libraries of retention time and mass spectra of compounds that can be shared among laboratories. This enables identification of sugars, sugar alcohols, amino acids and organic acids covering the central pathways of primary metabolism in plant samples. A comprehensive protocol on the application of GC-MS techniques for plant metabolomics is given by Shuman et al. [41]. However, the limitation as to the size and types of metabolites that can be analyzed along with extensive preparation and derivatization required is a big concern. Hence, metabolomics practitioners have resorted to liquid phase-based methods.

The LC-MS based approach allows for mild heating during ionization and is suitable for both thermolabile and thermostable metabolites. Thus, LC-MS approach gives a better coverage and is most suitable for non-targeted metabolic profiling. LC-MS also facilitates the usage of a wide variety of columns, such as those based on reverse phase, ion exchange and hydrophobic interactions principles [42]. This provides the potential to analyze a wide variety of metabolites in plants. Development of fast and efficient ultra-high performance liquid chromatography (UHPLC) has substantially increased chromatographic resolution and decreased the analysis time and peak capacity compared with HPLC. UHPLC has been shown to increase numbers of detectable compounds by 20 % compared with HPLC or LC method [43]. One of the major problems in knowledge extraction from LC-MS data is lack of reference databases. Variations in separation columns for HPLC and UHPLC render cross-comparison between different LCMS methods difficult and hence, a robust database for LC-MS has not yet been feasible. As discussed later (described in Subheadings 1.8, 3.1.6 and 3.2), large portion of plant metabolites are not represented in the databases. Hence, there is a need to determine chemical structures of novel metabolites using systematic approaches. Validation and identification of novel metabolites can be performed using NMR or tandem MS/MS and MRM (Multiple Reaction Monitoring) methods (for reviews on methods in LC-MS kindly refer to Bin et al. [44]).

The CE-MS is a powerful technique used for the separation of charged metabolites. It requires relatively smaller sample quantity than in LC-MS and has higher separation efficiency than LC. However, it has low-concentration sensitivity and thus requires enrichment of metabolites within the samples [45]. Furthermore, there is a lack of reference libraries for CE-MS.

Various ionization techniques, such as, electron ionization (EI), electrospray ionization (ESI), atmospheric pressure chemical ionization (APCI), chemical ionization (CI), MALDI, desorption ESI (DESI) and extractive ESI (EESI) can be used depending upon the range and nature of metabolites [46]. Thus, the goal of the metabolomics project, instrument costs, type and complexity of samples should be considered before choosing an instrument that is most likely to yield high-quality data for the research projects.

1.6 Experimental Design

Plants are able to survive in various environmental conditions by reconfiguration of their plant metabolic networks to maintain metabolic homeostasis and achieve a new steady state. Plant scientists often study effects of specific perturbations to understand the cause and effect relationships. Such experiments involve deliberate and controlled environmental perturbations to study the functional response of the system. Field sampling requires even a more robust experimental design to tackle the higher levels of variation inherent in environmental samples. Therefore, it is important to develop an experimental design that can provide insights into biological responses without confounding effects [47]. Limitations listed in Subheading 1.8 need to be considered so that the experimental design can be tuned to the biological question of interest. Care must be taken to ensure that the experiment is free of errors resulting from both biological and technical variability [48] (*see* **Note 1**). A pilot study to optimize various extraction and analytical procedures is ideal before embarking on a large-scale experiment. Some procedures are described in Subheading 3.1. The adequate number of biological and technical replicates can be determined by identifying the level of biological and nonbiological sources of variation ascertained from the pilot study. This will ensure reliability and reproducibility of the datasets. Factorial or randomized study designs can be considered to ensure minimal confounding effects due to machine, genotype, or the growth conditions and in separating out influences by the various data preprocessing steps.

The metabolomics research community has framed a set of minimal reporting standards for experiments to aid in setting up benchmarks for analysis and comparison of metabolomics datasets [49]. This requires diligently recording metadata such as temperature, growth conditions (minimal set of data of reporting standards and general guidelines) that will aid in biological interpretation and separating nonbiological sources of variation [49, 50]. A list of items to be considered for experimental design in the field of plant metabolomics has been nicely described in a review by Gibon et al. [47].

1.7 Data Analysis Strategies for High Throughput Omics Datasets

High-throughput metabolomics experiments produce large amounts of data, such as, measuring 10,000 features (ions) across 300 different conditions running into gigabytes of data (*see* **Note 2**). Handling and analyzing such complex data sets has a great impact on the ability to identify and quantify various metabolites that are different between samples/conditions, and influence the resulting biological interpretation. Extracting the relevant information from the overwhelming amount of data generated by these high throughput techniques is an important objective for knowledge discovery in this field [51].

In high-throughput omics datasets, systematic errors during experiments, misplaced samples and instrument errors can lead to serious misinformation during the analysis step. Hence, performing exploratory data analysis to quickly check data quality is a crucial step for checking experimental errors and can help in efficient analysis. Raw data obtained from the instruments has to be preprocessed to minimize noise and statistically robust analyses must be performed to capture real biological variations [52, 53]. A number of application-specific software pipelines and statistical techniques are available for handling and processing of experimental datasets to obtain meaningful biological information [54–57].

1.8 Challenges in Plant Metabolomics

The last decade has witnessed great advances in analytical techniques and instrumentation for the characterization of the complete complement of an organism's metabolome. However, a number of technical challenges from sample preparation to metabolite identification hinder comprehensive characterization of the metabolome.

Metabolites, being highly diverse, require specialized extraction protocols based on their spatiotemporal location and the genotype of the organism under study. The diverse chemical structures with their diverse chemical properties can cause large variations during extraction, leading to unequal stability, matrix effects, differences in detection limits and linearity ranges [46]. Thus, there is no single procedure that can be optimized to extract all metabolites as experimental conditions suitable for a given class of compounds will not be suitable for the extraction of other compounds or can interfere with their analysis. Therefore, it is necessary to tailor the extraction protocol to the metabolites to be profiled to accommodate variations in solubility, reactivity and other physicochemical properties [58]. This is further exacerbated while performing metabolomic profiling for characterizing novel metabolites or pathways where the nature of metabolites are largely unknown.

Another major challenge in sample preparation for metabolomics is to accurately capture the metabolite levels in the tissues when they are actively metabolized under specific conditions or at specific developmental time or location. Hence, first the metabolism has to be quenched (to obtain the metabolic snapshot of the sample at a particular time) and then cells/tissues should be homogenized followed by an extraction protocol (such as Solid Phase Extraction) to isolate the metabolites (*see* **Note 3**). Metabolites are also highly dynamic in size (ranging from 50 to 2,000 Da) and concentration, hence, the extraction protocol should be able to (1) separate out the extracellular and intracellular metabolites, (2) remove unwanted debris and (3) ensure that there are no further chemical changes or loss of metabolites [59]. For example, biosynthesis of different classes of metabolites occur in different regions of the plant such as hydrophilic substances stored in the vacuole or the lipophilic metabolites in latex, resin ducts, trichomes, or cuticle [5]. The extraction protocol will also vary depending upon the choice of instrument and study.

Metabolite identification is one of the key goals of high throughput metabolic profiling. A number of crucial metabolites remain to be unidentified. A 2009 survey by American Society for Mass Spectrometry (ASMS) revealed that majority of metabolomics users felt that the biggest bottleneck for this field is in metabolite identification and obtaining valid biological interpretations [60]. So far, metabolic profiling has been largely confined to a fairly limited list compounds that can be extracted together including major primary metabolites (sugars, sugar phosphates, amino acids, and organic acids) and certain secondary metabolites (e.g., phenylpropanoids, alkaloids). In high throughput analysis, metabolite identification is further complicated as there are a number of ambiguities present in direct matching of the m/z ratios to the databases (*see* **Note 4**). Furthermore, identification of metabolites is limited by the lack of an accurate and well-curated database for the structure of compounds [14, 58].

Hence, identification with high confidence relies on direct comparison to standards or structure elucidations using fragmentations patterns obtained using tandem MS and NMR approaches. This provides a possible list of structures based on the isotopic composition and fragmentation pattern of the compound. However, with the probable chemical space ranging from 600 million-determined using the seven golden rules, to 8 billion chemical formulas (theoretically possible C, H, N, S, O, P-formulas for compounds up to 2,000 Da), it becomes increasingly difficult to identify novel compounds when the reference compounds are not available [61]. Thus, structure determination and validation are increasingly becoming part of serious metabolomics programs.

As described earlier, LC-MS (described in Subheading 1.5) enables the detection of the diverse range of metabolites with varying polarity, magnitude and chemical nature. LC-MS can, therefore, be used for both targeted and non-targeted metabolomics. Here, we describe a user-friendly protocol for performing metabolomics experiments using LC-MS techniques.

2 Materials

2.1 Harvesting Samples

1. Recording equipment such as digital cameras for capturing high resolution images of the state of plant tissues or organs prior to harvesting.
2. A balance with a mass precision of ±0.1 mg to ensure accurate fresh weight determination of plant tissues.
3. Alcohol sterilized scissors or scalpels for separating the tissue from the plant and tweezers for picking up the collected tissue.
4. Round-bottomed 2 ml micro centrifuge tubes.
5. A notebook to record meta information such as sample identity numbers, developmental stage, weights, day, time, duration of harvesting, and physiological parameters like humidity, light fluence rate, light period (day length) and temperature at harvest.
6. Liquid nitrogen (Liq N_2).

2.2 Homogenization and Extraction

1. Homogenization is performed using a ball mill (e.g., available from Retsch, Germany).
2. Glass beads of size 0.5–2 mm can be used for homogenization.
3. Degassing devices, such as ultrasonic bath or pure argon gas bomb and a liquid cooling system are required to prepare the extraction mixture.
4. Ice bath to keep the samples at low temperature.
5. Micropipettes, which should be calibrated every 6 months to ensure accurate measurement.
6. Ice bath and liquid nitrogen flask for storing samples during extraction process.
7. A refrigerated centrifuge (20,800 × *g*), equipped with a rotor for microcentrifuge tubes, is used for fractionation.
8. A microtube shaker for extraction of metabolites.
9. In general, 80 % methanol is used as extraction buffer. Although different extraction buffers with varying percentages of

methanol, chloroform and water can also be used (*see* **Note 5**). All solvents used must be of the highest quality (e.g., >99 % ultrapure HPLC-MS gradient grade purity).

10. U-^{13}C-Sorbitol (200 ng per vial) to be added as an internal standard for normalization of metabolic profiles of different samples (*see* **Note 6**).

2.3 Mass Spectrometric Analysis

1. Ultra-Performance Liquid Chromatography (UPLC) C18 column with particle size 1.8 μm, 15 cm length. One may also use 10 cm column or higher particle size column based upon the complexity of the metabolic extract. Small particle size helps to improve resolution, while longer length helps to have better separation; hence, this setup will be best suited for highly complex metabolic samples. However, longer column length also means longer run time which adds sample running cost, whereas a smaller column may be used with high flow rate.
2. In-line guard columns, which are placed before UPLC column.
3. UPLC vials-total recovery screw cap glass vials.
4. LC-MS Auto-sampler and UPLC coupled to a TOF (Time-of-flight) mass spectrometer. All instruments should be controlled through a single software package for sample injection and for efficient data acquisition. Choose MS system based on type of metabolomics approach chosen for that study (described in Subheading 1.5). For instance, TOF is most suitable for non-targeted profiling type of application, while Q-trap or Triple quadrupole is most suitable for targeted approach, where fragmentation and high sensitivity is desired.
5. Acetonitrile and water acidified with 0.1 N acetic acid or formic acid to be used as LC mobile phase.
6. All MS manufacturers provide proprietary software such as MassHunter (Agilent Technologies), SIEVE (Thermo Scientific), MarkerView (AB SCIEX) to de-convolute raw data and Mass Profiler Professional (Agilent Technologies), SIMCA (UMETRICS) to perform statistical analysis to identify differential metabolites or to quantify its levels. Other freeware such as MetDAT, XCMS, Mzmine, MetaboAnalyst and datPAV [54–57, 62] are among several software options available which can be used to process raw data to generate useful information.
7. Metabolite databases such as AraCyc, PlantCyc, KEGG, METLIN, Chemspider and MassBank [3, 63–68] are required for metabolite identification and pathway mapping. Software such as ChemSketch (ACD/Labs), Mass Frontier (Thermo Scientific) helps to predict possible chemical structure of daughter ions from the structure of the parent ion. These help in mapping the fragmentation patterns resulting from MS2 to the chemical structure of the metabolite.

3 Methods

3.1 Experimental Design for Knowledge Extraction

Techniques used in sample preparation play a critical role in influencing the quality and detection of metabolites in MS. Hence, each sample preparation strategy should consider the type of plant, specificity and nature of metabolites from the sampling region, i.e., from tissues, stem, etc. and the nature of the study, i.e., qualitative or quantitative.

3.1.1 Harvesting Samples for Metabolite Extraction

1. Take plenty of high-resolution images of biological samples that are going to be harvested for metabolite extraction. This step is especially important for plant metabolomics, as this serves as a documentation and checkpoint to ensure that all samples harvested were at the same developmental stage.
2. Cut the organs of interest with alcohol sterilized scissors, scalpels or a cork borer. In some studies, one can collect entire plant or seedlings, which are used for metabolite extraction. Please note that the sampling step largely affects how much 'averaging' of metabolites will be done (*see* **Note** 7).
3. Harvested plant or its organ should be handled with tweezers carefully to avoid any unintended injury to sample and weigh the sample. In general, 100 mg fresh weight of sample is added with 1 ml of extraction solvent for metabolite extraction. Although one may increase or decrease this ratio depending upon application and studies, the ratio used should be strictly followed in all samples.
4. For studies at the cellular level, as a general target, extracts from one million cells are sufficient to provide a high quality profile with the modern mass spectrometers.
5. Collected samples should be stored prechilled in 2 ml centrifuge tube or 15–50 ml tube depending upon amount of samples harvested and must be snap freeze in liquid N_2. Harvested samples should be preserved at −80 °C till it is used for metabolite extraction.

3.1.2 Homogenization

1. Several methods are used for homogenization of biological samples. A good review on sample preparation is given by Fan [59]. Here, we are giving an example of the homogenization method using a ball mill.
2. Prechill sample holder and glass beads of size 1.0 or 2.5 mm.
3. Add the same number of beads to each tube containing samples.
4. Place tubes with glass beads into tube holders and balance them in ball mill. Set tube holders and mill at 35 s^{-1} for 30 s. Take out samples based on capacity of the ball mill and return

them into liquid N_2. All these steps should be performed quickly to avoid thawing of samples.

5. Repeat **steps 1–4** until all samples are homogenized (*see* **Note 8**).

3.1.3 Metabolite Extraction

1. Samples homogenized and kept in Liq N_2 should be taken out one-by-one and kept for few a seconds on ice before adding 80 % methanol as extraction buffer. In general, add approximately 1 ml of extraction buffer for 50–100 mg of homogenized samples, although one may change or optimize this proportion.
2. In general, for non-targeted metabolic profiling, 80 % methanol is used as extraction buffer, as it extracts both polar and organic metabolites. Selection of extraction buffer should be done based on type of metabolites being studied. Criteria for selecting extraction buffer are described in **Note 5**.
3. Add internal standards, e.g., U-^{13}C-Sorbitol (200 ng per vial) for normalization. Although adding internal standards is not essential, it is a good practice as this can be used for normalization as well as peak alignment (for other internal standards, *see* **Note 6**).
4. Incubate samples at 4 °C for 10 min (min) with regular vortexing of samples at every 2 min intervals. This is followed by incubating samples on ice till proceeding to the next step.
5. Centrifuge samples at high speed (20,800 × *g*) for 15 min at 4 °C.
6. Collect supernatant of each sample in a new microcentrifuge tube or a glass vial and store at −80 °C. The cell debris pellet can be discarded and glass balls can be reused after cleaning and autoclaving at 121 °C for 1 h.
7. Recentrifuge the collected supernatant at maximum speed for 10 min at 4 °C and transfer supernatant to a new micro centrifuge tube to ensure that there is no debris or particulate material present in extracted metabolites.
8. Repeat **steps 1–7** for all the samples till extraction is complete.
9. Metabolites extracted can be stored by directly keeping all sample vials in a liq N_2 tank. Another way of storing samples is by deoxygenating extracted metabolic samples using a gentle stream of nitrogen or argon gas for 1 min prior to tube closure. Tubes can then be stored in the dark at −80 °C for about 4 weeks.
10. Extracted metabolites can then be directly used for metabolic profiling using LC-MS approach. Although based on column used in LC, one can also change the solute by first freeze-drying the extracted metabolites and then dissolving them in desired solvent.

3.1.4 Chromatographic Separation

1. Using in-line chromatography with MS provides better resolution and helps to reduce signal to noise ratio compared to direct infusion MS approach. Here, we describe a protocol for developing an LC method.
2. Prepare fresh set of LC mobile phase and needle wash solution. For most purposes, acetonitrile with 0.1 N acetic acid as organic phase and water with 0.1 N acetic acid can be used as aqueous phase. Choice of organic phase ultimately depends upon column in use as well as metabolites under study. Methanol is another popular organic solvent. 50 % methanol is used as needle wash solution (*see* **Note 9**).
3. Prime the system, wash the auto-sampler needle and sample syringe using 100 % methanol to avoid any contamination.
4. Install a fresh, UPLC column and pre-line filter, and then condition the column by operating with 100 % mobile phase B (Acetonitrile with 0.1 N acetic acid) and starting flow at 50 μl/min for 5 min, increase flow rate to 100 μl/min for 5 min, again increase flow rate to 200 μl/min for 5 min before finally reaching flow rate to 300–400 μl/min as operating flow rate for at least 20 min. If using a previously used column, then wash it first by organic solvent and then by a gradient of organic solvent for a longer time to ensure that the column is clean before injecting the samples. It is a good practice to restrict the use of a column for each project to avoid contamination between samples from different projects.
5. Set column oven temperature to 50 °C. This helps to achieve high flow rate and better resolution without building high pressure across the column. Each LC column comes with a particular pressure tolerance mentioned by the manufacturer, and one should always ensure that it is below that value to avoid any damage to the column, which in turn can also affect the profiling spectra.
6. Change the mobile phase composition from 100 % B to 50 % B with operating flow rate and ensure that column pressure is not exceeding the set limit for the column. Column should be allowed to equilibrate for 30 min with this mobile phase composition.
7. Change the mobile phase to 5 % B and let pressure stabilize, this should be done to ensure that pressure across column is not exceeding the set limit. In case it exceeds the set value, one should check or change pre-line filter before running samples. In general, UPLC columns can operate at pressure close to 1,200 bars.
8. After cleaning LC/UPLC system, column and all lines, develop an LC gradient method for samples. A good LC/UPLC method helps to achieve better separation of metabolites and

improves resolution. To establish a LC/UPLC method, use a mix of standards with different degrees of hydrophobicity and mass, and build a gradient method which then can be tested on original samples. For untargeted profiling of samples, use a liner gradient to obtain proper separation of all ranges and properties of metabolites. Ensure that there is no carryover at the end of each sample run, by observing a chromatogram or TIC (total ion chromatogram as used here). If required, increase flow rate or run time to ensure complete separation with no carryover of the samples (*see* **Notes 10** and **11**).

9. Check that the manufacturer's recommended maintenance routines have all been carried out before submitting the batch. For LC-MS, chromatography column should be placed in-line, and ensure that there is no leakage through the UPLC system. Lines should be cleaned before starting batch process. Once LC/UPLC method is optimized, one can go for MS method optimization and setup before submitting the batch.
10. Change or clean the liner (tubing) of the LC and wash it thoroughly after every 50 samples.

3.1.5 Data Acquisition by Mass Spectrometer

1. After metabolite extraction, in-line chromatography with MS is used for data acquisition.
2. First step of data acquisition is to tune MS system to achieve optimal parameters for ion lenses, detector voltage and other settings. These parameters are integral parts of a MS instrument and are very crucial for achieving high sensitivity and resolution. Modern MS systems come with software, which allow users to perform auto tuning. This step is particularly important if the instrument has not been tuned for more than 3 months. MS systems from different manufactures come with different recommendations for tuning and one should perform this step based on recommendations.
3. Tuning of the instrument is followed by system calibration. Calibration of MS system is required to ensure that the system is providing high mass accuracy. MS systems these days (particularly for Q-TOF (quadrupole-time of flight)) offer software options which allow mass corrections based on calibration mixes, also provided by the manufacturers. Calibration mixes are mixtures of a set of compounds of known masses which are used to correct any mass deviation which is then applied to all masses profiled through the instrument. This step allows the user to achieve high mass accuracy.
4. Volume of sample injected per run depends upon sample loop, dilution factor, solvent and columns. Based on these one can inject sample volume from 2 μl to 50 ml depending upon application. Work with a particular injection volume with different

concentration or dilutions of sample and use it with developed LC method to check if good separation is achieved without any carryover. It is important to optimize dilution factor and injection volume depending on the metabolite concentrations and eventual signal-to-noise ratios in the MS profiles before applying this to all batches (*see* **Note 12**).

5. Injection programs have to include syringe washing steps before and after the injection, a sample pumping step for removal of small air bubbles, and an air buffer for complete sample removal during injection. This helps to avoid any sample carryover or cross contamination.
6. Different MS systems name their respective parameters differently; hence, one should follow manufacturer recommendation to set temperature of source, nebuliser gas flow rate, sheath gas flow rate, voltage of the source, scan rate, and saturation point. These values also depend upon flow rate and manufacturers often provide a rough estimation of these values for a particular range of LC flow rate. In general, source temperature of 150–250 °C and voltage 3,500–4,500 eV is used for LC-MS applications. At least two scans per second should be recorded in full scan mode, although based on sample complexity, one can increase scan rate to four or six scans per second. These scans although increase resolution, but compromise sensitivity. Hence, it should be optimized to obtain better results. The collision energies should also be tried at two to three levels during optimizations.
7. While setting-up a MS method, decide the mass range that has to be scanned. In general, for metabolomics based analysis, mass range of 50–1,200 m/z value is widely used, which cover metabolites including larger ones such as terpenoids (although this class is better suited for GC-MS).
8. For targeted analysis, such as MRM-based quantification or MS2 fragmentation, pure standards are required to optimize for the collision energy at which metabolites will be fragmented in the collision chamber. In many cases, obtaining pure standards is not possible for all compounds and hence ramp collision energy can be used to perform fragmentation to decide on the actual collision energy for the setup based on molecular mass. Although it should still be ensured that fragmentation of metabolites occurs, collision energy values obtained by this method may give a better estimate of the values to be used in the setup.
9. MS method for a particular type of analysis should be optimized and the samples should be submitted in a batch. Samples should be randomized in a batch before its acquisition to remove any possibility of biasness. Assign each sample a particular code. Samples should be submitted within a batch in a completely randomized form (*see* **Note 13**).

3.1.6 Data Processing and Analysis

Mass spectral data can be acquired using the instrument manufacture's software. The data is typically in the range of gigabytes resulting from elaborate experimental setups having several replicates, time-series studies and different types of treatment parameters for a single sample. It is highly recommended to perform Exploratory Data Analysis (EDA) on the raw data to check for any experimental errors.

1. EDA can be used identify systematic errors in data generation/extraction and quickly check the health of the data. Non biological sources of deviations can be easily identified using simple statistics and visualization techniques. We routinely perform simple statistical analyses to check for the presence of systematic errors. Here we showcase an example of application of EDA for checking data quality.

Raw data in the form of mzxml (converted using ProteoWizard [69]) is read into R using the XCMS package [55, 70]. Metabolic profiles are generated using the in-built plotting functions in R ("Plot" command) to visualize the TIC profiles. The TIC profiles (Fig. 1a) show that the third biological replicate has a completely different profile compared to the first two biological replicates. The raw matrix is then subjected to retention time correction and resulting data matrix is exported as a tab delimited file. We can use heat maps to analyze the variability between samples. Any deviation from the normal trend of the replicates can be easily visualized using this approach as the erroneous sample will exhibit different range of colors (Fig. 1b). For further analysis we can use datPAV [62], a Web-based tool providing various statistical and visualization options for exploratory data analysis. In order to compare the *m/z* features and their counts (intensities), scatter plots can be used. Figure 1c, d shows the differences between samples having similar and different *m/z* profiles. The similarity in the *m/z* features between biological replicates is demonstrated in the first and third figure which shows close clustering of the features. The difference in *m/z* profile of the third biological replicate can be easily seen from the second and fourth figures.

Following these simple measures enable generation of good quality data and allow the user to validate their metabolomics experimental setup before proceeding on an ambitious large scale analysis.

2. *Data analysis*: Analysis of high throughput metabolomics experiments is a resource-intensive process that includes preprocessing, pretreatment, and post-processing at each level of experimental hierarchy [54]. For a typical metabolomics experiment, data analysis can be performed using a number of statistical packages. Based on the nature of the experiment, the

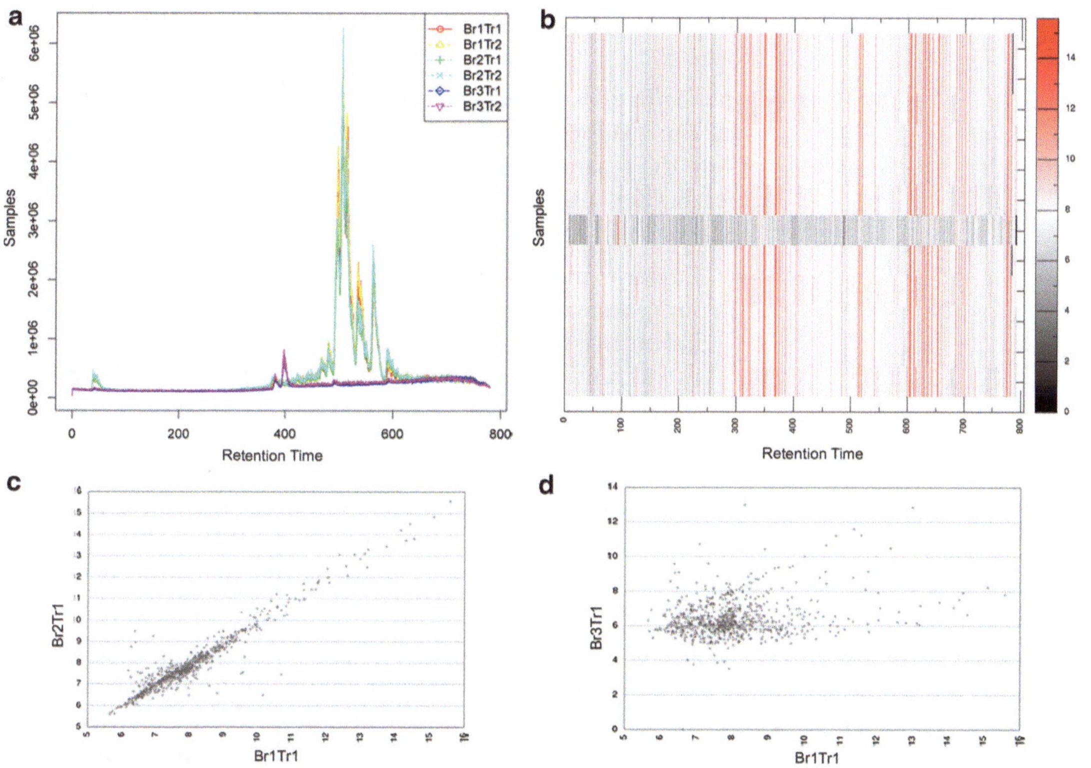

Fig. 1 Exploratory data analysis of metabolomics data. (**a**) Shows the total ion chromatograms (TIC's) between biological replicates. (**b**) Shows a heat map comparing metabolic profiles of different samples and their replicates. (**c**, **d**) Shows scatter plots comparing biological replicates from a sample

raw data can be analyzed using manufacturers' software such as Mass Profiler Professional, Sieve, and Mass Hunter or using freeware such as MetDAT, XCMS, Mzmine, and MetaboAnalyst [54–57, 71]. Raw data has to be cleaned using preprocessing steps such as baseline correction, filtering, noise removal and normalization to minimize instrument errors and remove noise. This is followed by advanced univariate and multivariate statistics using software listed above, to identify significant m/z features that can provide insights into the biological and technical variations. Multiple testing correction methods should be performed to avoid erroneous results leading to false positives.

3. Raw data can also be extracted in a standard formats such as MZXML, MZML, NETCDF, or ASCII for performing statistical analysis using open source software. This can be performed using tools such as ProteoWizard for converting the raw data from the proprietary format (such as RAW, .D files) [69]. Converting the data into MZXML, MZML, or NETCDF

formats is advised as the metadata can be preserved in these formats.

4. *Metabolite identification*: In a typical metabolomics study, users will aim to identify a set of significantly varying features or ions from their samples. These ions can be used for identifying metabolites using some but not limited to the following approaches:
   - Putative metabolite identification using databases: In its simplest form if the data has been acquired from high resolution mass spectrometers, then their significant *m/z* features can then be identified by matching the *m/z* value to different databases such as AraCyc, PlantCyc, KEGG, METLIN, and Chemspider [3, 63, 64, 66, 67] . However as the number of ions detected is not directly related to the number of metabolites, database matching produces only a list of putative metabolites based on the *m/z* ratio and needs to be confirmed using the analytical techniques (*see* **Note 4**).
   - Analytical techniques: Ambiguous metabolites can be confirmed through MS2 fragmentation and accurate differences in metabolite can be quantitatively measures using MRM methods. Fragmentation patterns of the daughter ions can then be matched using databases such as MassBank [68].
   - Isotopic pattern-based identification: Several software packages allow users to compare isotope patterns against pure standards of known compounds. Labeled compounds can be added at the earliest possible stages of analysis. This technique can also aid in absolute quantification by adding known quantity of labeled standards. Metabolic labeling approaches where compounds such as $^{13}CO_2$ or $^{15}NO_3$ can be used for labeling molecules [58]. Relative and absolute quantification can be easily accomplished by mixing labeled and unlabeled metabolites for each sample and then measuring changes in metabolite levels. This approach can also be used for tracking the partitioning of specific metabolites at various branch points of a metabolic pathway.
   - Pathway mapping and visualization: Mapping metabolites onto pathways is an important step in the biological interpretation of data. Once the metabolite has been identified by mapping onto databases or different analytical techniques, user can obtain the chemical identifier [72, 73]. This can then be mapped onto metabolic pathway databases such as KEGG, AraCyc, or PlantCyc [3, 65]. These can be performed using tools such as KaPPA-view, MetDAT, VANTED, iPath, and MetPA [54, 74–77].

3.2 A Typical Workflow for Metabolite Identification from a Complex Matrix of Plant Sample

In this scenario, we compared a glycosylation defective mutant line in Arabidopsis with its wild type (wt). The aim of this metabolomics analysis was to identify metabolites that significantly varied between the two lines. Conjugation by glycosylation adds sugar molecules to core structures of metabolites such as quercetin, kaempferol or other flavonoids. Previous studies on glycosylation of metabolites at different developmental stages showed that the glycosylation process starts in 3-day-old seedlings and reaches a maximum on day 5 with the amount of glycosylated metabolites being at their maximum levels. Then, it starts decreasing to two thirds of the maximum amount by the seventh day and goes on further decreasing. Hence, we selected 6-day-old seedling for harvesting as the glycosylation process would be highly active at this stage.

3.2.1 Harvesting Arabidopsis Samples for Metabolite Extraction for Non-targeted Metabolomics

1. Arabidopsis seeds of the mutant line and wt were grown on MS agar plate following seed surface sterilization and stratification at 4 °C in 16 h of light/8 h of dark cycle. 6-day-old seedlings for both lines were harvested in 2 ml clean centrifuge tube (45 seedlings for each biological replicate).
2. We ensured that all lines were from the same developmental stage and recorded fresh weight of each sample. For each line, five biological replicates were harvested and immediately snap frozen in liq N_2.

3.2.2 Homogenization and Extraction

1. Since the objective here was to perform non-targeted metabolic profiling of tissue extracts, we used 80 % methanol as the extraction buffer.
2. To each sample, precooled 10 mg of 2 mm sized glass beads were added and plant tissues homogenized in the extraction buffer as described in Subheadings 2.2 and 3.1.2.
3. For 50 mg of plant samples, 1 ml of extraction buffer was added and samples were incubated at 4 °C for 10 min followed by centrifugation at 20,800 × *g* for 15 min. Extracted metabolites were again centrifuged at 4 °C to remove any debris and were used for non-targeted metabolic profiling.

3.2.3 Chromatographic Separation

UPLC-MS approach was used with ESI-based soft ionization method.

1. For, UPLC separation, we used C18 column with 1.8 μm particle size and 10 cm column length and used chromatography conditions as described in Subheading 3.1.4 with total run time of 35 min.
2. Non-targeted metabolic profiling for 6-day-old seedlings of wt and mutant line was done using UPLC-MS-TOF in negative mode. Settings described in Subheading 3.1.5 were used with Q-TOF. High sensitivity of TOF coupled with UPLC helped us to readily identify the differential ions.

3.2.4 Data Acquisition and Analysis

1. In order to use the raw UPLC-MS-TOF data for comparative analysis, we performed EDA using datPAV [62]. Then preprocessing steps such as filtering, baseline correction and normalization using default values were used. Filtering was done to remove all the noise and contaminants from raw data, while preserving the sharpness of peaks in the data. For baseline correction, we observed the low-frequency baseline and then subtracted that baseline from the raw data. Normalization was used for reducing the systemic variation from the raw data.
2. TIC of wt and mutant lines were overlapped to visualize differential peaks between wt and mutant lines. As shown in Fig. 2a, TIC of mutant line showed absence of several peaks especially between 6 and 11 fractions of the total UPLC run. These fractions in wt showed high concentration of metabolites which were negligible in mutant line. Please note that this is only one type of non-targeted approach, where we start from LC profile differences. Users can alternatively compile complete list of significantly varying ions using statistical software and proceed to **step 2** in Subheading 3.2.5.

3.2.5 Metabolite Identification

1. *We chose* peaks labeled 6 and 7 (Fig. 2a, indicated using a horizontal line) for further analysis. Their MS spectra were extracted from the MS instrument's software. Total ion chromatogram (TIC) of mutant line from the selected region was subtracted from that of the wild type extract to identify differential ions as shown at the top right panel of Fig. 2a. Comparing the spectra from the samples of the wt and mutant lines, respectively, showed ions such as 265.312, 283.401, 297.22, 325.274, 311.239, 447.116, and 463.097 m/z to be significantly reduced in mutant line compared to wt. Most software that comes with the MS instrument allows these comparisons to be performed readily. Users can devise their own methods of

Fig. 2 Metabolomics workflow analysis, from non-targeted to targeted MS2 approach. (**a**) Shows an example of chromatography peaks present in the extract from wild type Arabidopsis but absent in that from the mutant line. Some of the differential peaks are indicated here using a *horizontal line*. Total ion chromatogram (TIC) was obtained by integrating MS spectra over the chromatography time period during which, the selected differential peaks had appeared. TIC profiles of extracts from mutant line were subtracted from those of the wild type extract. The *top right panel* shows the subtracted spectra. Two differential ions (m/z 463.097 and 447.116) are shown by *asterisk* and *filled circle* respectively. (**b**) Shows MS2 fragmentation of parent ion 463.097 and MS2 fragmentation of its most prominent daughter ion 301.1. (**c**) Similarly, shows MS2 fragmentation of parent ion 447.116 and MS2 fragmentation of its most prominent daughter ion 285.0. (**d**, **e**) Show further fragmentation patterns of the prominent daughter ions 301.1 from (**b**) and 285.0 from (**c**) which match those of quercetin and kaempferol, respectively. The fragments of quercetin and kaempferol are mapped to different parts of their respective molecular structures using freeware tools, such as ChemSketch (ACD/Labs) or commercial ones, such as Mass Frontier (Thermo Scientific)

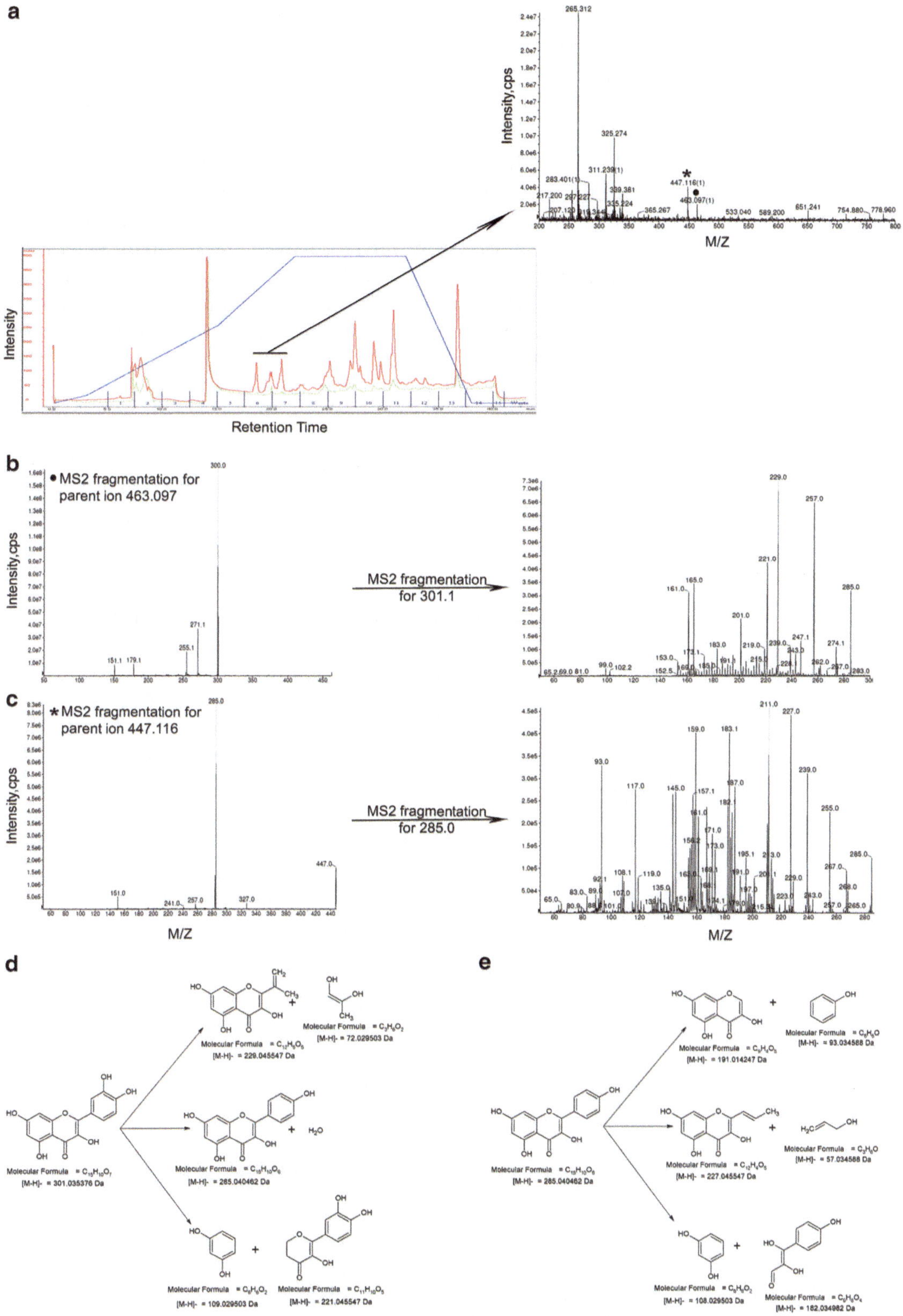
a
Intensity,cps
M/Z
Intensity
Retention Time
b
MS2 fragmentation for parent ion 463.097
MS2 fragmentation for 301.1
c
MS2 fragmentation for parent ion 447.116
MS2 fragmentation for 285.0
d
e
Molecular Formula = $C_{15}H_{10}O_7$
[M-H]- = 301.035376 Da
Molecular Formula = $C_{15}H_{10}O_6$
[M-H]- = 285.040462 Da

comparisons that are a variation of one described here. For example, users can scan sets of spectra from windows of LC separation. This example only shows how to pick the most prominent differential ions. The units "*m/z*" have not been included after the ions in the following sections for simplicity.

2. We used these *m/z* values to search for metabolites having similar *m/z* values against Aracyc [3] database using MetDAT [54]. Metabolites having molecular weights within ±50 ppm of the *m/z* ions were retrieved from the database as putative metabolites (Table 2). All the ions, except 265.312, matched with more than one metabolite. For instance, 447.116 matched with glycosylated forms of kaempferol-3-glucoside, kaempferol-7-O-glucoside and quercetin-3-O-rhamnoside, while 463.097 matched with quercetin-3-glucoside and quercetin 7-O-glucoside.

From non-targeted metabolic profiling, we observed that glycosylated forms of kaempferol and quercetin were down-regulated or absent in mutant line compared to wt. However, this approach

Table 2
Differential *m/z* values corresponding to putative metabolites

m/z	Possible metabolite
265.312	1,3-Diphosphateglycerate
325.274	*trans*-beta-D-Glucosyl-2-hydroxycinnamate
325.274	Coumarinic acid-beta-D-glucoside
325.274	2-Succinyl-5-enolpyruvyl-6-hydroxy-3-cyclohexene-1-carboxylate
325.274	6,7-Dimethyl-8-(1-D-ribityl)lumazine
311.239	13(S)-Hydroperoxyoctadeca-9,11-dienoate
311.239	Octadec-9-ene-1,18-dioic-acid
283.401	Stearate
283.401	Xanthosine
297.220	18-Hydroxyoleate
297.220	9,10-Epoxystearate
447.116	Kaempferol-3-glucoside
447.116	Kaempferol-7-O-glucoside
447.116	Quercetin-3-O-rhamnoside
463.097	Quercetin-3-glucoside
463.097	Quercetin-7-O-glucoside

generated ambiguous metabolite identities for each *m/z* value. This is expected as the matches are done using primary ions. Although non-targeted profiling showed defects in glycosylation of mutant line, it is was not possible to confirm if these differential glycones were derivatives of quercetin or kaempferol. To identify the backbone structure class, we performed MS2 fragmentation.

3.2.6 Targeted MS2 Fragmentation Approach

In order to ascertain identification of these differential metabolites, we performed MS2 analysis for 463.097 *m/z* and 447.116 *m/z*.

1. For the ion 463.097, putative metabolites matched were quercetin-3-glucoside and quercetin-7-O-glucoside. Among daughter ions obtained through MS2 fragmentation for 463.097, 300.0, and 301.1 (next to 300.0 which is seen by expanding spectra using raw data file) were dominating ions which matches with double and single negatively charged quercetin (Fig. 2b). To confirm if 301.1 was indeed quercetin, we performed fragmentation of 301.1 and fragmentation pattern matched with MS2 profile of quercetin standards and from MassBank database [68]. Daughter ions characteristic of quercetin, such as 201.0, 221.0, 229.0, 247.1, 274.1, 257.0, and 285.0 were present in MS2 fragmentation for 301.1 (Fig. 2b). We then used chemical structure of quercetin to see if these daughter ions may relate to fragmented structure of quercetin. As shown in Fig. 2d, daughter ions 285.0, 229.0, 221.0, 72.0, and 109.0 result from fragmentation of quercetin, thus confirming 463.097 as derivative of quercetin. Fragmentation of other daughter ions for 463.097 will be required to ascertain position of sugar molecule. Comparing MS2 profile for 463.097 from MassBank database can also help in determining the identity of this metabolic ion.
2. Similarly, ion 447.116 matched with glycosylated forms of kaempferol-3-glucoside, kaempferol-7-O-glucoside, and quercetin-3-O-rhamnoside. To ascertain if it was the derivative of kaempferol or quercetin, we performed MS2 fragmentation for 447.116. This showed 285.0 as the dominating daughter ion, matching with that of kaempferol, which has *m/z* value 285.0 in negative mode (Fig. 2c). Other daughter ions for 447.116 were 151.0, 241.0, 257.0, 327.0, and 447.0. Here 447.0 ion refers to the incomplete fragmentation of parent metabolite. To confirm if 285.0 is indeed kaempferol, we performed MS2 of 285.0 and the fragmentation pattern was matched with that of the kaempferol standard and fragmentation pattern in MassBank database. Fragmentation pattern of 285.0 is matched with the characteristic daughter ions of kaempferol such as 57.0, 93.0, 108.1, 151.0, 182.1, 191.0, 211.0, 227.0, and 255.0. We then used chemical structure of kaempferol to test if these daughter ions could be from frag-

mentation of kaempferol. Indeed, several of these daughter ions were shown to be related with different probable fragmentation structures of kaempferol (Fig. 2e). This matching of daughter ions of the corresponding parent ions and MS2 fragment ions thus confirmed that 447.116 is the glycosylated form of kaempferol, removing the possibility of quercetin-3-O-rhamnoside. However, we were left with two choices, kaempferol-3-glucoside or kaempferol-7-O-glucoside. We observed that MS2 profile of 447.116 matched with that of kaempferol-7-O-glucoside. This helped us confirm 447.116 as kaempferol-7-O-glucoside. Hence, availability of standards or previously characterized parent ions for their MS2 fragments helped in exact identification of the ion.

Thus, using both non-targeted and targeted MS2 approaches, we were able to identify two glucosylated metabolites that were down-regulated in mutant line compared with wt.

4 Notes

1. Generally, nested and factorial experimental designs are good, when a large number of parameters are monitored as treatments or growth conditions. Latin square design can be used when there are relatively fewer parameters to be tested. Plant scientists can also choose from a number of random-block and factorial experimental designs for growing plants and controlling the physiological conditions. It is good to consult field biologists or statisticians for large scale studies on experimental designs suitable for the project. Always include technical replicates for extractions and samples for the autosamplers of the MS. This is in addition to biological replicates, when performing metabolomics experiments.
2. Here is a caution on setting extraction thresholds for raw data. The threshold for the count data should be set after observing the TIC (total ion chromatogram) values at different retention times to establish the level of background noise. Very high threshold may result in loss of information, whereas low thresholds might make data analysis extremely complex with confounding effects due to the addition of noise. Be careful, as the default threshold values in many commercial software are generally set on the higher side based on our experience. For example, a metabolomics data matrix containing ten samples each with three biological replicates and two technical replicates each can result in 100 megabytes of data when the counts threshold is set as 150 CPS (counts per second). Capturing lower signals is important as many important biological molecules have low abundance. However, extracting

raw data without any thresholds will result in each run producing 1 gigabyte data (total 60 gigabytes) requiring high performance computers along with advanced statistics for the same data set.

3. Solid phase extraction (SPE) method has several advantages, such as greater reproducibility compared to other techniques, cleaner extracts and no emulsion during sample preparation. For plant metabolic extracts, one can use SPE to obtain a cleaner matrix.
4. Be careful in assigning metabolites to pathways, as a single *m/z* value can be matched onto a number of metabolites, and each metabolite can belong to multiple metabolic pathways. So, do not rely solely on automatic assignments of metabolites to pathways. As many metabolites are shared, keep large metabolic pathway maps to see different connections, when going from metabolite lists to pathways.
5. The more organic extraction buffer for more organic metabolites and more aqueous extraction buffer to extract polar metabolites should be used. Extraction methods are generally designed for specific classes of metabolites and include conditions targeting only that class, such as extreme pH values. Alternatives are neutral extraction agents such as chloroform, boiling ethanol or methanol, or a ratio of methanol and chloroform which determines the extent of polar compounds in the final extracted samples. Extraction buffers that extract more hydrophilic metabolites include pH 7 phosphate buffer saline, perchloric acid extraction, chloroform–water extraction, acetonitrile–water (1:1, v/v), acetonitrile–methanol–water mixture (2:2:1, v/v), and water–isopropanol–methanol (2:2:5, v/v) or 80 % methanol. Hence, samples have to be optimized for the extraction buffer and tested for the presence of desired metabolites before embarking on the actual study.
6. A fixed ratio of volume of extraction buffer to tissue weight should be used within a project and must be consistent for all samples. It is a good practice to add internal quantitative control such as ribitol and dinotefuran during the homogenization process. Internal standards, such as isotopic labeled metabolites are used to spike metabolite extracts. These aids in data preprocessing steps, such as normalization and peak alignments.
7. The more the number of cell types present in the tissues that are sampled, the higher will be the "averaging" of metabolite signals for these cell types. Hence, one should try to sample homogeneous groups of cells and check for any remaining cellular debris using a microscope when optimizing the methods.
8. While it seems trivial, a large number of samples in typical metabolomics projects require good sample labeling techniques,

with clear notes and cross references to avoid any potential labeling errors.

9. UPLC or HPLC buffer systems can consist of one eluent or a mixture of two solvent systems, consisting of various ratios of organic and aqueous phase, respectively. In this example, aqueous phase, termed as buffer A is ultra pure, double autoclaved and degassed water, while organic phase, termed as buffer B is HPLC-grade acetonitrile, which could also be substituted by 100 % v/v methanol. Both phases are acidified using acetic acid for better ionization of metabolites. Good UPLC separation of metabolites helps to achieve high MS resolution and sensitivity; therefore, optimization of gradient method is highly desirable.
10. If a metabolic extract is known to contain more organic compounds, then gradient should have steep-slope at less organic phase, while shallow-slope for more organic phase of the gradient. For non-targeted metabolic profiling of complex extracts, one can use a linear gradient starting with 5 % of buffer B to 100 % of buffer B. This state is maintained for a while, following which it is changed to 5 % of buffer B to equilibrate the column before next injection. Time for attaining 5–100 % of buffer B in a gradient and holding time for 100 and 5 % of buffer B at different stages of gradient run requires optimization, to ensure good separation of peaks and no carryover. Flow rate also helps to improve in LC separation, although it should be within acceptable range for column and LC, as well as for the MS. Generally, the flow rates ranges between 0.2 and 0.5 ml/min for UPLC columns.
11. Based on the complexity of the metabolite mixture, the LC run times can vary from 5 to 20 min. Remember, MS is also a separation technique; hence, it is not highly critical to separate each peak in the LC.
12. Each injection volume should be tested at different concentration or dilutions of sample to check if separation can be achieved without any carryover. It is important to optimize dilution factor and injection volume depending on the metabolite concentrations and eventual signal-to-noise ratios in the MS profiles before applying this setup on a global scale.
13. While arranging samples in the autosampler, it is a good practice to include blanks and calibrants after every five vials. Blanks in most cases are the extraction solvent followed using the same protocol as the samples. Calibrants can be in the blank solutions or spiked into the matrix. If following these methods, it should be ensured during optimization that the metabolites do not react with the calibrants. Both blanks and calibrant profiles should be extracted for further analysis to ensure that there are no carryovers in LC eluates from one sample to another.

Acknowledgements

We acknowledge the financial support from the Singapore-Peking-Oxford Research Enterprise, COY-15-EWI-RCFSA/N197-1. We gratefully acknowledge Agilent Technologies, Singapore for their support in acquiring and analyzing the mass spectrometry data for the differential analysis of metabolites.

References

1. Hartmann T (2007) From waste products to ecochemicals: fifty years research of plant secondary metabolism. Phytochemistry 68: 2831–2846
2. Vranova E, Coman D, Gruissem W (2012) Structure and dynamics of the isoprenoid pathway network. Mol Plant 5:318–333
3. Zhang P, Foerster H, Tissier CP, Mueller L, Paley S, Karp PD, Rhee SY (2005) MetaCyc and AraCyc. Metabolic pathway databases for plant research. Plant Physiol 138:27–37
4. Plant Metabolic Network. http://www.plantcyc.org/release_notes/content_statistics.faces. Accessed 02-Feb-2013
5. Wink M (2010) Introduction: biochemistry, physiology and ecological functions of secondary metabolites. Annu Plant Rev 40: Biochemistry of Plant Secondary Metabolism
6. Rhodes M (1994) Physiological roles for secondary metabolites in plants: some progress, many outstanding problems. Plant Mol Biol 24:1–20
7. Sakakibara H (2006) Cytokinins: activity, biosynthesis, and translocation. Annu Rev Plant Biol 57:431–449
8. Fluck RA, Leber PA, Lieser JD, Szczerbicki SK, Varnes JG, Vitale MA, Wolfe EE (2000) Choline conjugates of auxins. I. Direct evidence for the hydrolysis of choline-auxin conjugates by pea cholinesterase. Plant Physiol Biochem 38:301–308
9. Winkel-Shirley B (2001) Flavonoid biosynthesis. A colorful model for genetics, biochemistry, cell biology, and biotechnology. Plant Physiol 126:485–493
10. Winkel-Shirley B (2002) Biosynthesis of flavonoids and effects of stress. Curr Opin Plant Biol 5:218–223
11. Winkel-Shirley B (2001) It takes a garden. How work on diverse plant species has contributed to an understanding of flavonoid metabolism. Plant Physiol 127:1399–1404
12. Narasimhan K, Basheer C, Bajic VB, Swarup S (2003) Enhancement of plant-microbe interactions using a rhizosphere metabolomics-driven approach and its application in the removal of polychlorinated biphenyls. Plant Physiol 132: 146–153
13. Rasmussen S, Parsons AJ, Jones CS (2012) Metabolomics of forage plants: a review. Ann Bot 110:1281–1290
14. Reuben S, Bhinu VS, Swarup S (2008) Rhizosphere metabolomics: methods and applications. In: Karlovsky P (ed) Secondary metabolites in soil ecology. Springer, Berlin, pp 37–68
15. Schenk PM, Kazan K, Wilson I, Anderson JP, Richmond T, Somerville SC, Manners JM (2000) Coordinated plant defense responses in Arabidopsis revealed by microarray analysis. Proc Natl Acad Sci U S A 97:11655–11660
16. Dalmolin RJ, Castro MA, Rybarczyk Filho JL, Souza LH, de Almeida RM, Moreira JC (2011) Evolutionary plasticity determination by orthologous groups distribution. Biol Direct 6:22
17. Hart Y, Mayo AE, Milo R, Alon U (2011) Robust control of PEP formation rate in the carbon fixation pathway of C(4) plants by a bi-functional enzyme. BMC Syst Biol 5:171
18. Angelovici R, Fait A, Zhu X, Szymanski J, Feldmesser E, Fernie AR, Galili G (2009) Deciphering transcriptional and metabolic networks associated with lysine metabolism during Arabidopsis seed development. Plant Physiol 151:2058–2072
19. Brown PD, Tokuhisa JG, Reichelt M, Gershenzon J (2003) Variation of glucosinolate accumulation among different organs and developmental stages of Arabidopsis thaliana. Phytochemistry 62:471–481
20. Desbrosses GG, Kopka J, Udvardi MK (2005) Lotus japonicus metabolic profiling. Development of gas chromatography-mass spectrometry resources for the study of plant-microbe interactions. Plant Physiol 137: 1302–1318
21. Bhalla R, Narasimhan K, Swarup S (2005) Metabolomics and its role in understanding

cellular responses in plants. Plant Cell Rep 24:562–571

22. Alisdair RF, Richard NT, Arno JK, Lothar W (2004) Innovation: metabolite profiling: from diagnostics to systems biology. Nat Rev Mol Cell Biol 5
23. De Vos R, Moco S, Lommen A, Keurentjes J, Bino R, Hall R (2007) Untargeted large-scale plant metabolomics using liquid chromatography coupled to mass spectrometry. Nat Protoc 2:778–791
24. Hall R (2006) Plant metabolomics: from holistic hope, to hype, to hot topic. New Phytol 169:453–468
25. Glinski M, Weckwerth W (2006) The role of mass spectrometry in plant systems biology. Mass Spectrom Rev 25:173–214
26. Patti G (2011) Separation strategies for untargeted metabolomics. J Sep Sci 34:3460–3469
27. Werner E, Croixmarie V, Umbdenstock T, Ezan E, Chaminade P, Tabet JC, Junot C (2008) Mass spectrometry-based metabolomics: accelerating the characterization of discriminating signals by combining statistical correlations and ultrahigh resolution. Anal Chem 80:4918–4932
28. Kueger S, Steinhauser D, Willmitzer L, Giavalisco P (2012) High-resolution plant metabolomics: from mass spectral features to metabolites and from whole-cell analysis to subcellular metabolite distributions. Plant J 70:39–50
29. Kim HK, Choi YH, Verpoorte R (2011) NMR-based plant metabolomics: where do we stand, where do we go? Trends Biotechnol 29: 267–275
30. Dunn WB, Bailey NJ, Johnson HE (2005) Measuring the metabolome: current analytical technologies. Analyst 130:606–625
31. Morita A, Horie H, Fujii Y, Takatsu S, Watanabe N, Yagi A, Yokota H (2004) Chemical forms of aluminum in xylem sap of tea plants (Camellia sinensis L.). Phytochemistry 65:2775–2780
32. Veenstra T (2012) Metabolomics: the final frontier? Genome Med 4:40
33. David SW (2008) Quantitative metabolomics using NMR. TrAC Trends Anal Chem 27
34. Bedair M, Sumner LW (2008) Current and emerging mass-spectrometry technologies for metabolomics. Trac-Trend Anal Chem 27: 238–250
35. Dettmer K, Aronov P, Hammock B (2007) Mass spectrometry-based metabolomics. Mass Spectrom Rev 26:51–78
36. Parab GS, Rao R, Lakshminarayanan S, Bing YV, Moochhala SM, Swarup S (2009) Data-driven optimization of metabolomics methods using rat liver samples. Anal Chem 81:1315–1323
37. Broeckling CD, Huhman DV, Farag MA, Smith JT, May GD, Mendes P, Dixon RA, Sumner LW (2005) Metabolic profiling of Medicago truncatula cell cultures reveals the effects of biotic and abiotic elicitors on metabolism. J Exp Bot 56:323–336
38. Hall R, Beale M, Fiehn O, Hardy N, Sumner L, Bino R (2002) Plant metabolomics: the missing link in functional genomics strategies. Plant Cell 14:1437–1440
39. Bleeker PM, Diergaarde PJ, Ament K, Guerra J, Weidner M, Schutz S, de Both MT, Haring MA, Schuurink RC (2009) The role of specific tomato volatiles in tomato-whitefly interaction. Plant Physiol 151:925–935
40. Mayer F, Takeoka GR, Buttery RG, Whitehand LC, Naim M, Rabinowitch HD (2008) Studies on the aroma of five fresh tomato cultivars and the precursors of cis- and trans-4,5-epoxy-(E)-2-decenals and methional. J Agric Food Chem 56:3749–3757
41. Shuman JL, Cortes DF, Armenta JM, Pokrzywa RM, Mendes P, Shulaev V (2011) Plant metabolomics by GC-MS and differential analysis. Methods Mol Biol 678:229–246
42. Allwood J, Goodacre R (2010) An introduction to liquid chromatography-mass spectrometry instrumentation applied in plant metabolomic analyses. Phytochem Anal 21: 33–47
43. Nordstrom A, Want E, Northen T, Lehtio J, Siuzdak G (2008) Multiple ionization mass spectrometry strategy used to reveal the complexity of metabolomics. Anal Chem 80: 421–429
44. Bin Z, Jun Feng X, Leepika T, Habtom WR (2012) LC-MS-based metabolomics. Mol Biosyst 8
45. Obata T, Fernie A (2012) The use of metabolomics to dissect plant responses to abiotic stresses. Cell Mol Life Sci 69:3225–3243
46. Lei Z, Huhman D, Sumner L (2011) Mass spectrometry strategies in metabolomics. J Biol Chem 286:25435–25442
47. Gibon Y, Rolin D (2012) Aspects of experimental design for plant metabolomics experiments and guidelines for growth of plant material. Methods Mol Biol (Clifton, NJ) 860: 13–30
48. Leek JT, Scharpf RB, Bravo HC, Simcha D, Langmead B, Johnson WE, Geman D, Baggerly K, Irizarry RA (2010) Tackling the widespread and critical impact of batch effects in high-throughput data. Nat Rev Genet 11:733–739
49. Fiehn O, Wohlgemuth G, Scholz M, Kind T, Lee do Y, Lu Y, Moon S, Nikolau B (2008)

Quality control for plant metabolomics: reporting MSI-compliant studies. Plant J 53:691–704
50. Members MSIB, Sansone SA, Fan T, Goodacre R, Griffin JL, Hardy NW, Kaddurah-Daouk R, Kristal BS, Lindon J, Mendes P, Morrison N, Nikolau B, Robertson D, Sumner LW, Taylor C, van der Werf M, van Ommen B, Fiehn O (2007) The metabolomics standards initiative. Nat Biotechnol 25:846–848
51. Boccard J, Veuthey JL, Rudaz S (2010) Knowledge discovery in metabolomics: an overview of MS data handling. J Sep Sci 33: 290–304
52. Hendriks MMWB, van Eeuwijk FA, Jellema RH, Westerhuis JA, Reijmers TH, Hoefsloot HCJ, Smilde AK (2011) Data-processing strategies for metabolomics studies. Trac-Trend Anal Chem 30:1685–1698
53. Eliasson M, Rannar S, Trygg J (2011) From data processing to multivariate validation–essential steps in extracting interpretable information from metabolomics data. Curr Pharm Biotechnol 12:996–1004
54. Biswas A, Mynampati KC, Umashankar S, Reuben S, Parab G, Rao R, Kannan VS, Swarup S (2010) MetDAT: a modular and workflow-based free online pipeline for mass spectrometry data processing, analysis and interpretation. Bioinformatics 26:2639–2640
55. Smith CA, Want EJ, O'Maille G, Abagyan R, Siuzdak G (2006) XCMS: processing mass spectrometry data for metabolite profiling using nonlinear peak alignment, matching, and identification. Anal Chem 78:779–787
56. Pluskal T, Castillo S, Villar-Briones A, Oresic M (2010) MZmine 2: modular framework for processing, visualizing, and analyzing mass spectrometry-based molecular profile data. BMC Bioinformatics 11:395
57. Xia J, Mandal R, Sinelnikov IV, Broadhurst D, Wishart DS (2012) MetaboAnalyst 2.0—a comprehensive server for metabolomic data analysis. Nucleic Acids Res 40:W127–W133
58. Hegeman A (2010) Plant metabolomics—meeting the analytical challenges of comprehensive metabolite analysis. Brief Funct Genomics 9:139–148
59. Fan TW-M (2012) Considerations of sample preparation for metabolomics investigation. In: Fan TW-M, Lane AN, Higashi RM (eds) The handbook of metabolomics, vol 17, Methods in pharmacology and toxicology. Humana, Totowa, NJ, pp 7–27
60. American Society for Mass Spectrometry (2009) Metabolomics ASMS Workshop Survey 2009. http://fiehnlab.ucdavis.edu/staff/kind/Metabolomics -Survey-2009
61. Kind T, Fiehn O (2007) Seven Golden Rules for heuristic filtering of molecular formulas obtained by accurate mass spectrometry. BMC Bioinformatics 8:105
62. Biswas A, Rao R, Umashankar S, Mynampati KC, Reuben S, Parab G, Swarup S (2011) datPAV—an online processing, analysis and visualization tool for exploratory investigation of experimental data. Bioinformatics 27: 1585–1586
63. Kanehisa M, Goto S (2000) KEGG: kyoto encyclopedia of genes and genomes. Nucleic Acids Res 28:27–30
64. Kanehisa M, Goto S, Sato Y, Furumichi M, Tanabe M (2012) KEGG for integration and interpretation of large-scale molecular data sets. Nucleic Acids Res 40:D109–D114
65. Okuda S, Yamada T, Hamajima M, Itoh M, Katayama T, Bork P, Goto S, Kanehisa M (2008) KEGG Atlas mapping for global analysis of metabolic pathways. Nucleic Acids Res 36: W423–W426
66. Sana TR, Roark JC, Li X, Waddell K, Fischer SM (2008) Molecular formula and METLIN Personal Metabolite Database matching applied to the identification of compounds generated by LC/TOF-MS. J Biomol Technol 19:258–266
67. ChemSpider. http://www.chemspider.com/
68. Horai H, Arita M, Kanaya S, Nihei Y, Ikeda T, Suwa K, Ojima Y, Tanaka K, Tanaka S, Aoshima K, Oda Y, Kakazu Y, Kusano M, Tohge T, Matsuda F, Sawada Y, Hirai MY, Nakanishi H, Ikeda K, Akimoto N, Maoka T, Takahashi H, Ara T, Sakurai N, Suzuki H, Shibata D, Neumann S, Iida T, Tanaka K, Funatsu K, Matsuura F, Soga T, Taguchi R, Saito K, Nishioka T (2010) MassBank: a public repository for sharing mass spectral data for life sciences. J Mass Spectrom 45:703–714
69. Kessner D, Chambers M, Burke R, Agus D, Mallick P (2008) ProteoWizard: open source software for rapid proteomics tools development. Bioinformatics 24:2534–2536
70. R Core Team (2012) R: A language and environment for statistical computing. R Foundation for Statistical Computing, Vienna, Austria
71. Sugimoto M, Kawakami M, Robert M, Soga T, Tomita M (2012) Bioinformatics tools for mass spectroscopy-based metabolomic data processing and analysis. Curr Bioinformatics 7:96–108
72. Mochida K, Shinozaki K (2010) Genomics and bioinformatics resources for crop improvement. Plant Cell Physiol 51:497–523
73. Chagoyen M, Pazos F (2012) Tools for the functional interpretation of metabolomic experiments. Brief Bioinformatics. doi:10.1093/bib/bbs055

74. Tokimatsu T, Sakurai N, Suzuki H, Ohta H, Nishitani K, Koyama T, Umezawa T, Misawa N, Saito K, Shibata D (2005) KaPPA-view: a web-based analysis tool for integration of transcript and metabolite data on plant metabolic pathway maps. Plant Physiol 138:1289–1300
75. Junker BH, Klukas C, Schreiber F (2006) VANTED: a system for advanced data analysis and visualization in the context of biological networks. BMC Bioinformatics 7:109
76. Letunic I, Yamada T, Kanehisa M, Bork P (2008) iPath: interactive exploration of biochemical pathways and networks. Trends Biochem Sci 33:101–103
77. Xia J, Wishart DS (2010) MetPA: a web-based metabolomics tool for pathway analysis and visualization. Bioinformatics 26:2342–2344

Index

Ray J. Rose (ed.), *Legume Genomics: Methods and Protocols*, Methods in Molecular Biology, vol. 1069, DOI 10.1007/978-1-62703-613-9, © Springer Science+Business Media, LLC 2013

N

O

P

MIX
Papier aus verantwortungsvollen Quellen
Paper from responsible sources
FSC® C105338

If you have any concerns about our products,
you can contact us on
ProductSafety@springernature.com

In case Publisher is established outside the EU,
the EU authorized representative is:
Springer Nature Customer Service Center GmbH
Europaplatz 3, 69115 Heidelberg, Germany

Printed by Libri Plureos GmbH
in Hamburg, Germany